DYNAMIC MODI
MUSCULOSKELETAL MOTION

A Vectorized Approach for Biomechanical Analysis in Three Dimensions

DYNAMIC MODELING OF MUSCULOSKELETAL MOTION

A Vectorized Approach for Biomechanical Analysis in Three Dimensions

Gary Tad Yamaguchi, PH.D.
Department of Bioengineering
College of Engineering and Applied Sciences
Arizona State University
Tempe, AZ 85287

Library of Congress Cataloging-in-Publication Data

Yamaguchi, Gary Tad
Dynamic modeling of musculoskeletal motion : a vectorized approach for biochemical analysis in three dimensions / Gary Tad Yamaguchi
p. cm.
Includes bibliographical references and index.
1. Musculoskeletal system—Mathematical Models. 2. Biomechanics—Mathematical models. I. Title

ISBN10 0-7923-7430-4 (HC) ISBN 10 0-387-28704-3 (SC)
ISBN 13 9780792374305 ISBN 13 9780387287041

E-ISBN 0-387-28750-7

Printed on acid-free paper

First softcover printing, 2006

Printed in the United States of America.

9 8 7 6 5 4 3 2 1 SPIN 11544906

springeronline.com

for Janet, Jonathan,
Nathan, and Karissa

Contents

Acknowledgments

Much of the information and insights provided within have come about via the teaching of three biomechanics courses at Arizona State University over a 12 year period. Materials have been tried and tested on hundreds of senior undergraduate and graduate students in various forms. For their patience and willingness to learn from "half baked" course materials, and for their constant feedback, the author wishes to acknowledge their major contributions.

Artist and graduate student Dawn Taylor, and undergraduate student Amy Loveless, contributed their artistic talents voluntarily. Each of their illustrations is worth a thousand words. Byron Olson refined the mathematical presentation and added new insights. Technical editing was provided by graduate students Sherry Di Jorio, James Egan, and Alana Smith. Each gave me new insights on how to word things better, how to present ideas in more logical fashion, and to define each and every term.

The appendices containing the latest word in musculoskeletal modeling parameters was compiled by Michael R. Carhart. Other colleagues and former graduate students who have taken the time and effort to write meaningful comments include Neil Crawford, B. J. Fregley, Karin Gerritsen, Jonathan Green, Jiping He, Richard Hinrichs, He Huang, Ali Kakavand, Eric Mallett, Anthony Marsh, Philip Martin, Scott McLean, Daniel Moran, May Movafagh, Akinori Nagano, Joseph Peles, Darrin Richards, Jennie Si, Didik Soetanto, Brian Umberger, Peter Vint, Qingjun Wang, and Jack Winters.

Scott "Scooter" Delp and J. Peter Loan of Musculographics, Inc. have been very supportive through their generous academic developer pricing of Software for Interactive Musculoskeletal Modeling (SIMM). Without SIMM, many of the musculoskeletal models developed at ASU would have never come to be.

The author also wishes to acknowledge the support of National Science Foundation Young Investigator Grant BES9257395 which helped in part to more fully develop the theories and examples included in this text.

Foreword

When I arrived at Stanford University in the fall of 1984, Felix E. Zajac, III was a new researcher and professor studying the biomechanics of muscle and movement coordination within the Design Division, Department of Mechanical Engineering. After taking a quick look at my transcripts, he "strongly recommended" that I enroll for the graduate level dynamics course taught by Professor Thomas Kane within the Applied Mechanics Division. After explaining to Professor Zajac that I had already completed several courses in dynamics as an undergraduate physics major at Occidental College, an advanced dynamics class in my master's program at MIT, and had worked with the dynamics of atomic particles for several years at the Lawrence Livermore National Laboratory, he was resolute. Again, he strongly recommended that I cancel my plans to take a different course, and sign up for Professor Kane's course instead.

My first reaction was to be offended. After all, I had not only taken advanced courses in dynamics, I had worked on difficult problems in the field of magnetic fusion energy. I was familiar with what I could and could not do with classical dynamics techniques. But in a weak moment, I decided to swallow my pride and follow his advice. In my ignorance, I could not have realized how little I knew about dynamics, nor how limited was the set of problems that I could actually solve using classical techniques.

Within the first few weeks of the course, my eyes were opened. Armed with new tools which made working simultaneously with multiple reference frames in three dimensions almost trivial, I eagerly solved one of the perplexing problems that had stumped myself and a Ph.D. engineer at the Livermore laboratory for over a year. I remember it well – it only took a few minutes to solve it using Kane's systematic, vectorized approach.

Over the next several years, almost all of Zajac's students took Professor Kane's courses until Kane retired in the mid 1990's. Though there were many fine instructors and mentors at Stanford during those years, I consider it one of my greatest privileges to have been trained by both these men – a leading biomechanist and a pathbreaking dynamicist. The biodynamic models made by the students of Zajac and Kane are among the most highly sophisticated and realistic ever made. But curiously, Kane's Method has only found its way into a few biomechanical classrooms scattered across the country.

This textbook was written in hopes of stimulating further interest in Kane's Method and its applications to the biomechanical modeling and analysis. Some of the material is appropriate for senior level undergraduates who seek to enter a graduate program and for whom an early introduction to vector methods would be advantageous. Most of the material is presented to an audience of graduate students and to persons proficient in classical biomechanical methodologies. If the reader is not familiar with the field, he or she would benefit by having a basic physiology book which illustrates muscle pathways and joints. Additionally, a reader unfamiliar with matrix algebra, numerical analysis, and differential equations would be well served by a suitable reference book.

At Arizona State University, this material has been presented in a two semester course sequence to graduate students from the Department of Bioengineering and the Department of Exercise Science and Physical Education. It would be best to proceed through the material sequentially. However, because our student pool is divided almost equally in thirds between those desiring to learn 3-D experimental analysis, 3-D dynamic simulation, or both, the material is selectively presented according to the theme of each semester.

The first semester emphasizes vector kinematics and the inverse dynamic method, and only introduces the concept of Kane's Method. Thus, Chapters 3, and 4 are covered in detail, and portions of the remaining chapters are selected based on relevance to the topic of experimental analysis. To enable students to perform an inverse dynamic project without having to learn much of the material in Chapter 6, Gordon's method for writing dynamic equations is used (Section 6.10) instead of Kane's Method. Students are assigned the task of collecting data in the laboratory, processing the data, determining the segmental or joint torques, and finally, performing a distribution of forces amongst a redundant muscle set.

The second semester focuses upon Kane's Method and developing forward dynamic musculoskeletal models. Course material begins by reviewing the formulas for the velocity and acceleration of two points on a rigid body (Section 4.3.3), and proceeds to the end of the book. The topic of modeling musculotendon actuators (Chapter 1) is also included as part of the second semester. A dynamic modeling and simulation project is required. To facilitate the derivation and coding of the project, the *AUTOLEV* software program (OnLine Dynamics, Inc., Sunnyvale, CA) has proven useful.

Except for the previous statement, I have avoided mentioning or endorsing commercially available software products. The serious student should learn, *and diligently practice*, the techniques by hand before using a computer program. Many of these programs are "black boxes" which perform dynamic analyses internally and require very little knowledge to get running. The catch is, it becomes too easy to generate streams of numbers that may or may not have any relevance to the physical system being modeled. With patience, practice, and hard work, the learning will come.

In the literature, the material contained in this book is often summarized in a rather terse manner, such as, "... the model was developed and the dynamic equations were derived using Kane's Method." Descriptions of the actual methodology are rare in the biomechanical literature. This book was written to "fill in the blanks" for those not yet exposed to Kane's Method. The book is intended as a compliment to the literature, not as a compilation of the latest summary of research results. As much as possible, references to other works have been made to review articles and scholarly books, and to journal articles when necessary. Ph.D. dissertations are referenced when they are the only available source of additional information. With this in mind, the interested reader is encouraged to seek out the latest publications so the basic knowledge obtained here can be augmented with more current information.

I have devoted much of my career to developing musculoskeletal models that can be used to evaluate bold new designs and treatments designed to aid the disabled, and reestablish the control of paralyzed extremities. In doing so, I have learned how much more there is to know – and in seeking this knowledge I have been continually reminded of how masterfully and wonderfully we were made.

Best wishes to you on your journey!

Gary T. Yamaguchi
Tempe, Arizona
January 2001

I

INTRODUCTION

Chapter 1

OVERVIEW OF DYNAMIC MUSCULOSKELETAL MODELING

1.1 INTRODUCTION

It is now the new millenium, and 3-D is where it's at! The past few decades have seen an endless parade of analyses and modeling efforts confined to a two dimensional plane. In the past, limitations in computational power required such models to be simplified as far as one dared. Small and simple was good in many ways. Results could be easily interpreted, and generalities of the system were easily illuminated by elegant models, analyses, and interpretations. However, times have changed! With the advent of the inexpensive and powerful desktop computer, the new focus is upon creating models that are ever more realistic and accurate in terms of their predictive power.

Unfortunately, despite the recent advances in technology, most universities and educational centers still teach the old, classical methods of deriving dynamic equations. While there is nothing wrong with these methods, they are very difficult to implement for three dimensional (3-D) systems and even planar (2-D) systems having more than a few degrees of freedom. On the other hand, the computing revolution has brought about rapid advances in software as well as hardware. Computer software programs can now be purchased that do all of the hard work of deriving and programming the dynamic equations. The reader could stop reading now, return this book for a refund, and use the money to make a down payment on the appropriate software package if the simulation code is all that is desired. But, if the reader desires to know how the package works, and have confidence in its ability to produce accurate results, she must become intimately familiar with its inner workings and the mechanical principles upon which it is based.

Presently, this means anyone who seeks a real understanding of 3-D biomechanical systems must learn *vector kinematics* and *Kane's Method*. This book was written to introduce biomechanists at all levels of expertise to modern methods of modeling and analyzing dynamic biomechanical systems in three dimensions.

So be it. What's the big deal about yet another book on biomechanics? The younger generation might wonder why they should learn biomechanical analysis and modeling *this way*, while more traditional biomechanists might still be wondering what was wrong with the old approach.

The big difference is the way in which the development of the dynamic equations – the heart of every dynamic simulation and every dynamic analysis – are derived. At the very core of this book is a practical introduction to one of the most exciting and revolutionary developments of the modern era: Kane's Method. For over a decade, Professor Thomas Kane's students at Stanford University have used Kane's Method to solve difficult problems in the fields of aeronautics, spacecraft dynamics, robotics, and more recently in biomechanics (Radetsky, 1986). The vectorized approach is easy to learn, and is used in exactly the same way to solve every problem. No longer is intuition and judgement, developed by long years of tutelage under a dynamics master, a prerequisite. And with the advent of computerized methods of deriving the equations of motion, a single person can quickly and easily derive and program a complex set of dynamic equations in a fraction of the time needed by an entire *team* of classical dynamicists. This is not an exaggeration!

For example, it used to take months for a person trained in Kane's Method to derive, program, and debug a dynamic simulation program for a three dimensional system having more than a few degrees of freedom. Only a master classical dynamicist would be able to arrive at a solution. The debugging phase alone was likely to take half the time. If the simulation was found to be erroneous, it was difficult to find the error amid the myriad of complex symbols and equations. It was also hard to determine whether the error was made in deriving the equations or in transcribing them to computer code. Now, however, the situation has changed. Using Kane's Method and one of the associated symbolic computer tools available, just about anybody can derive the equations and create a bug-free computer simulation code for systems having up to 20 degrees of freedom – all in a day's work.

What this means for the biomechanist is more time can be spent on biomechanics, and much less time must be expended tediously deriving equations of motion. By the same token, the tools afforded by Kane's Method will enable one to analyze and model complex structures with relative ease. Because greater complexity can be handled, biomechanical models have the potential to become much better representations of the physiological system, and will not be plagued by simplifications that in the past had to be made for reasons of mathematics.

If you are one of the many experienced biomechanists who have yet to discover the power of Kane's Method, this book is for you. This book is intended to let traditional biomechanists "in" on techniques that make modeling in three dimensions as easy as doing it in two dimensions. Or if you are a new modeler and wish to be trained with the latest and most powerful techniques of 3-D dynamic modeling, this book has been designed for you as well. In fact, the outline of material forms a syllabus for a one year graduate sequence, in which Kane's Method is taught in conjunction with computational movement biomechanics.

It is important to note that this book cannot hope to explain Kane's Method in total, and the interested reader is referred to the excellent text by Kane and Levinson (1985) for a more complete treatise. This book seeks only to extract the core methodology, to apply it to biomechanics, and to explain it in more understandable terms.

Support for the methodologies herein has come from the performance of graduate students in performing their exercises and required projects. Virtually all of the students have been able to formulate a problem of interest to themselves, generate state-of-the-art models of musculoskeletal motion (four to six degrees-of-freedom in three dimensions), and answer a research question in the field of movement biomechanics. Some of these class projects have been presented as research findings at national and international biomechanics meetings, and have formed the basis for masters and doctoral theses. From the author's experience, it is indeed possible to teach students the basic elements of Kane's Method along with musculoskeletal biomechanics. These techniques are powerful, empower students with a lifelong ability to work comfortably in three dimensions, and provide a basis for continual learning in this fascinating field.

1.2 OVERVIEW

Mechanics is a broad subfield of physics in which forces and torques, and their effects, are studied from a variety of viewpoints and with a multiplicity of methods. It should be helpful to define the coverage of this book within this greater context.

- *Statics*, a subfield of mechanics, refers to the study of systems that are in static equilibrium, *i.e.,* systems that have zero accelerations. The methods presented in this book are well suited to performing static analysis in two and in particular, three dimensions.

- *Kinematics* is the study of systems and their motions, including all possible motions that can be exhibited by the system. As such, kinematics lies outside of the broader definition of mechanics, as no attempt is made to relate forces and torques to motions. However, the ability to perform rigorous kinematical analyses in two and three dimensions is critical to predicting the forces arising from inertial sources. A significant emphasis is placed upon *vector* kinematics in this text, as vector kinematics is an essential foundation of movement biomechanics.

- *Dynamics* can be used to predict which of the infinitely many possible movements of a given system will occur, given sufficient information about the initial conditions, forces, and torques acting upon the system. Dynamic equations simply relate the forces and torques to the positions, velocities, and accelerations of the system.

- *Kinetics* focuses upon the measurement and calculation of forces, moments, and torques in both static and dynamic situations.

- *Biomechanics* refers to the study of forces, torques, and the effects of their actions upon biological structures and materials. Possible effects include the initiation of relative and whole body motions (a subfield called *movement biomechanics*), fluid flows (*biological fluid mechanics*), and distributed stresses upon the tissues and potential tissue failures (*orthopaedic biomechanics*).[1]

This book is devoted to the development and usage of dynamic musculoskeletal models. Vector methods in kinematics are emphasized over the standard trigonometric approaches, and Kane's Method of formulating the dynamical equations of motion is emphasized over more traditional Newton-Euler and Lagrangian formulations. The reader who masters these modern techniques will be able to formulate and use models that are far more realistic and complex than can otherwise be accomplished – and for this reason the book is aimed at both the traditional biomechanist seeking to expand his or her analytical skills and at graduate students in movement biomechanics.

In Chapter 2, the macroscopic behavior of muscle and tendon is presented, with an emphasis on modeling the most important characteristics of the musculotendon actuator. It is important to start with a discussion of muscles and tendons because these apply the actuating forces without which there would be no movement. A literature review on the subject of muscle modeling is included.

---

[1]This book focuses upon the relatively large movements exhibited across joints of virtually rigid subsystems. Thus this text does not contain much in the way of orthopaedic biomechanics, which emphasizes the study of the stresses and strains within biological structures that are deformable, but from a macroscopic point of view are primarily rigid. However, the study of movement often interfaces with orthopaedics, as the movement models are useful for predicting and analyzing the time dependent joint loads that cause the stresses within the biological tissues. Muscles, too, are deformable tissues that are strongly related to the creation of large joint movements and hence are included in this book. Except for a brief mention of viscosity, this book is generally not directed at the study of biomechanical fluid flows.

Chapters 3 and 4 develop the fundamental skills in managing rigid body reference frames and the kinematics of vectors. It is essential that this material be mastered – not simply understood – to carry out two and three dimensional analyses of biomechanical problems! It is recommended that the techniques introduced in these chapters be practiced until they become second nature. Many examples are given which the reader may emulate in order to practice his or her technique.

Chapter 5 presents methods to transform models that are mechanical or robotic in nature (*e.g.*, torque actuated linkage structures with joints that can rotate more than 360 degrees) into biologically relevant models (*e.g.*, musculotendon force actuated structures with kinematically realistic joints). Joint models are presented in Chapter 5, including a case study describing a simple model of the knee joint.

In Chapter 6, the dynamic equations of motion are developed via Kane's method. This material would serve as an appropriate introduction to Kane's Methods for students studying robotics or prosthetics. It is recommended that the reader intending to seriously pursue these fields of study purchase the textbook by Kane and Levinson (1985) , as the material contained in Chapter 6 is only an introduction to the methods more fully explained in Kane and Levinson's original text. For instance, some of the formulas contained herein are presented without proof. The interested reader desiring more is therefore referred to Kane and Levinson (1985).

Biological organisms typically have many more muscles than needed to actuate and coordinate movements. Therefore, the subject of the "redundant problem in biomechanics" is covered in Chapter 7. An infinite variety of muscle actuations can produce the same outward movement, but which control strategy is utilized by the central nervous system? Via optimal control techniques, the best mathematical solutions can be determined and compared to experimental measurements of the biological movement. Through these comparisons, some idea of the control strategies employed to coordinate movement can be found. A brief literature review describes the development of this aspect of biomechanical modeling.

The Appendices present the parameters used for a recently developed model of human gait. This model was developed to evaluate paraplegic stepping for the purpose of dynamic balance recovery. The Appendices provide all the important parameter values for the lower extremity and will enable readers to practice their developing technique using real data. Additional appendices including parameters for an upper extremity model of both humans and primates is under development and is not included in this First Edition. It is hoped that models similar to these will be increasingly employed to explore current questions in the fields of neuroscience and motor control.

An overview of Kane's Method will now be given.

1.3 KANE'S METHOD VERSUS THE NEWTON-EULER APPROACH

Other methods of deriving dynamic equations have been in existence since Newton first postulated that the resultant force (the vector sum of p forces $\vec{F}_i$, $(i = 1, 2, 3, \ldots, p)$ acting on body B) on an object of mass m was proportional to its acceleration,

$$\sum_{i=1}^{p} \vec{F}_i = m\vec{a}\,. \tag{1.1}$$

This was revolutionary because it *enabled the motion of the object to be predicted.* In other words, from the object's initial position and velocity, and knowing the vector sum of the forces acting upon it, the acceleration of the object could be computed and integrated to find its new positions at future times. This is the essence of a dynamical equation – a mathematical formula which makes it possible to compute the motions of an object in response to applied forces and torques. Usually, a dynamic equation simply relates these forces and torques to the positional variables (such as distance from an origin or an angular orientation) and their first and second time derivatives.

D'Alembert rewrote the very same equation in a different form,[2]

$$\sum_{i=1}^{p} \vec{F}_i + \vec{F}^* = 0 \tag{1.2}$$

where the inertial force acting on the body is defined as,

$$\vec{F}^* = -m\vec{a}\,. \tag{1.3}$$

While it may not seem that this "refinement" was anything new, it was indeed profound because it allowed a dynamic problem to be solved using static methods. In other words, D'Alembert treated the inertial force $\vec{F}^*$ in the same way as any other force. All one had to do to solve a dynamic problem was to determine the inertial force and sum it along with all the other external forces acting on a body.

Other methods for determining the dynamic equations were developed which were useful for specific types of problems. For example, Lagrange's Method was useful whenever the potential and kinetic energies of a system could be readily identified (Andrews, 1995).[3] However, unless one became a master of

[2] Strictly speaking, the right-hand side of Equation 1.2 should be written as a zero vector, $\vec{0}$, but the usual practice is to dispense with the vector overbar.

[3] There are hundreds of books covering the subjects of mechanics and dynamics which describe these methods. A few that provide background material on this subject from the biomechanical point of view include Allard *et al.* (1995), Hay (1993), Özkaya and Nordin (1999), Tözeren (2000), and Winter (1990). See the references at the end of the chapter for additional listings.

the art of dynamics it was difficult to determine which method was the best to use to derive the equations for a particular problem. Newtonian Methods, D'Alembert's Method, energy methods based on the works of Hamilton and Lagrange, *etc.*, were all capable and available, but with the right choice a seemingly complex problem could be solved rather simply. Secondly, some of the methods required imaginary perturbations of exactly the right size to be introduced. One had to carry the perturbations along in the mathematics and know from intuition or experience when the effects of the perturbations could be neglected, and when they had to be accounted for. Finally, most of these methods required the usage of differential and integral calculus. While not always difficult when problems were simple, the geometry of more complex systems in which the objects exhibited strong interactions made the calculus tedious, difficult, and error prone.

Kane's Method differs from most of these classical methods in the following ways:

- The method is vector-based, and therefore is extremely well suited for 3-D analyses;
- Vector cross and dot products are used to determine velocities and accelerations rather than calculus;
- The dynamic equations for an n degree of freedom system are formed out of simplified forms of the forces, moments, and torques, called *generalized forces* F_r $(r = 1, 2, \ldots, n)$, rather than the actual forces, moments, and torques;
- Inertial forces and inertial torques are also incorporated in simplified form as *generalized inertia forces* F_r^*;
- Generalized forces and generalized inertia forces are simply added together to create n dynamic equations describing the motions of the n degrees of freedom,

$$F_r + F_r^* = 0\,. \tag{1.4}$$

Though other authors emphasize vector methods to determine velocities and accelerations (for example, Merriam & Kraige, 1997), Kane invented the concepts of the *generalized speed, partial velocity, partial angular velocity, generalized active force,* and *generalized inertia force.* These provided the key simplifications enabling forces and torques having no influence on the dynamic equations to be eliminated early in the analysis. Early elimination of these "noncontributing forces and torques" greatly simplified the mathematics and enabled problems having greater complexity to be handled.

The generalized forces are scalars formed by taking dot products of the vector forces and particular portions of the velocity vectors called "partial velocities." As we shall see in Chapter 6, the use of generalized forces greatly simplifies the equations themselves. Forces and moments that do not contribute to the dynamic equations of motion are eliminated prior to the formulation of the dynamic equations. In biomechanical models, the joint reaction forces are generally considered to be noncontributing, and thus can be ignored when using Kane's Method. On the other hand, if knowledge of the joint reaction forces is desired, they can still be found rather easily.

Whether or not the reader chooses to ultimately use Kane's Method to derive the dynamic equations of motion, the vector-based methods for determining the system kinematics are invaluable and may be used to define the crucial accelerations necessary for other dynamic methods.

1.3.1 EXAMPLE: DERIVATIONS OF THE EQUATIONS OF MOTION FOR A TWO-LINK PLANAR LINKAGE

The following example illustrates the methodological differences between the more standard Newton-Euler approach and Kane's Method. It is directed at the reader who possesses some knowledge of dynamic systems. At this point, the more knowledgable reader is encouraged to examine this section only to gain an overview of what will be covered in detail later. Many of the minute details of nomenclature and methodology are purposely left out so that the focus will remain upon the advantages and disadvantages posed by Kane's approach. A less knowledgable reader can skim over the remainder of this chapter, and return later when more familiar with kinematic and dynamic systems.

Figure 1.1 shows the arrangement of the system having two degrees of freedom, represented by angles q_1 and q_2. Rigid link A is connected to the ground reference frame N with a frictionless pin joint at point A_o. Likewise, link A is connected to rigid link B at point B_o. A torque, $\vec{\tau}_{N/A}$ is exerted by N on segment A, and torque $\vec{\tau}_{A/B}$ is exerted by A on segment B. The torques exerted by one segment on the other across the joints are equal and opposite, so that the torque exerted by B on A is $\vec{\tau}_{B/A} = -\vec{\tau}_{A/B}$. A force $\vec{F}$ of arbitrary magnitude and direction acts at the endpoint of the linkage.

With both the Newton-Euler approach and Kane's Method, the first step is to derive expressions for the angular velocity and angular acceleration of each rigid body comprising the system. Along with geometrical information, these angular quantities enable the velocities and accelerations of all the important points within the system to be computed.[4] These are called *kinematical equations* because they describe motions, but do not relate the motions to the forces and torques which determine the motions. Most persons even somewhat

[4]For now, consider the "important points" to be points having mass, and points where forces are applied.

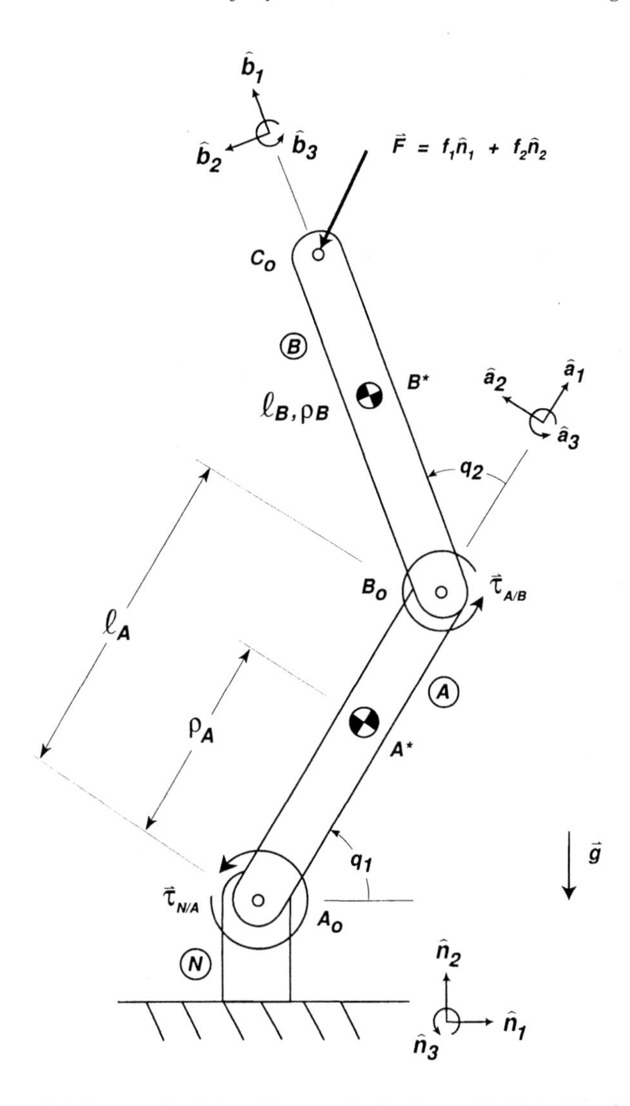

Figure 1.1. Planar two-link kinematic chain with an endpoint force. Rigid bodies A and B, and the ground reference frame N are joined together with frictionless pins at points A_o and B_o. Triads of mutually perpendicular unit vectors define vector component directions for each of these three rigid body reference frames. $\hat{a}_1$ remains parallel to $\overline{A_oB_o}$ (a line joining points A_o and B_o) as body A rotates by angle q_1, and $\hat{b}_1$ remains parallel to line $\overline{B_oC_o}$ as both bodies A and B rotate. Angle q_2 is defined to be the counterclockwise rotation angle from $\hat{a}_1$ to $\hat{b}_1$. Mass centroids A^* and B^* for the respective bodies are located at distances ρ_A and ρ_B from their proximal ends. A torque, $\vec{\tau}_{N/A}$, is exerted by N on A, and another torque, $\vec{\tau}_{A/B}$, is exerted by A on B. An endpoint force of arbitrary direction and magnitude, $\vec{F} = f_1\hat{n}_1 + f_2\hat{n}_2$ is exerted by an external influence at the endpoint of the linkage, C_o.

familar with physics or mechanics are used to seeing the kinematical equations expressed in terms of a global, Cartesian coordinate reference frame. For example, it is common to see velocities and accelerations expressed in terms of their horizontal and vertical components. However, it is much easier to describe the kinematics of the system using *local reference frames* affixed to each body. We can relate the reference frames N, A, and B by creating *tables of direction cosines,*

$$\begin{array}{c|ccc} {}^{N}R^{A} & \hat{a}_1 & \hat{a}_2 & \hat{a}_3 \\ \hline \hat{n}_1 & c_1 & -s_1 & 0 \\ \hat{n}_2 & s_1 & c_1 & 0 \\ \hat{n}_3 & 0 & 0 & 1 \end{array} \tag{1.5}$$

$$\begin{array}{c|ccc} {}^{A}R^{B} & \hat{b}_1 & \hat{b}_2 & \hat{b}_3 \\ \hline \hat{a}_1 & c_2 & -s_2 & 0 \\ \hat{a}_2 & s_2 & c_2 & 0 \\ \hat{a}_3 & 0 & 0 & 1 \end{array} \tag{1.6}$$

$$\begin{array}{c|ccc} {}^{N}R^{B} & \hat{b}_1 & \hat{b}_2 & \hat{b}_3 \\ \hline \hat{n}_1 & c_1c_2 - s_1s_2 & -c_1s_2 - s_1c_2 & 0 \\ \hat{n}_2 & s_1c_2 + c_1s_2 & -s_1s_2 + c_1c_2 & 0 \\ \hat{n}_3 & 0 & 0 & 1 \end{array} \tag{1.7}$$

where the abbreviations $c_i \equiv \cos(q_i)$ and $s_i \equiv \sin(q_i)$ for $i = 1, 2$. The component vectors of reference frame B, say, are represented by three mutually orthogonal unit vectors (indicated by the ˆ overstrike) with the corresponding lower case letter, $\hat{b}_1, \hat{b}_2, \hat{b}_3$. Simply stated, one can express any vector in terms of the component vectors of N, A, or B using these tables. For example, the unit vector $\hat{n}_2$ may be equivalently expressed in terms of components in the B reference frame as,

$$\hat{n}_2 = (s_1c_2 + c_1s_2)\,\hat{b}_1 + (-s_1s_2 + c_1c_2)\,\hat{b}_2 \tag{1.8}$$

by simply reading across the $\hat{n}_2$ row of the table in Equation 1.7. The third table is formed directly from the previous tables by substitution, for instance,

$$\hat{n}_1 = c_1(\hat{a}_1) - s_1(\hat{a}_2) \tag{1.9}$$

$$= c_1\left(c_2\hat{b}_1 - s_2\hat{b}_2\right) - s_1\left(s_2\hat{b}_1 + c_2\hat{b}_2\right) \tag{1.10}$$

$$= (c_1c_2 - s_1s_2)\,\hat{b}_1 - (c_1s_2 + s_1c_2)\,\hat{b}_2\,. \tag{1.11}$$

Direction cosine tables also provide the key to working simultaneously in multiple reference frames.[5] The elements in a direction cosine table contains the

[5] There is no need to continually transform position vectors, velocities, and accelerations into a common reference frame. Biomechanists who insist on converting everything to Cartesian coordinates, for instance,

dot products between unit vectors of different coordinate reference frames. For instance, the dot product of $\hat{n}_2$ and $\hat{b}_1$ can be obtained from Equation 1.8 above, but is more easily and directly read from the table contained in Equation 1.7. One needs only to find the element in the $\hat{n}_2$ row and the $\hat{b}_1$ column and read off the result,

$$\hat{n}_2 \cdot \hat{b}_1 = s_1 c_2 + c_1 s_2 \,. \tag{1.12}$$

Kinematic analyses of all kinds will benefit from direction cosine tables. The tables are particularly well suited for biomechanical structures because they enable points and vectors to be tracked in the coordinate systems (defined by the bones) in which they are most easily defined. Kane's Method is greatly facilitated via direction cosine tables because of the regular need to compute dot products of vectors in different reference frames.

Using the methodologies explained in Chapter 3, the angular velocities of bodies A and B with respect to reference frame N are determined to be,

$$^{N}\vec{\omega}^{A} = \dot{q}_1 \hat{a}_3 \tag{1.13}$$
$$^{N}\vec{\omega}^{B} = (\dot{q}_1 + \dot{q}_2)\hat{b}_3 \,. \tag{1.14}$$

Likewise, the angular accelerations are,

$$^{N}\vec{\alpha}^{A} = \ddot{q}_1 \hat{a}_3 \tag{1.15}$$
$$^{N}\vec{\alpha}^{B} = (\ddot{q}_1 + \ddot{q}_2)\hat{b}_3 \,. \tag{1.16}$$

The astute reader may notice the angular velocity and acceleration associated with the A body are expressed with an $\hat{a}_3$ unit vector, while those quantities associated with the B body are given $\hat{b}_3$ vectors. This is done to facilitate the derivations of the linear velocities and accelerations which require vector cross products to be performed. While dot products of vectors in different reference frames can be found easily from direction cosine tables, unfortunately cross products require vectors to be expressed in a common reference frame. Since position vectors defined within bodies A and B are most conveniently expressed in reference frames A and B, it is advantageous to also express the angular velocities and accelerations in the same reference frame.

The kinematical equations are complete when the velocities of all points with forces acting through them are known, and all accelerations of mass centers are known. The velocity of point A_o in reference frame N is denoted as $^{N}\vec{v}^{A_o}$, which is of course, zero since it is the location of a point pinned rigidly to reference frame N. The velocities of points A^*, B_o, B^*, and C_o are found to

will be limited to performing planar and simple 3-D analyses simply because the mathematical expressions become unwieldy as the degrees of freedom increase.

be,

$$ {}^{N}\vec{v}^{A^*} = \rho_A \dot{q}_1 \hat{a}_2 \tag{1.17} $$
$$ {}^{N}\vec{v}^{B_o} = \ell_A \dot{q}_1 \hat{a}_2 \tag{1.18} $$
$$ {}^{N}\vec{v}^{B^*} = \ell_A \dot{q}_1 \hat{a}_2 + \rho_B \left(\dot{q}_1 + \dot{q}_2\right) \hat{b}_2 \tag{1.19} $$
$$ {}^{N}\vec{v}^{C_o} = \ell_A \dot{q}_1 \hat{a}_2 + \ell_B \left(\dot{q}_1 + \dot{q}_2\right) \hat{b}_2 \,. \tag{1.20} $$

Deriving dynamic equations often appears to be difficult because finding expressions for the accelerations of the mass centers can be arduous. However, in a "kinematic chain" of linked rigid bodies, a vectorized method of deriving the accelerations can be done using a step by step process that uses vector cross products instead of differential calculus. Vector cross products require the analyst to copy text strings and then to simplify their products using nothing more than algebra. After only a few lines of computation the accelerations ${}^{N}\vec{a}^{A*}$ and ${}^{N}\vec{a}^{B*}$ of mass locations A^* and B^*, with respect to reference frame N, can be derived as,

$$ {}^{N}\vec{a}^{A^*} = -\rho_A \dot{q}_1^2 \hat{a}_1 + \rho_A \ddot{q}_1 \hat{a}_2 \tag{1.21} $$

$$ {}^{N}\vec{a}^{B^*} = -\ell_A \dot{q}_1^2 \hat{a}_1 + \ell_A \ddot{q}_1 \hat{a}_2 - \rho_B \left(\dot{q}_1 + \dot{q}_2\right)^2 \hat{b}_1 + \rho_B \left(\ddot{q}_1 + \ddot{q}_2\right) \hat{b}_2 \,. \tag{1.22} $$

Now that the kinematics is complete, a comparison of the two methods can proceed. Kane's Method is presented first, as it refers to Figure 1.1.

1.3.1.1 KANE'S METHOD

In Kane's approach, quantities called *partial angular velocities* and *partial velocity vectors* must be generated directly from angular velocity expressions and velocity expressions. The partial angular velocities are important quantities to define for bodies that rotate in response to applied torques. Likewise, the partial velocities are important, and come from the velocities of points which change in response to forces acting through them. The "partial" quantities are picked out of the expressions of a body's angular velocity or a point's velocity. Doing so is easy to learn, and is much like learning how to pick cherries from a cherry tree. All one needs to do to become a successful cherry picker is to learn to differentiate the cherries from the leaves and branches!

As an example, the angular velocity of body A with respect to reference frame N has already been given in Equation 1.14, and can be factored into the following form,

$$ {}^{N}\vec{\omega}^{A} = \left(\hat{a}_3\right) u_1 + (\vec{0}) u_2 \,, \tag{1.23} $$

where the quantities $u_i \equiv \dot{q}_i$, $(i = 1,\, 2)$. u_1 is known as *the first generalized speed* of the system, and u_2 is called the *second generalized speed.* Most of the time, n generalized speeds will be defined for a system having n degrees of freedom. The *first partial angular velocity* of body A in reference frame N (the "cherry") is the term in parentheses that multiplies u_1 in the expression for ${}^N\vec{\omega}^A$, and is denoted as ${}^N\vec{\omega}_1^A$,

$$ {}^N\vec{\omega}_1^A = \hat{a}_3 \,. \tag{1.24} $$

The *second partial angular velocity* of body A in reference frame N is denoted as ${}^N\vec{\omega}_2^A$, which is,

$$ {}^N\vec{\omega}_2^A = \vec{0} \,, \tag{1.25} $$

because the second generalized speed u_2 is multiplied by the zero vector.[6] For body B, the first and second partial angular velocities turn out to be equal in this system,

$$ {}^N\vec{\omega}_1^B = {}^N\vec{\omega}_2^B = \hat{b}_3 \,. \tag{1.26} $$

The partial velocity vectors ${}^N\vec{v}_r^{A^*}$, ${}^N\vec{v}_r^{B^*}$, and ${}^N\vec{v}_r^{C_o}$ for $r = 1, 2$ are determined just as easily as the partial angular velocities. For instance, ${}^N\vec{v}_r^{B^*}$ from Equation 1.19 can be factored into a form which emphasizes the generalized speeds,

$$ {}^N\vec{v}^{B^*} = \left(\ell_A \hat{a}_2 + \rho_B \hat{b}_2\right) u_1 + \left(\rho_B \hat{b}_2\right) u_2 \,. \tag{1.27} $$

From this equation the partial velocities are again the terms in parentheses that multiply the generalized speeds,

$$ {}^N\vec{v}_1^{B^*} = \ell_A \hat{a}_2 + \rho_B \hat{b}_2 \,; \qquad {}^N\vec{v}_2^{B^*} = \rho_B \hat{b}_2 \,. \tag{1.28} $$

Similarly, the partial velocities for the other mass center A^* and for the endpoint of the linkage C_o are found,

$$ {}^N\vec{v}_1^{A^*} = \rho_A \hat{a}_2 \,; \qquad {}^N\vec{v}_2^{A^*} = 0 \,; \tag{1.29} $$

$$ {}^N\vec{v}_1^{C_o} = \ell_A \hat{a}_2 + \ell_B \hat{b}_2 \,; \qquad {}^N\vec{v}_2^{C_o} = \ell_B \hat{b}_2 \,. \tag{1.30} $$

It turns out that in using Kane's Method there is no need to compute partial velocities at the joints (points A_o and B_o) because the joint interaction forces are noncontributory.

What exactly are the partial angular velocities and partial velocities? For now, it is sufficient to say that the partial velocities are vector quantities extracted from the angular velocity and velocity expressions. In very simple systems, they are often parallel to the velocities but of different magnitudes. Beyond this, they have no physical meaning, and are simply very convenient quantities to pluck from the kinematic equations.

[6] Usually, the vector notation for the zero vector is replaced by a simple scalar 0 to simplify writing. But here Equation 1.25 is written as a vector equation to emphasize the vector nature of partial angular velocities and partial velocities.

Generalized active forces are then formulated for the system A and B considered together as a whole. To form them, vector dot products between the partial velocities of points and the forces acting at those points are computed and added together. Additionally, dot products between partial angular velocities and torques are added together and summed together with the previous result. For instance, the *first and second generalized active forces* are formulated by performing the following dot products,

$$\begin{aligned} F_1 = & \left({}^N\vec{v}_1^{A^*} \cdot -m_A g\hat{n}_2\right) + \left({}^N\vec{v}_1^{B^*} \cdot -m_B g\hat{n}_2\right) \\ & + \left({}^N\vec{v}_1^{C_o} \cdot \vec{F}\right) \\ & + \left({}^N\vec{\omega}_1^A \cdot (\vec{\tau}_{N/A} - \vec{\tau}_{A/B})\right) + \left({}^N\vec{\omega}_1^B \cdot \vec{\tau}_{A/B}\right) ; \end{aligned} \tag{1.31}$$

$$\begin{aligned} F_2 = & \left({}^N\vec{v}_2^{A^*} \cdot -m_A g\hat{n}_2\right) + \left({}^N\vec{v}_2^{B^*} \cdot -m_B g\hat{n}_2\right) \\ & + \left({}^N\vec{v}_2^{C_o} \cdot \vec{F}\right) \\ & + \left({}^N\vec{\omega}_2^A \cdot (\vec{\tau}_{N/A} - \vec{\tau}_{A/B})\right) + \left({}^N\vec{\omega}_2^B \cdot \vec{\tau}_{A/B}\right) . \end{aligned} \tag{1.32}$$

The first line of Equation 1.31 contains the dot products between the first partial velocities of A^* and B^* with the gravitational forces acting at those points. The second line is the dot product of the endpoint force $\vec{F}$ with the first partial velocity of C_o in reference frame N. The last line contains the dot products of the first partial angular velocities and the total torques acting on segments A and B. Note that body A has torque $\vec{\tau}_{N/A}$ exerted by N on A, and another torque $-\vec{\tau}_{A/B}$ acting at its distal end. $-\vec{\tau}_{A/B}$ is nothing more than an *equal and opposite* torque to $\vec{\tau}_{A/B}$, the torque exerted by A on body B. Equation 1.32 contains the same forces and torques, but the *second* partial velocities and partial angular velocities are used instead of the first.

One of the main features to recognize at this point is that Kane's Method does not typically require dot products to be formulated between partial velocities and joint reaction forces. Because the systems A and B are considered together as a whole, it turns out that forces having equal and opposite counterparts within the system cancel during the summation of dot products. Also, many contact forces with bodies outside the system are eliminated if the points of contact have either zero velocity or velocities perpendicular to the contact force. The act of taking dot products between the partial velocities and the forces guarantees that any perpendicular combination of force and velocity will be *noncontributing* and need not be expressed within the dynamic equations of motion.

After computing the first and second generalized active forces, the *generalized inertia forces* are calculated next. These are composed of the dot products between the partial velocities of the mass centers and the inertial forces there, as well as the dot products between the partial angular velocities and the inertial torques. For example, the *first and second generalized inertia forces* are formulated as,

$$\begin{aligned} F_1^* &= \left({}^N\vec{v}_1^{A^*} \cdot (-m_A\,{}^N\vec{a}^{A^*})\right) + \left({}^N\vec{v}_1^{B^*} \cdot (-m_B\,{}^N\vec{a}^{B^*})\right) \\ &\quad + \left({}^N\vec{\omega}_1^{A} \cdot (-I_{A^*}\,{}^N\vec{\alpha}^{A})\right) + \left({}^N\vec{\omega}_1^{B} \cdot (-I_{B^*}\,{}^N\vec{\alpha}^{B})\right)\,; \end{aligned} \tag{1.33}$$

$$\begin{aligned} F_2^* &= \left({}^N\vec{v}_2^{A^*} \cdot (-m_A\,{}^N\vec{a}^{A^*})\right) + \left({}^N\vec{v}_2^{B^*} \cdot (-m_B\,{}^N\vec{a}^{B^*})\right) \\ &\quad + \left({}^N\vec{\omega}_2^{A} \cdot (-I_{A^*}\,{}^N\vec{\alpha}^{A})\right) + \left({}^N\vec{\omega}_2^{B} \cdot (-I_{B^*}\,{}^N\vec{\alpha}^{B})\right)\,. \end{aligned} \tag{1.34}$$

Once the generalized active forces and the generalized inertia forces are computed, the dynamic equations of motion are obtained simply by adding them together. If there are two degrees of freedom as in this example, then there will be two dynamic equations because two and only two variables require their accelerations to be predicted. The equations are,

$$F_1 + F_1^* = 0\,; \tag{1.35}$$

$$F_2 + F_2^* = 0\,. \tag{1.36}$$

These coupled equations describe the relationships between q_1, q_2, $\dot{q}_1$, $\dot{q}_2$, $\ddot{q}_1$, and $\ddot{q}_2$. Typically, they are solved simultaneously to obtain the two accelerations as functions of the angular positions and angular speeds.

To summarize this section, Kane's Method views the system together as a whole, creates auxiliary quantities called partial angular velocities and partial velocities, and uses them to form dot products with the forces and torques acting from external and inertial forces. The dot products form quantities called the generalized active forces and the generalized inertia forces, which are *simplified forms of the forces and moments used to write the dynamic equations of motion.* It is in forming these dot products that the main advantage of Kane's Method becomes apparent, because only the forces and torques that actually create motion changes will survive.

1.3.1.2 NEWTON-EULER METHOD

Free body diagrams are the keys to the Newton-Euler method. A dynamic system is broken up into isolated components which communicate with each other via interfacing forces and torques. Figure 1.2 contains the definitions and free body diagrams of links A and B.

The first step after making the free body diagram is to solve for the joint interaction force $\vec{F}_{A/B}$. If there were more than $n = 2$ joints as in this simple example, then expressions for $n-1$ joint reaction forces need to be determined.

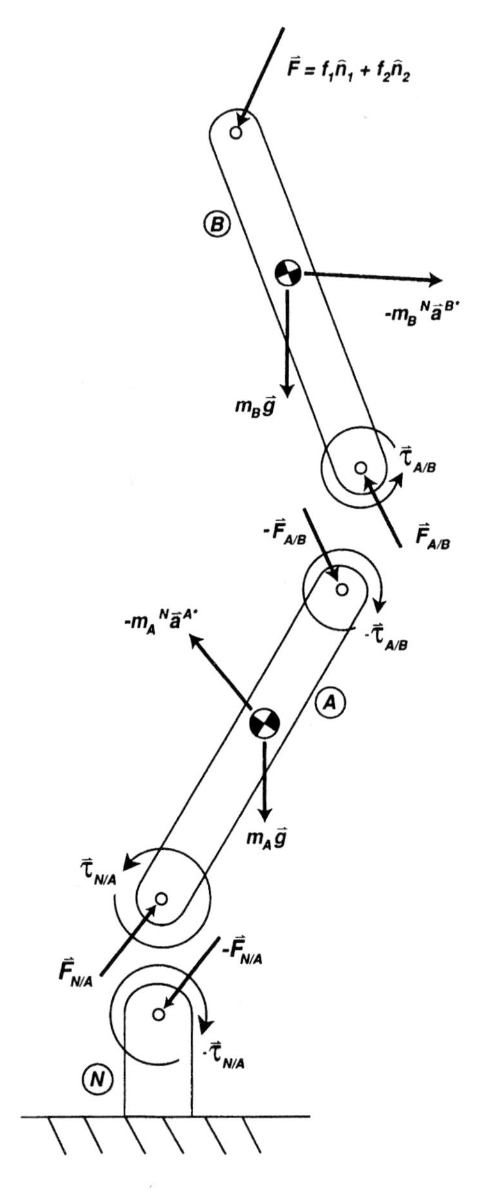

Figure 1.2. Free body diagram of the model depicted in Figure 1.1. Rigid bodies A and B are now drawn separately, with forces $\vec{F}_{A/B}$ and $\vec{\tau}_{A/B}$ depicting the joint interaction force and torque exerted by body A on body B. Likewise, $\vec{F}_{N/A}$ and $\vec{\tau}_{N/A}$ are the force and torque exerted by the ground frame N on body A. The directions of these forces and torques are unimportant, as they depend upon the variables of the system. Inertial forces $-m_A g\hat{n}_2$ and $-m_B g\hat{n}_2$ are also drawn as acting upon the centers of mass according to D'Alembert's Principle. All other definitions are consistent with Figure 1.1.

This is begun by summing together *all* forces acting upon the outermost link B,

$$\vec{F}_B + \vec{F}^{B^*} = 0\,. \tag{1.37}$$

$\vec{F}^{B^*}$ is the inertial force which resists the acceleration, and must be summed together with the endpoint force $\vec{F}$ and the joint reaction force $\vec{F}_{A/B}$ according to *D'Alembert's Principle* contained in Equation 1.2. Equation 1.37 becomes

$$\vec{F} + \vec{F}_{A/B} - m_B g\hat{n}_2 - m_B\,{}^N\vec{a}^{B^*} = 0\,, \tag{1.38}$$

which allows the unknown joint reaction force to be solved,

$$\vec{F}_{A/B} = -\vec{F} + m_B g\hat{n}_2 + m_B\,{}^N\vec{a}^{B^*}\,. \tag{1.39}$$

If there were more than two links in the system, one would proceed inward using *D'Alembert's Principle* to compute the rest of the joint reaction forces. For the next link, the joint reaction force $-\vec{F}_{A/B}$ would be summed together with the other forces acting upon it. This force is equal and opposite to the force acting on the outermost link because it is a reactive force. The force between the ground and the first link (here, $\vec{F}_{N/A}$) does not need to be determined because the moment of the ground reaction force can always be computed about the point of ground contact (here, point A_o) and thus be made zero.

Once the joint reaction forces are determined, the analysis proceeds from the ground outward. Moments about the proximal (inward) ends are summed together with torques acting upon each free body of the system. For body A, torque $\vec{\tau}_{N/A}$ acts at its lower end and torque $-\vec{\tau}_{A/B}$ acts at its upper end. The moments are cross products of position vectors and forces. For body A, the moments are computed from a common point, A_o and summed together with the torques,

$$\vec{M}_{A_o} = I_A\,{}^N\vec{\alpha}^A\,, \tag{1.40}$$

or,

$$\vec{\tau}_{N/A} - \vec{\tau}_{A/B} + \left(\vec{p}^{A_oA^*} \times m_A(-g\hat{n}_2 - {}^N\vec{a}^{A^*})\right) + \left(\vec{p}^{A_oB_o} \times (-\vec{F}_{A/B})\right) = I_A\,{}^N\vec{\alpha}^A\,. \tag{1.41}$$

Equation 1.40 is the first dynamic equation, as ${}^N\vec{\alpha}^A$ contains the second time derivative of angle q_1.

The second dynamic equation is determined in the same manner, this time summing moments of forces acting on body B about point B_o,

$$\vec{M}_{B_o} = I_B\,{}^N\vec{\alpha}^B\,; \tag{1.42}$$

$$\vec{\tau}_{A/B} + \left(\vec{p}^{B_oB^*} \times m_B(-g\vec{n}_2 - {}^N\vec{a}^{B^*})\right) + \left(\vec{p}^{B_oC_o} \times \vec{F}\right) = I_B\,{}^N\vec{\alpha}^B\,. \tag{1.43}$$

In contrast to Kane's Method, the components of the system are isolated, and the forces of interaction between them are computed, usually working inward from the outermost link. Once completed, the analysis proceeds in the opposite direction (outward), summing moments and torques together to create the dynamic equations. One should not underestimate the complexity of having to first compute expressions for the joint interaction forces, and second to compute moments of the joint reactions. Moment calculations require the vectors being crossed to be expressed in the same reference frame. Therefore, it becomes an arduous task to continually have to convert the joint reaction forces into the same reference frames as the position vectors. Kane's Method is much easier because the joint reactions are eliminated altogether. If one wishes to *solve for* the joint reaction forces, it is an easy matter to introduce them. When three dimensional analyses are performed, there simply is no easier nor better way than Kane's Method!

1.4 DYNAMIC MUSCULOSKELETAL MODELS AND MODELING

Musculoskeletal models can be made to represent virtually every portion of the body having skeletal segments and muscles to actuate them. They can be formulated to explore primarily kinematic (referring to motions) or kinetic (referring to forces) problems in nature. Furthermore, kinetic models can be static (no accelerations) or dynamic (having nonzero accelerations) . The kinetic models solve for the forces and moments in evidence at different parts of the structure in situations that are both static and dynamic, because statics is simply a special case of dynamics. Kinematic models move also, but not in response to forces and moments. Hence kinematic models are typically used to define all the motion possibilties. For example, in human factors research, kinematic models are often used to define ranges of motion, operator workspaces, comfortable operating positions and postures, *etc.*, for specific population groups.

The appeal of a dynamic musculoskeletal model is enhanced by its utility for both static and dynamic situations. These models are responsive to forces and moments from internal (*e.g.,* musculotendon forces, joint reaction forces, *etc.*) and external sources (*e.g.,* ground reaction forces, gravitational forces, *etc.*). Therefore dynamic models can be used to predict which motion among the infinite number of possible motions will actually occur. As later explained in Chapter 7, the dynamic equations which define the mathematical relations between the forces, moments, and motions can be exploited to yield either predicted motions (via a discrete time process called forward dynamic simulation) or internal forces and moments (via an instantaneous process called inverse dynamics). Dynamic models can also be used with optimal control methods and an optimal criterion or strategy to predict the optimal forces and moments needed to create a specific movement. For example, an optimal criterion function might minimize total muscle power while still achieving the desired motions of the limbs. Though this application is far from fully developed, it affords musculoskeletal modelers a non-invasive means of exploring the movement control strategies utilized by the central nervous system to coordinate multijoint movements.

1.5 EXERCISES

1. The dynamic equations for the two link planar linkage determined by the Newton-Euler method were given symbolically in Equations 1.40 and 1.42. Complete the derivations of the two dynamic equations of motion. Use the following expressions for the *joint* torques,

$$\tau_{N/A} = T_1 \hat{a}_3 \tag{1.44}$$

$$\tau_{A/B} = T_2 \hat{b}_3 \,. \tag{1.45}$$

   The answers are given below.

$$\begin{aligned}\Big(I_A + \rho_A^2 m_A + \ell_A^2 m_B + \ell_A \rho_B m_B c_2\Big)\, \ddot{q}_1 + \Big(\ell_A \rho_B m_B c_2\Big) \ddot{q}_2 \\ = \Big(T_1 - T_2\Big) - g\Big(\rho_A m_A c_1 + \ell_A m_B c_1\Big) \\ + \ell_A \rho_B m_B s_2 \Big(\dot{q}_1 + \dot{q}_2\Big)^2 + \ell_A \Big(f_2 c_1 - f_1 s_1\Big)\end{aligned} \tag{1.46}$$

$$\begin{aligned}\Big(I_B + \rho_B^2 m_B + \ell_A \rho_B m_B c_2\Big)\, \ddot{q}_1 + \Big(I_B + \rho_B^2 m_B\Big)\, \ddot{q}_2 \\ = T_2 + g \rho_B m_B \Big(s_1 s_2 - c_1 c_2\Big) - \Big(\ell_A \rho_B m_B s_2\Big) \dot{q}_1^2 \\ + \ell_B \Big(- f_1 (c_1 s_2 + s_1 c_2) + f_2 (-s_1 s_2 + c_1 c_2)\Big)\end{aligned} \tag{1.47}$$

2. In Figure 1.2, the free body diagram for the two link planar linkage is shown. The angle of the second link, q_2, is defined as a *joint angle* between the $\hat{a}_1$ and $\hat{b}_1$ axes, as shown in Figure 1.1. If, instead, angle q_2 was defined as a *segmental angle* between the $\hat{n}_1$ and $\hat{b}_1$ axes, the dynamic equations of motion will change.

   Using the Newton-Euler approach, and using the same torque expressions given in Problem 1, find the dynamic equations for the two link planar linkage when segmental angles are employed for segments A and B. The answers are given on the next page.

$$\begin{aligned}\Big(I_A + \rho_A^2 m_A + \ell_A^2 m_B\Big)\,\ddot{q}_1 + \Big(\ell_A \rho_B m_B c_{21}\Big)\ddot{q}_2 \\ = \Big(T_1 - T_2\Big) - g\Big(\ell_A m_B + \rho_A m_A\Big)c_1 \\ +\Big(\ell_A \rho_B m_B s_{21}\Big)\dot{q}_2{}^2 + \ell_A\Big(f_2 c_1 - f_1 s_1\Big)\end{aligned} \tag{1.48}$$

$$\begin{aligned}\left(\ell_A \rho_B m_B c_{21}\right)\ddot{q}_1 + \Big(I_B + \rho_B^2 m_B\Big)\,\ddot{q}_2 \\ = T_2 - g\rho_B m_B c_2 - \Big(\ell_A \rho_B m_B s_{21}\Big)\dot{q}_1^2 \\ +\ell_B\Big(f_2 c_2 - f_1 s_2)\Big)\end{aligned} \tag{1.49}$$

where $c_{21} = cos(q_2 - q_1)$ and $s_{21} = sin(q_2 - q_1)$.

References

Allard, Paul, Stokes, Ian A. F., and Blanchi, Jean-Pierre (1995) *Three-Dimensional Analysis of Human Movement*. Human Kinetics, Champaign, IL.

Andrews, James G. (1995) "Euler's and Lagrange's Equations for linked Rigid-Body models of Three-Dimensional Human Motion." Chapter 8 in *Three-Dimensional Analysis of Human Movement*, Allard, Paul, Stokes, Ian A. F., and Blanchi, Jean-Pierre (eds.) Human Kinetics, Champaign, IL.

Hay, J. G. (1993) *The Biomechanics of Sports Techniques*. Prentice-Hall, Englewood Cliffs, NJ.

Kane, T. R., and Levinson, D. A. (1985) *Dynamics: Theory and Applications*. McGraw-Hill, New York, NY.

Meriam, J. L., and L. G. Kraige (1997) *Engineering Mechanics Volume 2 – Dynamics*. John Wiley & Sons, New York, NY.

Özkaya, N., and Nordin, M. (1999) *Fundamentals of Biomechanics*. Springer-Verlag, New York, NY.

Radetsky, Peter. (1986) "The man who mastered motion," *Science 86*, V. 7, n. 4, pp. 52-60.

Tözeren, Aydin (2000) *Human Body Dynamics*. Springer-Verlag, New York, NY.

Winter, D. A. (1990) *Biomechanics and Motor Control of Human Movement*. John Wiley & Sons, New York, NY.

Winters, J. M. and Crago, P. E., eds. (2000) *Biomechanics and Neural Control of Posture and Movement.* Springer-Verlag, New York, NY.

Winters, J. M., and Woo, S. L.-Y., eds. (1990) *Multiple Muscle Systems: Biomechanics and Movement Organization*. Springer-Verlag, New York.

Chapter 2

AN INTRODUCTION TO MODELING MUSCLE AND TENDON

Objective – To apply mathematically predictable driving forces to the segments of the body using lumped-parameter models of the muscles and tendons.

The musculoskeletal system is one of the most amazing works of physiological machinery ever studied. Simple acts like standing and walking are such commonplace activities that we rarely contemplate the wonders of producing coordinated movements. And, when one considers truly extraordinary movements – such as leaping into the air to catch a ball thrown at high velocity and landing in perfect balance – it is truly awe inspiring to imagine all of the physiological processes and electrochemical events going on beneath the skin. Even a cursory study of this topic, called *motor control*, requires some knowledge of nerve and muscle physiology, the central nervous system and the senses, control theory, and psychology – enough material to fill an entire book on musculoskeletal systems. In a single chapter, the reader can only be introduced to the topic, and only in a very limited way. The goal of this chapter is rather modest – to give enough information to the reader so simple musculotendon models can be made and used to create the moments which in turn create joint movements. In this chapter, the reader is introduced to the subject by describing only the macroscopic force producing properties of skeletal muscle. Because biomechanics is the study of forces and their actions on and within living organisms, this chapter is devoted to the biomechanics of skeletal muscle and tendons.

The reader will be introduced to this topic by a quest. If a model of skeletal muscle was desired, how would one make it? The first step would be to determine the overall properties of actual muscles. Once we develop an understanding of these properties, we should be able to construct a model to predict the force that a muscle could produce under a given set of conditions. Idealized representations of actual mechanical devices such as springs, dashpots, and force generators could then be used to construct a mathematical model of muscle and tendon. If a quantitative model could be created, it could be used to analyze and better understand the innerworkings of the musculotendon actuator. Readers who develop an interest in the subject of musculotendon modeling, and who wish to understand this topic more thoroughly are directed to the research literature and additional coursework. At a minimum, the muscle modeling techniques introduced here should give the reader a glimpse of the many ways in which muscle models can be used together with dynamic skeletal models to numerically analyze and simulate movements.

There are virtually unlimited applications, because quantitative models incorporating the salient features of the skeletal system and the musculotendon actuators can be used to predict the interactions of man and machines *even if those machines cannot yet be built.* For instance, a mechanical designer can learn much about the way a proposed new tool or product will interact with the human body given a certain set of mechanical properties. A rehabilitation engineer might study the limitations of applying a new technology proposed in healthcare *before* the risks are fully known and human subjects are recruited.

So much for philosophy! Let's bring this discussion back down to earth and see how this topic fits into the progression of this book. In the following chapter, the topics of rigid bodies and reference frames are introduced. Because the emphasis of this book is upon learning how to analyze and simulate skeletal motions in three dimensions, it might be easy to miss the fact that the muscles, tendons, and ligaments are the primary structures within the body that actually create these motions. At the same time, it is the motions of the rigid bodies within – the skeletal segments – which determine the lengths and velocities of the musculotendon actuators. These two scalar quantities either enhance or undermine the ability of the muscles to produce contractile force. It follows that the positions and velocities of the muscles' attachments to the bones, called *origin* and *insertion points*, must be precisely tracked. Therefore, the material contained in this chapter provides the need for the material in Chapter 3.

Chapter 4 describes the vector based kinematical methods which allow the motions of the rigid bodies to be precisely defined. In Chapter 4 are examples of muscle moments, and it is a prerequisite for the reader to have a good understanding of how the forces produced by the muscles vary in response to length and velocity. Chapter 5 discusses the modeling of joints, and how moments are created by muscles and ligaments spanning the joints. Deriving the dynamic equations of motion is the subject of Chapter 6, and an entire section is devoted to showing how torques are developed in the segments by single muscles that cross one, two, and three joints. Finally, Chapter 7 concerns the problem of distributing forces among a redundant set of muscles, and introduces the reader to methods of finding the set of muscle forces needed to drive the musculoskeletal models along a desired motion trajectory. In summary, the entire book relates in some way to the topic of muscle and a macroscopic understanding of how muscle tension is produced.

The mechanical properties of the tendons are also very important, since the production of force by muscle is critically dependent on the lengths of the muscle fibers, which in turn is influenced by the amount of stretch in the tendons. Even though tendons themselves are not very elastic on an absolute scale (tendons break in tension at about 10% strain as opposed to a synthetic fiber such as nylon which has greater than 50% strain at failure), many of the tendons in the distal extremities have lengths an order of magnitude longer than the muscle fibers. When a long tendon is stretched by the contraction of a muscle having short muscle fibers, the alteration in muscle force caused by the total tendon stretch can be appreciable. Including tendon as a separate mechanical element from the muscle model brings out the interplay between muscle and tendon. To develop these ideas further, the subject of muscle modeling is addressed first.

2.1 MACROSCOPIC PROPERTIES OF MUSCLE

Early ideas of the muscle mechanics go back to Erasistratus in the third century B.C., who thought the nerves were hollow, fluid-filled tubes through which an animal spirit could flow from the head to the muscles (McMahon, 1984). Inflated with this mysterious fluid-like spirit called *pneuma*, the muscles were envisioned to shrink lengthwise as they expanded in volume and breadth. Sim-

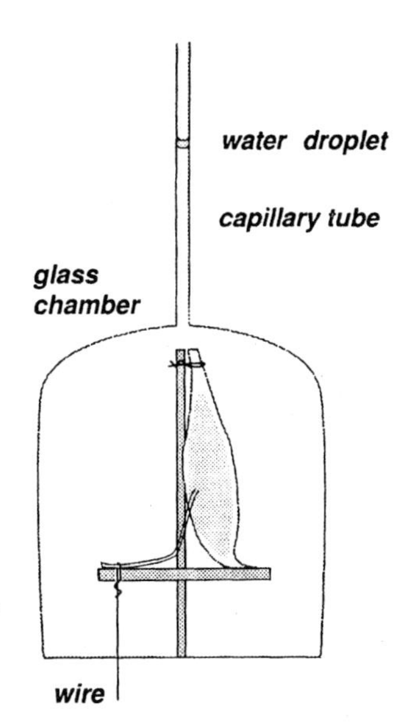

Figure 2.1. Schematic of the experimental apparatus used by Swammerdam in 1663 to prove that muscle contracts while maintaining a nearly constant volume.

ple observations must have tended to confirm this view, as muscles obviously increase in circumference as they contract.

This idea of muscle being a bag which was somehow inflated to cause simultaneous circumferential increase and lengthwise shortening persisted for centuries until an experiment was performed by Swammerdam in 1663. His experiment elegantly disproved this idea, by showing that muscle maintains a constant volume as it contracts (Figure 2.1). A frog's muscle was sealed within a glass chamber, which was airtight except for an open capillary tube extending upward. A fine wire was pulled, which pinched the nerve and caused the muscle to contract. Observation of a small droplet of water placed in the tube was indicated whether the volume of air and tissue inside the chamber increased, decreased, or remained the same. If the hypothesis of the day were true, the volume of the muscle would increase during a contraction, and the total volume of air and muscle would also increase leading to a rise in the position of the water droplet. When Swammerdam pulled a fine wire which pinched the nerve and caused the muscle to contract, the small droplet of water placed in the tube was observed to remain fixed in place or even to fall slightly, demonstrating that the total volume of the muscle remained relatively constant during contraction.

2.1.1 TWITCH, TETANUS

More recent experiments have further developed our current understanding of muscle twitch, tetanus, tension-length and force-velocity dependence (Hill, 1938). When an electrical stimulus of sufficient magnitude is applied to an isolated muscle preparation, the muscle membrane depolarizes and a short-duration contraction known as a twitch develops.[1] Figure 2.2 shows a common current pulse which is "charge balanced" so the total amount of positive and negative charges injected in the vicinity of the nerve are equal. It is thought that balancing these charges will minimize any electrochemical byproducts which might be harmful.

Two twitches separated by a suitably long time interval develop identical records of force development and relaxation. But, if the second twitch is initiated before the first twitch has completely decayed, a slightly higher peak force will be reached (Ritchie and Wilkie, 1955). This is called a "doublet." Presumably the higher level of force can be accounted for with superposition (Figure 2.3). This effect is magnified greatly when the muscles fibers pull

[1] The stimulus can also be applied to the nerve which innervates the muscle, or to the nerve-muscle junction. The resulting action potential propagates to the muscle via a fascinating electrochemical process involving neural transmitters, sodium and potassium channels, ionic currents and differences in electrical potentials. This lies outside of the scope of this book, but the reader is strongly urged to study the physiology of nerve and muscle.

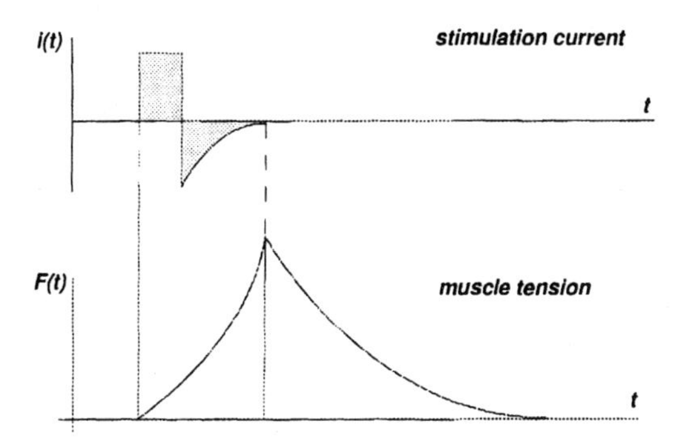

Figure 2.2. An idealized sketch of a balanced biphasic current pulse (top) and the resulting muscle tension developed by the twitch it produces (bottom).

against compliant, rather than unyielding, structures. Three twitches are called a "triplet," and develop slightly higher forces. The increase in force going from the doublet to the triplet appears to be smaller than the increase in force going from the single pulse to the doublet. The force multiplying effect thus becomes progressively smaller as the number of pulses increases. Applying a series of stimuli closely spaced in time produces a plateau of force exhibiting a ripple

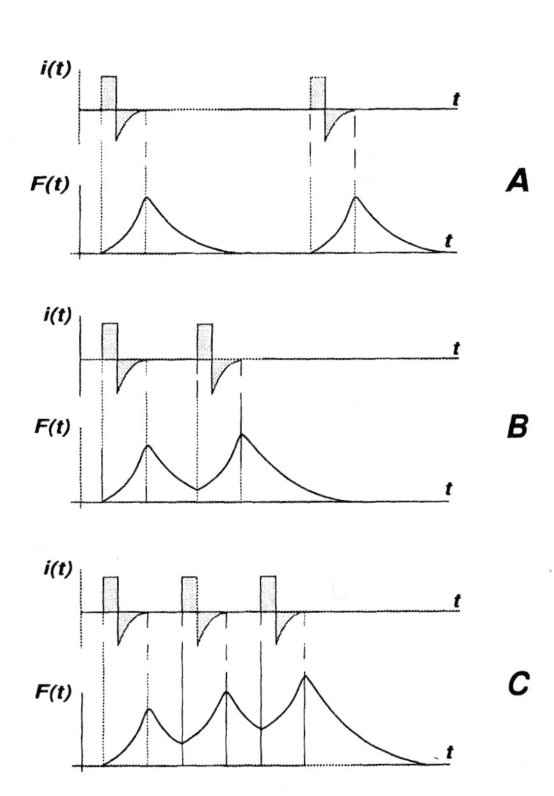

Figure 2.3. A. Two single twitches occur when the second stimulus is applied after the first twitch has died out. B. A doublet is obtained when the second stimulus is applied when there is a residual force left over from the first contraction. The force level attained in the second contraction exceeds that of the first. C. A triplet achieves a greater force than a doublet.

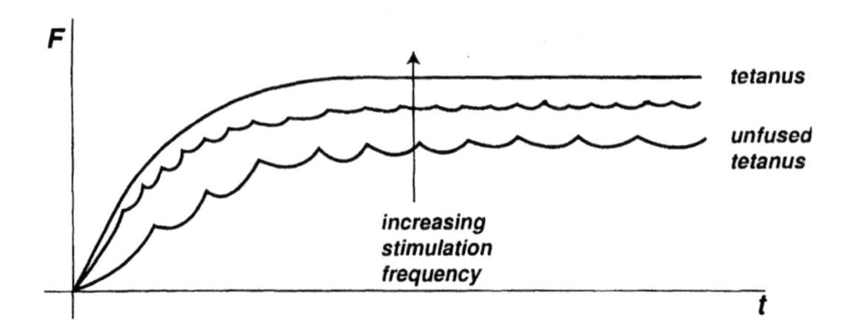

Figure 2.4. Unfused tetanus and tetanus.

at the frequency of stimulation (Figure 2.4). This is called an *unfused tetanus.* As the frequency increases, the plateau reaches greater levels of force while the magnitude of the ripple decreases. At the limit, force ceases to rise with further increases in stimulation frequency, and the ripple becomes very small. This limit is effectively reached at frequencies of 50 to 60 Hz in mammalian muscle at ambient body temperatures, and is referred to as *tetanus.*

2.1.2 RECRUITMENT AND ACTIVATION

Increasing the frequency of stimulation is not the only way to increase muscle force. A good rule of thumb is that only about half of muscle force development occurs by increasing the frequency of stimulation. Muscle also builds up force through a process called recruitment. One can think about individual muscle fibers like soldiers in an army. When a country goes to war, the first soldiers sent to the front lines are likely to be the career soldiers who volunteered for duty well before the war started. When additional soldiers are needed, new soldiers are recruited voluntarily, trained, and shipped to the front lines. If the war is going badly and still more are required, a draft is implemented to bring additional (non-voluntary) troops to battle.

Recruitment of muscle fibers happens in much the same way during voluntary contractions (Figure 2.5). Slow oxidative (SO) muscle fibers are like the career soldiers – they are used for almost every contraction because oxygen is taken directly from the bloodstream to fuel the contraction. Fast oxidative glycolytic (FOG) fibers are like the soldiers who volunteer at the start of the war – these add to the pool of active fibers when additional force or contractile speed is needed. And, fast glycolytic (FG) fibers are like the draftees – they are added to the pool when either high forces or high contraction velocities are required. The FG fibers are anaerobic, and use glycogen stored within the cells to drive the contractions. The FOG fibers represent a continuum of fiber types which exhibit some of the properties of both SO and FG fibers. It is also known that the SO fibers are somewhat smaller in diameter, and have smaller diameter nerves that innervate the muscle fibers. The FG fibers are the largest in diameter, and have the largest diameter nerves. Thus, the order of recruitment in voluntary contractions goes from slow oxidative to fast glycolytic, and in terms of size, from small diameter to large diameter. The latter principle has become known as the *size principle* (Henneman, 1957).

It is interesting to note that the high rate of fatigue in electrically stimulated muscle can be directly related to the size of the muscle and nerve fibers, which reach threshold depolarization levels at smaller applied currents than the smaller fibers. Hence, in functional neuromuscular stimulation (FNS), the highly fatigueable FG fibers are recruited first, and supplemented by FOG and SO fibers as the speed and force of contraction increase. This is a *reverse recruitment order* from the natural voluntary process. As recently as a decade ago, reverse recruitment was a discouraging factor for researchers who were trying to apply FNS to regain function in paralyzed limbs. Recent work on electrically stimulated conditioning protocols has suggested that the muscle fiber type population within a muscle can be changed by applying the right kind of stimulation during an exercise regime for several hours each day. For all practical purposes, it appears that the muscle fibers can adapt and change their properties in response to a changing demand.

The simplest model of activation $a(t)$ defines activation as the percentage of muscle fibers that are instantaneously active at a particular time t. Figure 2.6 shows schematically how both increasing the numbers of fibers actively contracting (recruitment) and increasing the stimulation rate of each active fiber (stimulation frequency) increases the activation when $a(t)$ is defined in this way. This definition will be used here with full knowledge that this definition is only used for mathematical convenience. It is hoped that continued development in this area will enable muscle models to better reflect the effects of recruitment, fiber type, and stimulation frequency.

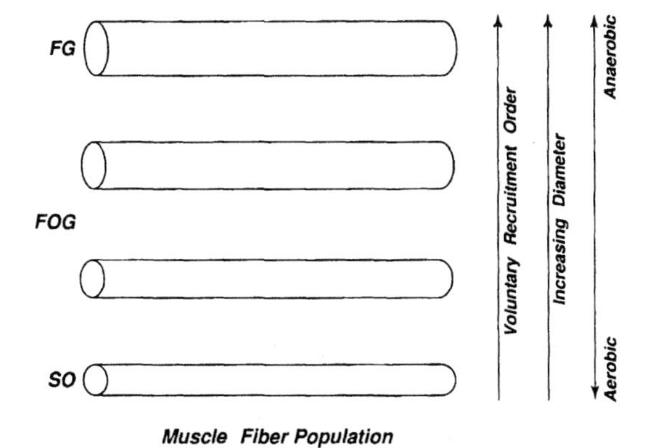

Figure 2.5. The normal process of recruiting muscle fibers in voluntary contractions starts from small (slow oxidative or SO), and proceeds to large (fast glycolytic or FG). Under conditions of artifically imposed electrical stimulation, the order of recruitment reverses, and goes from FG to SO.

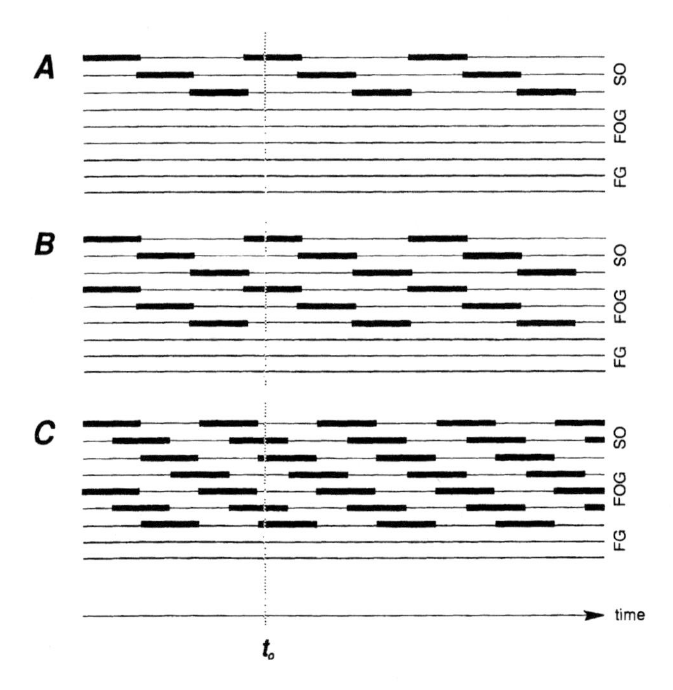

Figure 2.6. In the simplest model of activation, the percentage of active fibers increases with both recruitment and stimulation frequency.

2.1.3 HE-ZAJAC-LEVINE BILINEAR ACTIVATION MODEL

Let's assume we wish to approximate the "neural" excitation of the muscle fibers in a simple way, such that a control signal $c(t)(0 \leq c(t) \leq 1)$ is used as a step input to elicit an activation that rises in time as a first-order step response (Figure 2.7). We cannot pretend that the control $c(t)$ is anything but a mathematically convenient form of control signal, as it does not pretend to encode neural stimulus frequencies or anything remotely physiological.[2] Furthermore, we'll assume the muscle tension, length, and velocity do not affect the activation $a(t)$, even though, for example in FNS of muscle these parameters have been shown to affect the activation achieved.

[2]However, it is hoped here that a reasonable neuromuscular model will be developed in the future that will remedy these shortfalls.

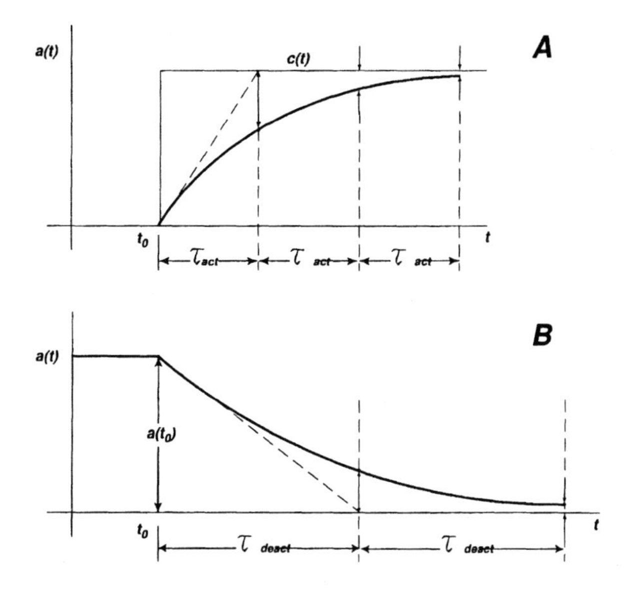

Figure 2.7. First order activation and deactivation model. A. A first order rise from zero activation approaches the control value $c(t)$. At $t = 0$, the initial slope of the activation curve is $c(t)/\tau_{act}$. After n time constants, $a(t)$ approaches $c(t)$ with an error of $c(t)/e^n$. B. A first order decay from activation $a(t)$. The time constant τ_{deact} is larger than the time constant for activation due to the slower rate of calcium diffusion out of the muscle fibers.

Then, a convenient first-order model for both activation and deactivation of muscle is given by He, Zajac, and Levine (1991) as,[3]

$$\dot{a}(t) + \frac{1}{\tau_{act}} \left[\beta + (1-\beta)c(t)\right] a(t) = \frac{1}{\tau_{act}} c(t) , \qquad (2.1)$$

where β is the ratio of the activation time constant τ_{act} to the deactivation time constant τ_{deact},

$$\beta \equiv \frac{\tau_{act}}{\tau_{deact}} , \qquad (2.2)$$

where $0 \leq \beta \leq 1$. Note that the term $\left[\beta + (1-\beta)c(t)\right] / \tau_{act}$ is a *rate constant* which depends upon the control input $c(t)$, and therefore increases with larger values of $c(t)$. This implies the activation dynamics is fastest when $c(t) = 1$ (rate constant $= 1/\tau_{act}$), and slowest at $c(t) = 0$ (rate constant $= \beta/\tau_{act} = 1/\tau_{deact}$). The reader is cautioned that the *He-Zajac-Levine* first order model is good for effecting a time delayed activation response, but that the activation approaches the control value only under the simplest conditions. As $t \to \infty$, $a(t) \to c(t)$ only for activations starting from the resting state ($a(t) = 0$) and rising to the specified value of $c(t)$, or for complete deactivations where $c(t)$ is set to zero.

[3]The model was developed by this team, but reported within a more comprehensive work by He, Levine, and Loeb (1991).

2.1.3.1 EXAMPLES

Case 1. A first order rise from an inactive state. Let $\beta = 0.5$, activation $a(t) = 0$ at time $t = 0$, and control $c(t) = 1$ at time $t = 0$. $a(t)$ should thus be a first order step response, with $a(t)$ approaching $c(t)$ as $t \to \infty$. With these values, Equation 2.1 becomes

$$\dot{a}(t) + \frac{1}{\tau_{act}} a(t) = \frac{1}{\tau_{act}}. \tag{2.3}$$

One approach to finding a solution is to guess the form of the solution,

$$a(t) = Ae^{\sigma t} + B, \tag{2.4}$$

and solve for the unknown coefficients A, B, and σ. This form is used because only an exponential function, when added to a multiple of its first derivative, can satisfy a first order differential equation in the form of Equation 2.3. $a(t)$ and

$$\dot{a}(t) = A\sigma e^{\sigma t} \tag{2.5}$$

are "plugged into" Equation 2.3 to yield

$$A\sigma e^{\sigma t} + \frac{1}{\tau_{act}} \left(Ae^{\sigma t} + B\right) = \frac{1}{\tau_{act}}, \tag{2.6}$$

which can be factored into form

$$\left(\sigma + \frac{1}{\tau_{act}}\right) Ae^{\sigma t} + \frac{B}{\tau_{act}} = \frac{1}{\tau_{act}}. \tag{2.7}$$

The first portion of the left hand side of Equation 2.7 is *time varying*, because it is multiplied by the term $e^{\sigma t}$. The rest of Equation 2.7 is constant in time. In order that this equation be satisfied *for all times t*, it is necessary for the time varying portion to be self canceling, or

$$\sigma = -\frac{1}{\tau_{act}}. \tag{2.8}$$

Also, the constant portion of Equation 2.7 is trivially satisfied when

$$\frac{B}{\tau_{act}} = \frac{1}{\tau_{act}} \tag{2.9}$$

or $B = 1$. At this point, we have two of the three unknown coefficients,

$$a(t) = Ae^{-t/\tau_{act}} + 1. \tag{2.10}$$

A is found from the initial condition $a(t) = 0$ at $t = 0$, which implies that

$$0 = A + 1 \tag{2.11}$$

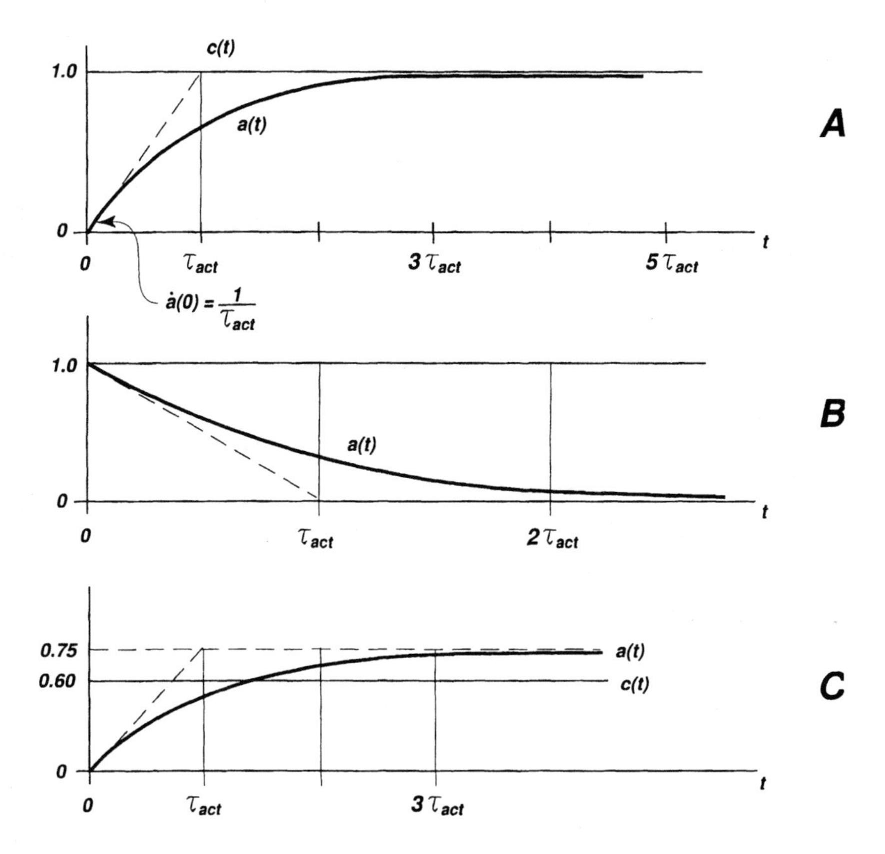

Figure 2.8. A. Numerical plot of activation for Case I, a first order rise from an inactive state to a control value of $c(t) = 1.0$, with time constant τ_{act}. B. Case II, a first order decay from an active state to an inactive state. Note that the deactivation follows time constant τ_{deact}, which is about twice as long as τ_{act}. C. Case III. A first order rise from an inactive state to an active state other than 1.0. Note that the $He-Zajac-Levine$ approximation overshoots the control value $c(t)$.

at $t = 0$. Hence,

$$a(t) = 1 - e^{-t/\tau_{act}} \tag{2.12}$$

which does indeed approach the value of $c(t) = 1$ when $t \to \infty$.

A plot of $c(t)$ and $a(t)$ is shown in Figure 2.8A. τ_{act} is called the *activation time constant* because, when t is a multiple of τ_{act} the exponent of e becomes a whole number. Typical values used for muscle activation time constants range from about $8\,ms$ to $12\,ms$, with deactivation time constants being roughly double the activation time constant. $a(t)$ will come to within $1/e$ of its final value in one time constant, within $1/e^3$ of its final value in three time constants, and within $1/e^5$ of its final value after five time constants as shown in the following equations:

$$a(\tau_{act}) = 1 - e^{-\tau_{act}/\tau_{act}} = 1 - \frac{1}{e} = 0.632\,; \tag{2.13}$$

$$a(3\tau_{act}) = 1 - e^{-3\tau_{act}/\tau_{act}} = 1 - \frac{1}{e^3} = 0.950\,; \tag{2.14}$$

$$a(5\tau_{act}) = 1 - e^{-5\tau_{act}/\tau_{act}} = 1 - \frac{1}{e^5} = 0.993\,. \tag{2.15}$$

The extension to other values of time t is obvious. These particular values of time were chosen to yield the well known rules of thumb that a first order step response achieves 63.2% of its magnitude in a time equal to one time constant, 95% of its magnitude in three time constants, and better than 99% of its magnitude in five time constants.

An aid to drawing the first order step response accurately is also shown in Figure 2.8A. The slope of the $a(t)$ curve at the initial time is obtained from Equation 2.5,

$$\dot{a}(0) = A\sigma = -\frac{A}{\tau_{act}}\,. \tag{2.16}$$

In our specific example, $A = -1$, so that the initial slope of the activation curve is

$$\dot{a}(0) = \frac{1}{\tau_{act}}\,. \tag{2.17}$$

It is easy to draw the line tangent to $\dot{a}(0)$, shown in the figure as a dashed line, because it is simply the diagonal line of positive slope joining the corners of a rectangular box of width τ_{act} (one time constant) and height 1. For cases other than our example, the height of the box is the value that $a(t)$ approaches as $t \to \infty$.

Case 2. A first order decay from an active state to an inactive state. Let $\beta = 0.5$, activation $a(t) = 1$ at time $t = 0$, and control $c(t) = 0$ at time $t = 0$. $a(t)$ should thus be a first order decay, with $a(t)$ approaching $c(t)$ as $t \to \infty$. In this case, the differential equation (Equation 2.1) becomes

$$\dot{a}(t) + \frac{1}{\tau_{deact}} a(t) = 0\,, \tag{2.18}$$

which has the solution

$$a(t) = e^{-t/\tau_{deact}}\,. \tag{2.19}$$

The result is plotted in Figure 2.8B.

Case 3. A first order rise from an inactive state to an active state other than 1.0. Let $\beta = 0.5$, activation $a(t) = 0$ at time $t = 0$, and control $c(t) = 0.6$ at time $t = 0$. One would expect that $a(t)$ should rise as a first order step response to the value 0.6. As shown in Figure 2.8C, it does not, and this illustrates one of the weaknesses of the *He-Zajac-Levine* model. The numerical solution is obtained by plugging the parameter values into Equation 2.1, which becomes,

$$\dot{a}(t) + \frac{1}{\tau_{act}}\left[0.5 + (1 - 0.5)\, 0.6\right] a(t) = \frac{1}{\tau_{act}}\,(0.6)\,; \tag{2.20}$$

$$\dot{a}(t) + \frac{0.8}{\tau_{act}}\, a(t) = \frac{0.6}{\tau_{act}}\,. \tag{2.21}$$

Following the same procedure as before, we guess the form of the solution and plug Equations 2.4 and 2.5 into Equation 2.21 and solve for the coefficients,

$$A\sigma e^{\sigma t} + \frac{0.8}{\tau_{act}}\left(Ae^{\sigma t} + B\right) = \frac{0.6}{\tau_{act}}\,; \tag{2.22}$$

$$\left(\sigma + \frac{0.8}{\tau_{act}}\right) Ae^{\sigma t} + \frac{(0.8)\, B}{\tau_{act}} = \frac{0.6}{\tau_{act}}\,. \tag{2.23}$$

Therefore,

$$B = \frac{0.6}{0.8} = 0.75\,, \tag{2.24}$$

$$\sigma = -\frac{0.8}{\tau_{act}}\,, \tag{2.25}$$

and

$$a(t) = Ae^{-0.8t/\tau_{act}} + 0.75\,. \tag{2.26}$$

Solving for A from the initial conditions yields

$$a(t) = 0.75\left(1 - e^{-0.8t/\tau_{act}}\right) \tag{2.27}$$

which approaches the value 0.75, not 0.60, as $t \to \infty$. A closer examination reveals that as $t \to \infty$, $a(t)$ approaches the value,

$$a(t) \to \frac{c(t)}{[\beta + (1 - \beta)c(t)]}\,, \tag{2.28}$$

with a modified time constant τ' of,

$$\tau' = \frac{\tau_{act}}{[\beta + (1 - \beta)\, c(t)]}\,. \tag{2.29}$$

τ' is the time required for $a(t)$ to rise from $a(t) = 0$ to $a(t) = 0.75(1 - 1/e)$, which is within $1/e$ of its final value of 0.75.

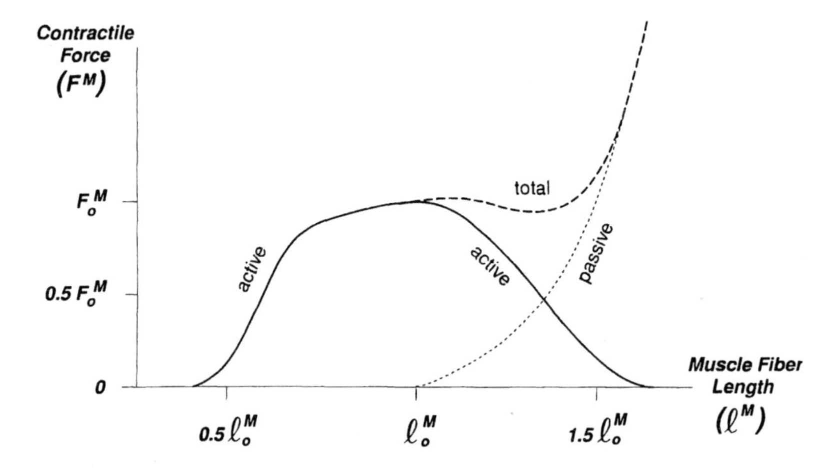

Figure 2.9. Isometric tension (F) produced by muscle fiber as a function of muscle fiber length (ℓ^M) and activation $a(t)$. The dashed line represents the tension produced by passive elastic structures within the muscle, and the solid line represents tension actively produced by the contractile proteins. The active and passive contributions are considered to be generated in parallel, and are additive.

2.1.4 FORCE-LENGTH PROPERTY

Let's begin a discussion of muscle behavior by reviewing the more obvious muscle properties. The first property states that muscle can only produce tensile (pulling) forces. These tensile forces are transmitted via tendons to the attachment locations of the tendon to bone, called the muscle's origin and insertion, respectively. No matter how much a muscle might try, it simply cannot push against the bones it is connected to because the tendons simply buckle. The second property is called the "force-length" property. A muscle at a constant activation level and shortening from its relaxed length develops progressively less and less force as it gets shorter. When muscle fibers reach half of their resting lengths, they can no longer produce any tensile force, and cannot shorten any further. If the muscle is stretched beyond its resting length, it does develop a tensile force and can sustain tension whether or not the muscle is actively contracting.

A plot of the tensile force produced by active muscle held at different fixed lengths is called the isometric length-tension curve, as shown in Figure 2.9 (solid lines). The figure shows the dependence of muscle force as a function of muscle fiber length for the case when all fibers are simultaneously active. The solid line shows the total force developed by the muscle for the case of 100% ($a(t) = 1$) activation. If the tension of an unstimulated muscle ($a(t) = 0$) is monitored instead, the muscle still resists being stretched beyond its relaxed, or resting length by developing a passive tensile force (dashed line,

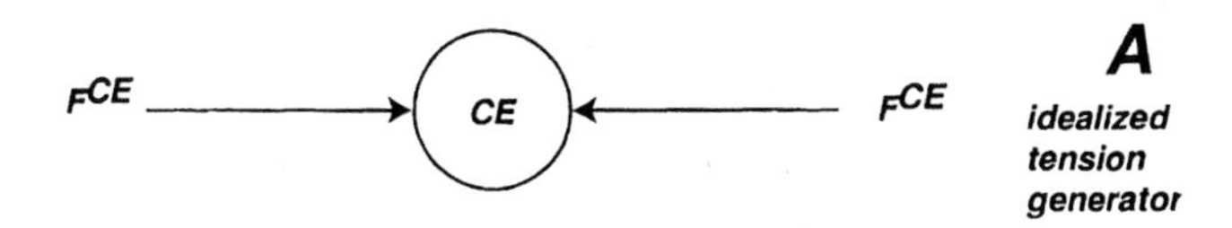

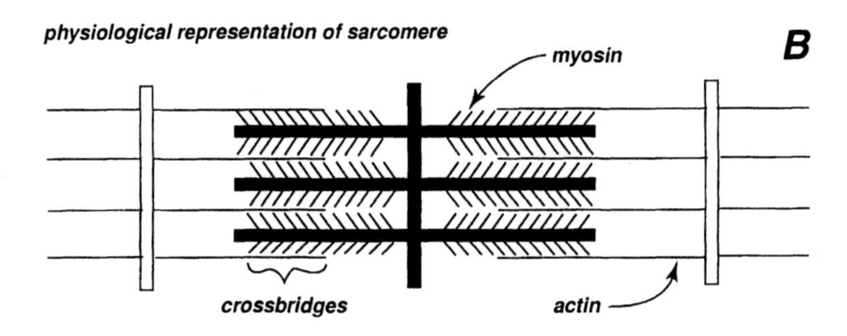

Figure 2.10. A. Schematic representation of the contractile element (CE), which generates tensile forces due to the interactions of actin and myosin protein filaments within the muscle cells. B. Physical explanation of the isometric length tension property shown in Figure 2.9. In the sliding-filament theory, the actin and myosin filaments within the sarcomere have fixed lengths, and slide past each other to shorten or lengthen the muscle. In the cross-bridge theory bonds between the sliding filaments are formed when the heads of the myosin filaments attach to binding sites on the actin filaments. According to the classical explanations, a maximal amount of active force (F_o^M) is generated when the number of crossbridges is maximized and the ends of the filaments are not compressed against the sarcomere walls.

Figure 2.9). The difference between the total force-length and the passive force-length curves results in the active force-length curve (dotted line) that apparently arises from forces generated by the contractile proteins actin and myosin within the muscle fibers.

The active portion of the force-length curve has a maximum value at the optimum fiber length, ℓ_o^M, where the M superscript refers to muscle, and the "ought" ($_o$) subscript refers to the optimum value of muscle length. The optimum muscle length and the relaxed, or resting length of the muscle fibers are very close together, and for the purposes of this text they are considered to be equal. ℓ_o^M is theorized by the sliding-filament theory to be the length at which a maximum number of cross-bridges between the actin and myosin protein filaments can be formed (Figure 2.10, Hill, 1953; Gordon, *et al.*, 1966; Huxley, 1974). The force developed at 100% activation and fiber length $\ell^M = \ell_o^M$ is called the optimal force F_o^M.[4] It can be easily estimated by multiplying a

[4]Beginning with Chapter 3, we will become much more precise regarding our designation of force as a vector quantity. In the context of muscle, force F refers to the magnitude of the vector force, or $|\vec{F}|$.

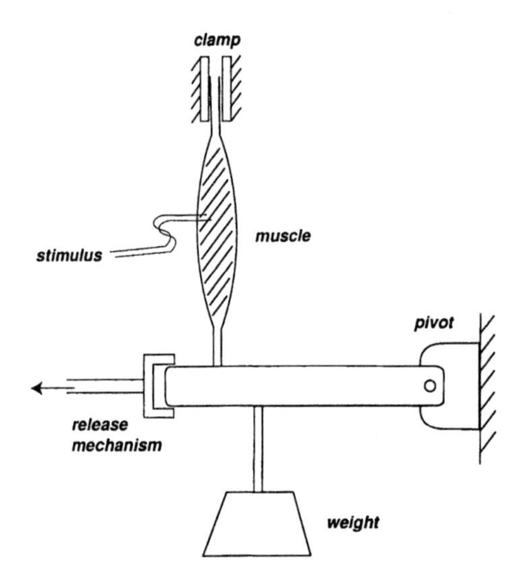

Figure 2.11. Schematic for a quick release experiment. The muscle is stimulated and allowed to come to force and length equilibrium. Maintaining a constant activation level, the rigid lower attachment is instantaneously removed by means of a sudden force, and the muscle is exposed to a sudden change in load. The muscle response includes an instantaneous change in its length.

muscle's average cross-sectional area by the maximal muscle specific tension of approximately $31.39\ N/cm^2$. This number is fairly constant across a wide range of muscles and mammalian species. While it is called the "optimal" muscle force, it is the maximum amount of force that can be *actively* generated. When muscle is stretched beyond ℓ_o^M, the passive and active tensions sum and develop total forces exceeding F_o^M.

The full explanation for how the sarcomeres which make up the muscle fibers actively develop contractile force and somehow shortens has yet to be discovered. The sliding-filament theory and the cross-bridge theory do not fully account for the shape of the length-tension curve in experiments on single muscle fibers (Pollack, 1990). A satisfying explanation for how the myosin heads swivel and cause the "ratcheting" action to shorten the sarcomeres has not yet been developed. Energy and entropy considerations suggest that it is possible for the myosin filaments to shorten *without swiveling the heads* (Pollack, 1990). These more recent criticisms of the classical theories have been controversial, but evidence is mounting. The biomechanist should keep an open mind. Many exciting developments are occurring in the field of muscle physiology, and much remains to be discovered.

At lengths less than half the optimal fiber length and greater than 1.5 times the optimal length, the muscle cannot generate much, if any, tension by active means. The range of active force generation is thus equal to the optimal fiber length. Generally speaking, it's usually a good rule of thumb for a muscle to be near maximally contracted ($\ell^M = 0.5\ell_o^M$) and near maximally extended ($\ell^M = 1.5\ell_o^M$) when the joint spanned is at the limits of joint flexion and extension. Taking this further, it is reasonable to assume that the active force generated isometrically will be maximal somewhere near the midrange of the joint's range of motion. Exercise machines that seek to fully exercise a muscle over the entire range of joint motion have taken advantage of this knowledge by adding a cam to the weight lifting mechanism. The purpose of the cam is to change the leverage of the weight in response to the expected length and leverage of the muscle. When the muscle is at its shortest, less force needs to be applied to raise the weight.

Note that the passive tension increases exponentially as the muscle is stretched. The force developed passively is independent of the level of activation. Resting skeletal muscle and other collagenous tissues such as tendon, skin, and mesentery have all been shown to exponentially rising passive force-length behavior (Fung, 1967). In muscle, this force was originally thought to arise from the muscle cell membranes and connective tissue surrounding the fibers, but there is some evidence to suggest that there are some tensile force contributions caused by cross-bridge interactions between the actin and myosin filaments producing the contractile force (D. K. Hill, 1968, see McMahon, 1984). Whatever the source of the contribution, in 1973 Pinto and Fung showed that the derivative of applied muscle stresses with respect to strain in inactive rabbit cardiac muscle was a linearly increasing function of the applied stress (see McMahon, 1984). By integrating this result, they obtained the stress ($\sigma = F/A$) as an exponentially increasing function of the extension, (ℓ^M/ℓ_o^M),

$$\sigma = \mu e^{\alpha \ell^M / \ell_o^M} - \beta\,. \tag{2.30}$$

μ and β are scalar constants used to fit the form of the equation to physiological data. Constant α determines the rate at which the slope changes, and should not be confused with pinnation angle. The"element" giving rise to the passive elastic property of muscle has become known as the *parallel elasticity* because of its spring-like, activation-independent characteristic.

Quick release experiments gave further insight into the nature of another elastic contributor to muscle tension, but this one is dependent upon activation. A muscle was initially held at a fixed length, and tetanically stimulated to develop a tensile force (Figure 2.11). If the length restriction was instantaneously removed, and replaced by a constant force, the length of the muscle was observed to shorten instantly by an amount proportional to the difference between the pre-release and the post-release tension. This force dependent, and velocity independent shortening is characteristic of an undamped mechanical spring. Thus, another spring-like, but activation dependent component was theorized to exist based on these experiments (Hill, 1938, 1950, 1953; Wilkie, 1956). This active series elasticity, together with the passive elasticity, comprise the elastic force producing elements of the muscle.

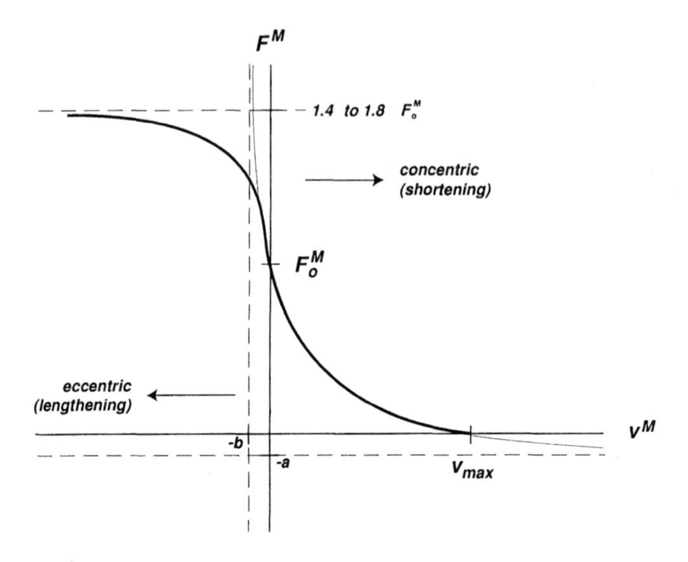

Figure 2.12. Contractile force development as a function of muscle fiber velocity, where positive velocity $v^M(t)$ is defined as shortening velocity $v^M \equiv -\dot{\ell}^M$. When $v^M \geq 0$, the "$f - v$" curve follows a hyperbola with asymptotes at $v^M = -b$ and $F^M = -a$. The left-hand side of the curve indicates that force production during eccentric (lengthening) activity exceeds that produced isometrically ($v^M = 0$), while the right-hand side shows that force diminishes during concentric contractions. Maximum shortening velocity $v_{max} \approx 10\ell_o^M$ / sec.

2.1.5 FORCE-VELOCITY PROPERTY

In addition to the dependence of active muscle force on length, the rate at which a muscle fiber changes length influences the magnitude of the active tension developed by the muscle (Figure 2.12). When activated muscle shortens (called a *concentric* contraction), the resulting force is less than that observed during isometric contractions. The opposite is true in cases where the muscle is lengthening *eccentrically*. It is as if the muscle's contractile machinery acts in parallel with a fluid "damper" which generates a force that is proportional to, but opposite in direction to, the velocity. An empirical relation by Hill (1938) relating the tensile force, F^M, to the muscle shortening velocity, v^M, was found that adequately matches data for cardiac and skeletal muscle,

$$\left(F^M + a\right)\left(v^M + b\right) = \left(F_o^M + a\right)b, \tag{2.31}$$

which is a hyperbola described by asymptotes at $F^M = -a$ and $v^M = -b$. F_o^M is the optimal muscle force as defined in Section 2.1.4.

2.1.5.1 ACTIVATION SCALING AND NORMALIZATION

Subtracting the passive contributions from the isometric tension-length curves yields a family of length-dependent curves of *active* force development. If we

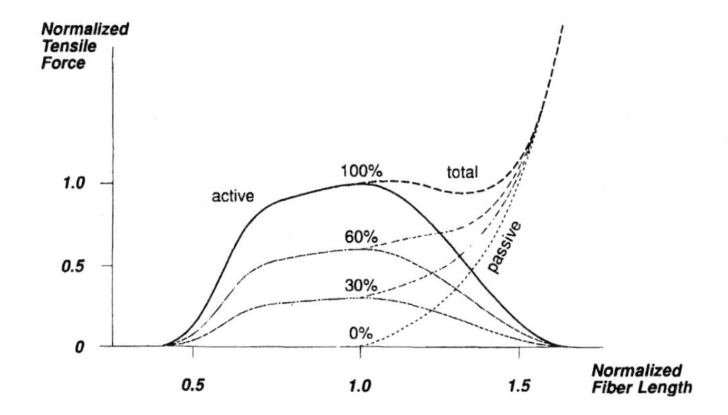

Figure 2.13. Normalized force-length curves, where tensile force has been scaled by $a(t)$ and normalized by F_o^M and length has been normalized by ℓ_o^M. Note that the same passive force is added to the various active force curves to produce the total isometric force length curves.

make the simplifying assumption that all active muscle fibers (regardless of diameter) have similar strengths, then this active portion of the force-length curve for activations $a(t) < 1$ can be linearly scaled by the activation level $a(t)$ of the muscle.[5]

[5] This is a big assumption. Future work in the field of muscle biomechanics should include the measurement of forces produced by fibers of different type and diameter.

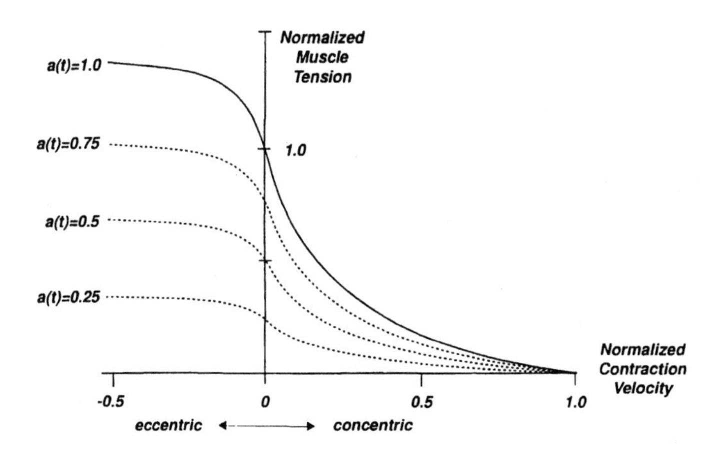

Figure 2.14. Normalized force-velocity curves, where tensile force has been scaled by $a(t)$ and normalized by F_o^M and velocity has been normalized by $v_{max} \approx -10\ell_o^M$ per second). Note that the same passive force is added to the various active force curves to produce the total isometric force length curves.

The maximum active force developed is thus $F_o^M a(t)$ when the muscle is at its optimal length (ℓ_o^M). According to this simplified force scaling approximation, muscular tension is zero when ℓ^M is less than half or more than 1.5 times resting length. Since similar tension-length curves are observed for most mammalian skeletal muscles, it is also convenient to normalize muscle length by (ℓ_o^M) and active force by F_o^M to obtain a generic, non-dimensional curve easily scaled to any skeletal muscle (Figures 2.13 and 2.14, Zajac, 1989).

2.2 MATHEMATICAL MODELS OF MUSCLE

In order to answer a particular problem involving movement, coordination, and muscle action, a *muscle model* is often created to yield quantitative information relating the actuator tensions and the resulting movements of the musculoskeletal system. The muscle model is but a portion of the larger musculoskeletal model, which is typically composed of a collection of mathematical representations of skeletal segments, joints, and the soft tissues that surround them. Only the essential mechanical properties of the muscles need to be included in the muscle model. We can include these properties by linking together idealized mechanical elements arranged to create a composite "lumped parameter" model that mimics the behavior of the actual physiological system under a set of well-defined conditions.

The strategy for building a muscle model is to first introduce the basic mechanical elements of a spring and a dashpot, and explain how series and parallel arrangements can be made to accurately model the viscoelastic behavior of soft tissues. The Maxwell, Voight, and Kelvin models are good for soft tissues under both compressive and tensile loads.

It is emphasized that the reader must master this material before he can fully understand the muscle model. After all, the basic Hill muscle model is nothing more than a soft tissue model with a length dependent tension generator. Following this section, the mechanical properties of tendon will be introduced, and finally the properties of a musculotendon actuator will be discussed. We will not proceed to models of muscle and tendon which have their own mass, because these are refinements that create additional problems in exchange for small increases in accuracy.

2.2.1 BASIC MECHANICAL ELEMENTS OF A MUSCLE MODEL

In biomechanical systems, the basic elements for modeling include idealized masses, inertias, springs, and dashpots (dampers, see Figure 2.15). These are all passive elements, which produce no power on their own and simply respond to forces and displacements. In addition to the passive elements, a model of muscle requires an active generator of tensile force. These elements can be

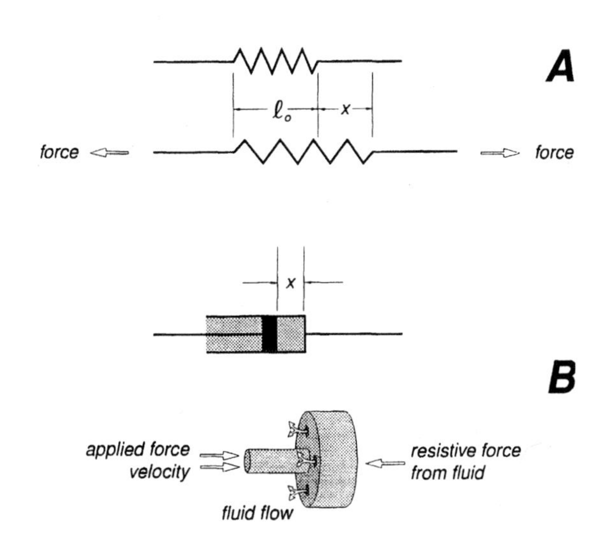

Figure 2.15. A. An idealized representation of a spring. The idealized spring is assumed to be massless, linear in its force-deflection characteristics, and infinitely deformable. B. An idealized dashpot. Its physical analog is represented as a fluid filled cylinder with a piston through which holes have been made to allow fluid to pass from one side of the piston to the other. The idealized dashpot is assumed to be massless, infinitely deformable, and to generate a resistive force which is opposite in direction and proportional to the piston velocity.

mixed and matched depending upon the behavior desired under the conditions imposed. For instance, the macroscopic properties of a tendon can be modeled accurately by a nonlinear spring during quick extensions and relaxations (those which occur within seconds). Damping must be added to the tendon model if the extension/relaxation occurs over a long period of time (minutes to hours), and mass must be added if inertial effects are important.

The most basic building block comprising a muscle model is the ideal spring (Figure 2.15A). A linear spring creates a force proportional to its deflection,

$$F_k = kx\,, \tag{2.32}$$

where F_k is the magnitude of the force exerted by the spring on its attachments, x is the extension of the spring beyond its relaxed length and k is the constant of proportionality. F_k can be expressed as a nonlinear function of x, or the coefficient k can alternatively be expressed as a nonlinear function of x to maintain the linear form of Equation 2.32. In this latter case, a mathematical function is created which increases the stiffness of the spring as x increases. In mechanical machines, x is allowed to be negative to model compression, but in physiological (tension only) muscle models, springs are not allowed to generate forces in compression. The essential property of an ideal linear spring is obvious – it must produce a force proportional to its displacement. Not so obvious, an ideal spring is considered to be massless. When subjected to a force *a spring produces an instantaneous deflection.* Also, unlike real springs, there is no limit to the amount of extension, no residual deformation when relaxed, and no dissipation of energy as heat.

An ideal dashpot or damper creates a force proportional to its velocity,

$$F_b = b\dot{x}, \tag{2.33}$$

where F_b is the magnitude of the force exerted by the dashpot on its attachments, $\dot{x}$ is the time rate of extension x, and b is the viscous damping constant. An ideal dashpot is considered to be massless and is immune from all forms of friction except viscous friction. In biomechanical systems, the dashpot is allowed to produce forces in either direction as long as it opposes the velocity. One can think of a dashpot as a massless piston being pushed through a cylinder of heavy oil. If there are small holes linking the two sides of the piston, it would take some time to allow the oil to flow through the holes in the piston to the other side (Figure 2.15). A dashpot is also like moving one's hand through a viscous fluid. The faster the hand is moved, the greater the shear drag and pressure drag become. One inobvious, but important principle of an ideal dashpot, is that *a dashpot cannot be displaced instantaneously* unless acted upon by an infinitely large force.

2.2.2 SERIES AND PARALLEL ARRANGEMENTS OF PASSIVE ELEMENTS

Before proceeding further with our discussion of muscle models, it is worthwhile to examine the properties of springs and dashpots in combination. Springs and dashpots can be connected end to end (in *series*), side by side (in *parallel*), or both to create mathematical models which describe the *viscoelastic* behavior of passive tissues when loaded or deformed.

When a ligament, for instance, is suddenly loaded by a constant tensile force, it immediately stretches by some amount, then continues stretching an additional amount slowly over time (Figure 2.16A). After the initial stretch, the *rate* of stretching is finite and slows down as time proceeds until the stretch rate becomes effectively zero. If modeled by a first-order process, the time constant for the tendon stretch is on the order of seconds to tens of seconds. Note that the ligament won't keep on stretching forever – the amount of stretch asymptotically approaches a limit as $t \to \infty$. A slightly different loading condition is created by suddenly stretching and holding a ligament at a constant length. Under these conditions, the tension within the ligament instantaneously rises and then falls asymptotically to a lower tension (Figure 2.16B). This second loading condition is created when an athlete stretches and holds a posture at an extreme joint position. Using a musical analogy, one can actually hear the pitch decrease as the tension falls in a newly installed guitar string.[6] The decrease in tension

[6] The actual process by which tension is created in a guitar string is to fix one end on the guitar face and wind the other end of the string around a gear driven post. Slowly turning the post increases the tension.

is created because the material is said to "creep" in response to the tensile force. In biological tissues the presence of viscous fluids helps to explain the time dependent nature of the constant load and constant deformation responses. Similar load and deformation patterns are exhibited by most physiological soft tissues, with the exception that some tissues, such as cartilage, are usually subjected to compressive loading rather than tensile loading.

The following examples will illustrate how a structural arrangement of linear springs and dashpots is affected by both the arrangement, and by the properties of its component parts. The reader should study these examples fully in order to lay the foundation for a deeper understanding of the muscle model to follow.

2.2.2.1 MAXWELL MODEL

A series arrangement of a spring (k) and a dashpot (b) is depicted in Figure 2.17A. In a series arrangement, the tensions in the spring (F_k) and dashpot (F_b) are equal, and the deformations x_k and x_b are additive. A first order differential equation is sought to relate the applied tension F, the total deformation

This does not impose a tensile force suddenly, nor does it a suddenly impose a deformation. So this musical analogy holds only *after* the desired initial tension is reached.

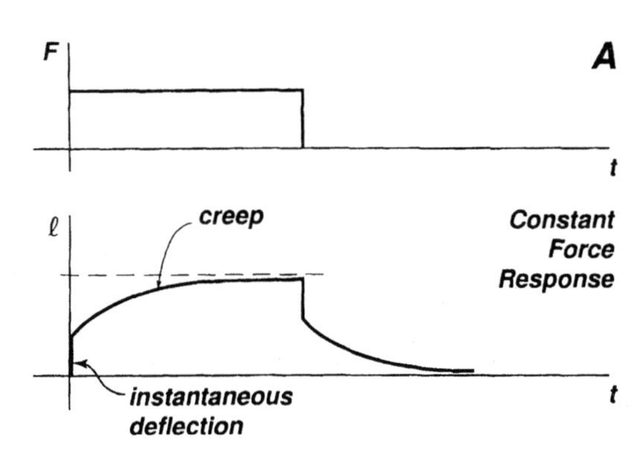

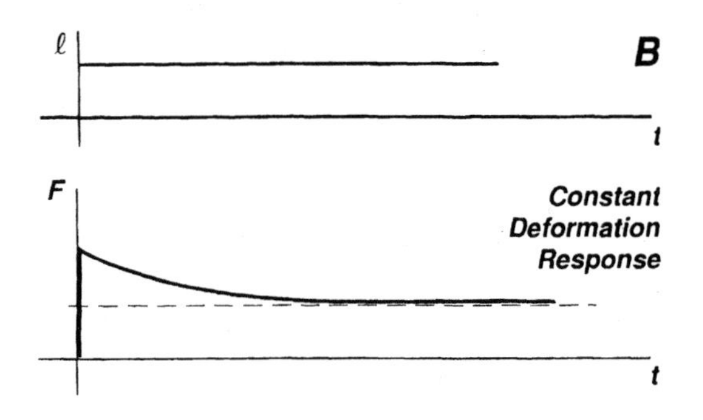

Figure 2.16. Responses of physiological tissues to a constant force and a constant deformation. A. When a constant force is applied, the tissue shows both an instantaneous elongation and a slower elongation which gradually approaches an asymptotic value. B. When the tissue is instantaneously stretched, the force in the tissue is high at first, but asymptotically decreases to a nonzero asymptotic value.

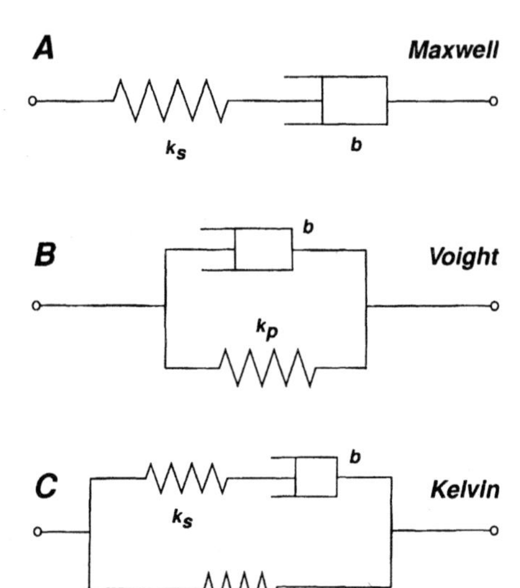

Figure 2.17. A. The Maxwell model of passive soft tissue consists of an ideal spring and an ideal dashpot in series. B. A Voight model contains the same elements in parallel. C. The Kelvin solid model combines the Voight model in series with an idealized spring.

x, and their first time derivatives. The mathematical solution of the equation will yield an expression for either the displacement $x(t)$, or the force $F(t)$, as a function of time t. If the mathematical behavior of the model matches that of the physiological system as shown in Figure 2.16, then the model is said to be successful. If the model does not match the physiological system studied, then the model is altered until it can successfully predict the behavior of real tissues under the specified set of conditions.

To begin,

$$F = F_k = F_b \tag{2.34}$$

$$x = x_k + x_b \,. \tag{2.35}$$

The forces in the spring and dashpot are given in Equations 2.32 and 2.33. Differentiating Equation 2.35 yields

$$\dot{x} = \dot{x}_k + \dot{x}_b \,, \tag{2.36}$$

and plugging in Equation 2.33 and $\dot{x}_k = \dot{F}/k$ yields a differential equation relating F, x (zero in this instance), and their first time derivatives $\dot{F}$ and $\dot{x}$,

$$\dot{F} + \frac{k}{b}F = k\dot{x} \,. \tag{2.37}$$

Case I. Maxwell Model Constant Load Response. Applying a constant force F at time $t = 0$ implies that $\dot{F} = 0$. Equation 2.37 becomes

$$\frac{k}{b}F = k\dot{x} \,, \tag{2.38}$$

or,

$$\dot{x} = \frac{F}{b}. \tag{2.39}$$

The solution for the displacement $x(t)$ as a function of time t is

$$x(t) = \frac{F}{b}t + x_o, \tag{2.40}$$

where x_o is the initial displacement at $t = 0$. The physical arrangement of the spring and dashpot, and the mechanical properties of these individual components allows us to determine x_o. Because the ideal dashpot cannot be displaced instantaneously by a finite force, the displacement of the dashpot must be zero at $t = 0$.[7] The idealized linear spring is considered to be massless, and thus responds instantaneously to the applied tension with displacement

$$x_o = \frac{F}{k}. \tag{2.41}$$

Therefore, the solution is

$$x(t) = \frac{F}{b}t + \frac{F}{k}, \tag{2.42}$$

which is graphed in Figure 2.18B. Note that the initial deformation is observed, like with the real tissue, but the linearly increasing deformation with time implies the tissue will continue to elongate forever. Also, when the force is removed, a residual deformation exists. These results suggest that the Maxwell model is not a good model of physiological tissue under constant load except when time t is small.

Case II. Maxwell Model Constant Deformation Response. A quick stretch at $t = 0$ to a constant deformation implies the ligament will have no elongation at $t > 0$, or $\dot{x} = 0$. The differential equation for the Maxwell Model, Equation 2.37, becomes

$$\dot{F} + \frac{k}{b}F = 0. \tag{2.43}$$

This is a first order equation in the variable F. Because functions of F and $\dot{F}$ are added together to produce a null result, F must be an exponential. Because the form of the solution is known *a-priori*, an easy way to solve for $F(t)$ is to formulate an expression for it including three unknown parameters A, B, and σ,

$$F(t) = Ae^{\sigma t} + B. \tag{2.44}$$

[7]More specifically, the displacement of the dashpot is considered at time $t = 0^+$ immediately after the application of the tension F at time $t = 0$.

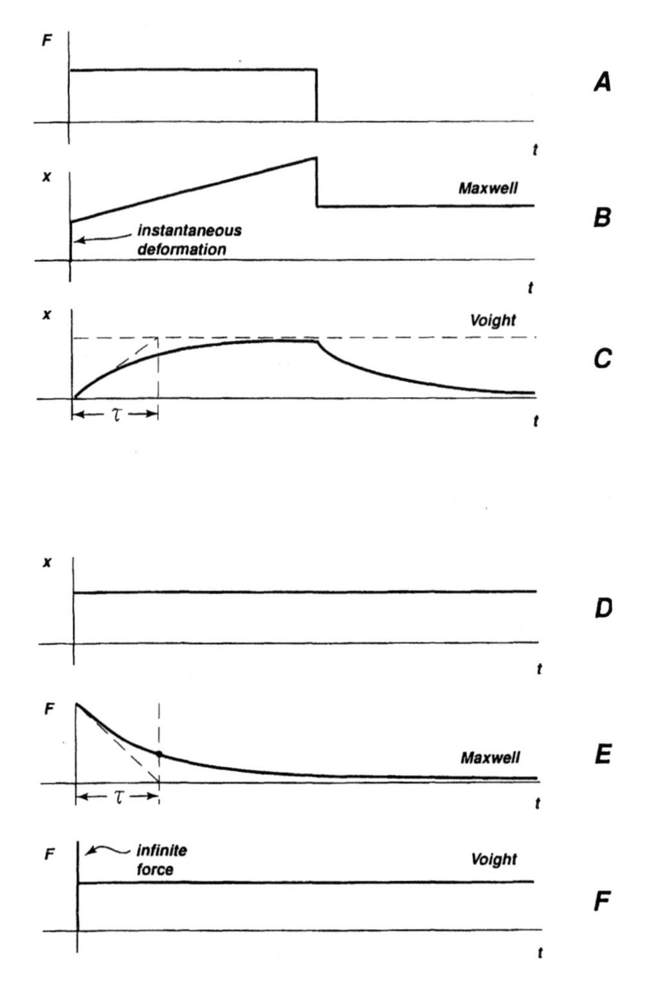

Figure 2.18. Mathematical responses of the Maxwell and Voight models to a constant applied force (A) and to a fixed deformation (B). A good model would mimic the responses of physiologic tissues as in Figure 2.16. The Maxwell model (B and E) appears to model the transient behavior well, but does not mimic the steady state behavior of physiological tissues very well. The Voight model (C and F) models the steady state behavior, but not the transient behavior.

By "guessing" the form of the solution it can be differentiated with respect to time,

$$\dot{F}(t) = A\sigma e^{\sigma t}\,, \tag{2.45}$$

and "plugged in" to Equation 2.43, yielding

$$A\sigma e^{\sigma t} + \frac{k}{b}\left(Ae^{\sigma t} + B\right) = 0\,. \tag{2.46}$$

This equation is factored by separating the time varying terms of the equation (every part multiplied by $e^{\sigma t}$) from the terms that are constant in time,

$$A\left(\sigma + \frac{k}{b}\right)e^{\sigma t} + \frac{k}{b}B = 0\,. \tag{2.47}$$

In order for this equation to hold at all values of time t, the time varying portion and the constant portion of the equation must be self cancelling, or

$$B = 0 \tag{2.48}$$

$$\sigma = -\frac{k}{b}\,. \tag{2.49}$$

This solves for two of the three parameters of Equation 2.44, which now takes the form

$$F(t) = Ae^{-t/\tau}\,, \tag{2.50}$$

where $\tau = b/k$. Only the parameter A remains undetermined. It is obtained by examining the initial conditions in light of the physical properties of the spring and dashpot. At $t = 0$, only the spring can deform instantaneously, so $x = x_k = F/k$, and $A = F(0) = kx$, which is constant because the applied deformation x is constant. The final solution is

$$F(t) = kxe^{-t/\tau}\,, \tag{2.51}$$

which is plotted in Figure 2.18E. An initial deformation is observed, as in the physiological response, but the longterm solution decays to zero and thus the Maxwell model does not provide a good match to the behavior of actual physiological tissue.

2.2.2.2 VOIGHT MODEL

We have seen that a series arrangement of a spring (k) and a dashpot (b) was unable to model the physiological situation well in the long term. One might try a parallel arrangement next, as depicted in Figure 2.17B. In a parallel arrangement, the "vertical" bars joining the horizontal lines extending from the spring and dashpot are constrained to remain vertical at all extensions, and thus the deformations x_k and x_b are equal in the two members while the tensions (F_k) and (F_b) are additive. As with the Maxwell model, a first order differential equation is sought that relates the total applied tension F, the common deformation x, and their first time derivatives. The mathematical solution of the equation will yield an expression for either the displacement $x(t)$, or the tension $F(t)$, as a function of time t.

To begin,

$$x = x_k = x_b \tag{2.52}$$

$$F = F_k + F_b\,. \tag{2.53}$$

The force levels in the spring and dashpot are given in Equations 2.32 and 2.33,

$$F_k = kx_k = kx \tag{2.54}$$
$$F_b = b\dot{x}_b = b\dot{x}\,. \tag{2.55}$$

Plugging these into Equation 2.53 yields our first order differential equation,

$$F = kx + b\dot{x}\,. \tag{2.56}$$

Case I. Voight Model Constant Load Response. Applying a constant force F at time $t = 0$ implies $\dot{F} = 0$. Equation 2.56 remains the same because $\dot{F}$ does not appear explicitly. Since F is constant, we seek the time dependent deformation $x(t)$. The form of $x(t)$ must be an exponential, since 2.56 adds functions of x and $\dot{x}$ together and sums to a constant. Therefore, we can *guess* the form of the solution as

$$x(t) = Ae^{\sigma t} + B\,. \tag{2.57}$$

Differentiation yields

$$\dot{x}(t) = A\sigma e^{\sigma t}\,, \tag{2.58}$$

and these expressions plugged into Equation 2.56 produce

$$F = k\left(Ae^{\sigma t} + B\right) + b\left(A\sigma e^{\sigma t}\right)\,. \tag{2.59}$$

Separating the *time varying* parts of the equation from the portions that are constant in time,

$$F - kB = (k + b\sigma)\,Ae^{\sigma t}\,. \tag{2.60}$$

As before with the Maxwell model example, this equation can only hold for all values of time t if the time varying portion on the right side is self cancelling,

$$\sigma = -k/b\,. \tag{2.61}$$

The constant B on the left side is also determined,

$$B = F/k\,, \tag{2.62}$$

so our expression for $x(t)$ becomes

$$x(t) = Ae^{-t/\tau} + \frac{F}{k}\,, \tag{2.63}$$

where $\tau = b/k$. The final unknown A is determined by initial conditions. Because the dashpot cannot deform instantaneously when acted upon by a finite force, $x(0)$ must be zero and thus the prior equation with $t = 0$ becomes

$$x(0) = A + \frac{F}{k} \tag{2.64}$$

$$A = -F/k\,, \tag{2.65}$$

and hence, the time varying solution is

$$x(t) = \frac{F}{k}\left(1 - e^{-t/\tau}\right)\,. \tag{2.66}$$

$x(t)$ is plotted in Figure 2.18C. Note that the longterm solution is good, in that $x(t)$ exponentially approaches an asymptote as t approaches ∞, but that the initial instantaneous deformation exhibited by the physiological tissue and seen in the Maxwell model is not present.

Case II. Voight model Constant Deformation Response. When $\dot{x} = 0$, the differential equation for the Voight model, Equation 2.56, becomes trivial,

$$F = kx\,. \tag{2.67}$$

This equation still expresses the correct relationships between F, $\dot{F}$, x, and $\dot{x}$, but the tensile force response $F(t)$ is trivial and uninteresting for $t > 0$. However, it is interesting to speculate on what the force is *at* $t = 0$, because with our parallel arrangement of spring and dashpot, an instantaneous deformation cannot take place with the application of a finite force. Therefore, in order to have instantaneously deformed the Voight model by a specified amount, *an infinite force must have been applied.* This leads to the discontinuous nature of the force response under a constant deformation at $t = 0$ (Figure 2.18F).

2.2.2.3 KELVIN MODEL

The previous examples have shown the following. *A spring in series with a dashpot* is needed to obtain instantaneous deformations, and *a spring in parallel with a dashpot* provides a long term response which gradually approaches an asymptote. The next structural arrangement, called the *Kelvin solid* adds a spring in series with a Voight model (Figure 2.17C). The spring in series with the dashpot is necessary to allow an instantaneous deformation, because the dashpot prevents anything in parallel with it from changing length instantaneously. With proper specification of parallel spring constant k_p, series spring constant k_s, and damping coefficient b, the Kelvin model can be made to effectively match the behavior under both short term and long term conditions. The interested reader should work out Homework Problem 2.5 to derive the differential equation governing the Kelvin model's response, and its solution under constant load and constant deformation conditions.

2.2.3 HILL MUSCLE MODEL

To review, elasticity in both the contractile machinery and passive portions of the muscle, as well as active force dependence upon muscle length and velocity, are properties desired in a useful model of muscle. Nearly all researchers

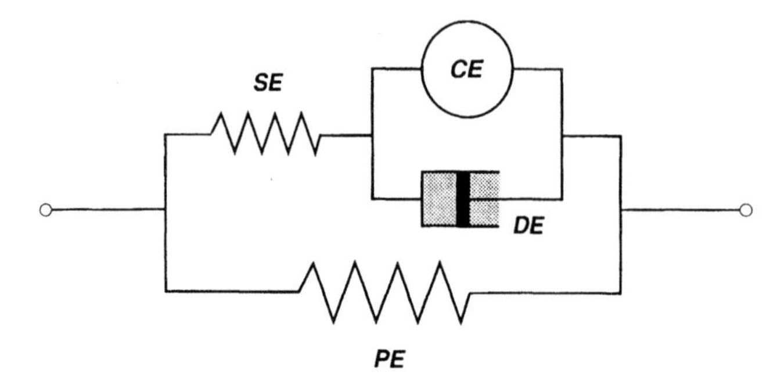

Figure 2.19. Hill type muscle model. The muscle model is a Kelvin solid with a tensile force generator (the contractile element, CE) added in parallel with the damping element (DE). The parallel elastic element (PE) creates passive elastic tension when muscle is stretched beyond its resting length. The series elasticity (SE) is required to account for the behavior of muscle during quick release experiments.

today use variations of the model attributed to Hill (1938) (Figure 2.19). The Hill type model is a Kelvin solid model with the addition of a force generator to account for the tension produced by the contractile proteins actin and myosin. As discussed in Section 2.1.4, the variation in active muscle tension as a function of muscle fiber length is classically explained by the number of crossbridges formed between actin and myosin filaments. Another difference between the muscle model and the Kelvin solid is that the idealized elements are not necessarily assumed to be linear in their mechanical properties. For instance, nonlinear springs are typically used to represent the parallel elastic (PE) and series elastic (SE) elements.

To represent the active force producing or "contractile component" of the muscle, Hill proposed a force generator, called the "active state", working in parallel with a velocity dependent damping element (DE). He theorized that stimulated muscle fibers always developed maximal forces, but that the viscous resistance to shortening reduced the aggregate muscle force during shortening. Though the concept of the active state has fallen into disfavor due to a number of reasons, the basic structure of Hill's model has endured for decades. Most, if not all, macroscopic muscle models continue to be variations of the original Hill model.

In any variation of the basic model, a damping element (DE) must be placed in parallel with the active tension producing element. The damping element must be placed in parallel so that the velocities of the active state and the DE maintain the same direction. Because this guarantees that the DE will shorten when the active state shortens, the force generated by the DE opposes the tensile force generated by the active state during concentric (shortening) contractions. On the other hand, the damping element will lengthen during eccentric contractions, and will create a force that augments the tension generated by the active state. The damping element therefore accounts for the force reduction in concentric contractions, and the force enhancement in eccentric contractions as explained by the force-velocity property (Section 2.1.5). The viscous force arises because of the presence of intracellular and extracellular fluid within the muscle.

The parallel elasticity (PE) accounts for the nonlinear passive elastic properties of the muscle. This was already introduced as the passive tension length property in Section 2.1.4, and is plotted as the dashed curve in Figure 2.9. Note that the slope of the passive tension length curve increases as the muscle length increases beyond ℓ_o^M, and therefore the stiffness of the PE increases with muscle stretch. It is thought that the parallel elastic (PE) properties come mainly from the connective tissues within the muscle.

An elasticity in series with the damping element *must be present* to account for the instantaneous length change observed during quick release experiments (see Section 2.1.4 and Figure 2.11). As discussed in Section 2.2.1, a damping element cannot change length instantaneously. Because muscle at constant activation does exhibit an instantaneous length change in response to a change in load, the SE element is required in any muscle model that is subject to transient forces. For instance, a musculoskeletal model developed to study stepping movements would need to include SE elements within the muscle models, as transient ground reaction forces are applied to the bottoms of the feet. It is thought that the series elasticity (SE) arises from the elasticity of the actin-myosin cross bridges.

2.3 MODELING TENDON PROPERTIES

Tendon is a complex structure composed of proteins, water, hydrophilic gels, and cementing substance. At a macroscopic level, groups of collagen fascicles, each composed of many subfilaments or fibrils, are usually bundled together in roughly parallel fashion to form the tendon. Water accounts for much of the total weight and volume, while collagen makes up approximately one third of the protein in the tendon or about 6% of its weight (see Butler *et al.*, 1979). Elastin, proteoglycans, and other noncollagenous proteins are also components, and influence the tissue's elastic, and time and history dependent viscoelastic properties.

Time and space do not permit going further with the subject of tendon physiology. However, it is important for the reader to realize that a wide range of tests and testing conditions, species, tendon preparations, and ages of subjects have been used to experimentally determine the mechanical properties of tendons. Tendons vary widely in geometrical shape, and have a wide variation in properties from subject to subject and even muscle to muscle. A

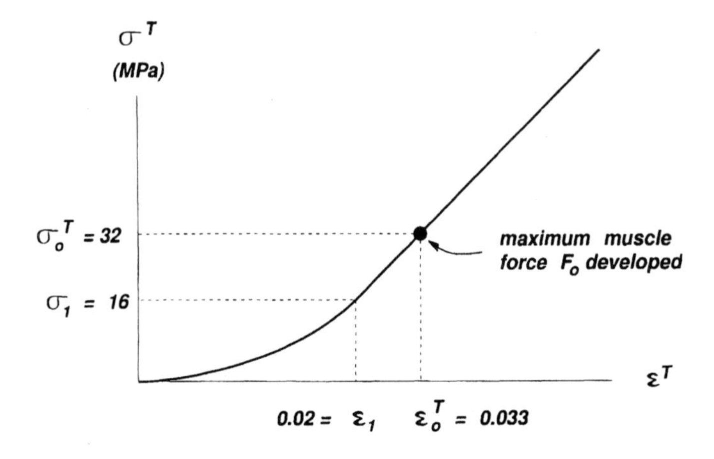

Figure 2.20. "Toe" and "linear" regions of the tendon stress-strain curve.

generic model of tendon, then, will be difficult to formulate from this data if high precision is desired.

Most of the time, a first order approximation is sufficient. Since the muscular forces expected during normal activities are not high relative to their maximum strengths, one would not anticipate a great deal of tendon stretch (relative to muscle-fiber length) for most of the muscles. Even for the muscles which have long tendons, only a moderate amount of stretch is expected. This stretch must be accounted for, but until our musculoskeletal models become more accurate, the stretch only needs to be estimated. As long as the stretch estimate is close, a

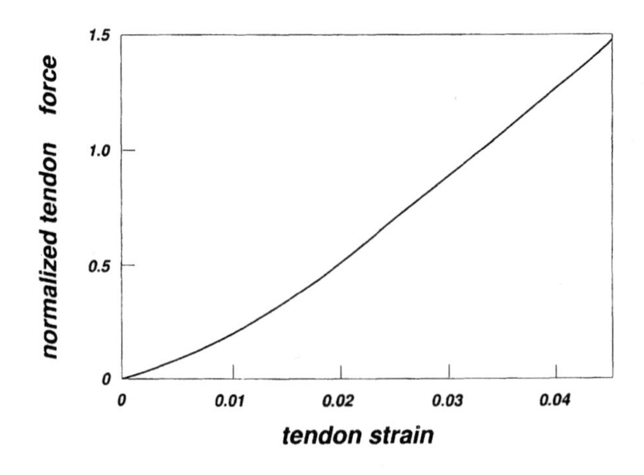

Figure 2.21. A tendon force-strain curve. Tendon force is normalized by optimal muscle force F_o^M (from Yamaguchi, 1989).

small percentile error out of a small amount of total stretch should not seriously jeopardize muscle length and muscle tension calculations.

As an example, let us define strain as the ratio of extension to relaxed length $\epsilon = \Delta\ell^T/\ell_r^T$, and carry out some calculations. $\Delta\ell^T$ is the change in tendon length ℓ^T, and ℓ_r^T is defined as "tendon resting length." Assuming a linear relationship between tension and strain, a maximum of 3.3% strain at F_o^M (ultimate tendon strain is about 10% according to Zajac, 1989), and activations of 50% or less during walking, the maximum change in length of tendon during normal walking should be less than 3% of its total length when the muscle contracts eccentrically. An error of 10% in this would be 0.3%, which would account for only 10 x 0.3% = 3% of muscle fiber length for muscles having the high ratio of tendon to muscle resting length of 10. Near the plateau region of the muscle tension length curve, this error would yield detectable, but insignificant errors in calculated force. In any event, it is highly unlikely for tendon to be stressed at levels remotely approaching ultimate levels during such a normal activity of daily living. Tendon rupture is relatively rare, occurring primarily in extreme cases of physical activity such as during active sports and weight lifting. Woo (1988) has noted that tendon strength decreases with age, but this might be at least partially explained by a lessening of activity.

The tendon model of Zajac (1989) is explained here as it is relatively simple and is easily incorporated into dynamic musculoskeletal models. It expresses the tendon stress-strain curve in terms of parameters averaged across a wide range of tendons, which is shown in Figure 2.20. Here only the lower range of stresses and strains is considered, which can be conveniently divided into a "toe" and a linear region. The strain in the toe region is modeled logarithmically as a function of stress until tendon force reaches $16\,MPa$, where strain is thereafter approximated by a linearly increasing function with slope equal to the elastic modulus $E_m = 1200\,MPa$. It is worth noting that the strain at maximum muscle force F_o^M is only about 3.3%, which is about a factor of three below ultimate tendon strain. We assume that tendon is perfectly elastic, energy conservative, and massless so that tendon force (F^T) is instantaneously and unambiguously given as a function of tendon length. In practice, it is often useful to convert the stress-strain curve into a tendon *force*-strain curve (Figure 2.21) to make it compatible with muscle models.

2.4 THE MUSCULOTENDON ACTUATOR

So far, we have carried out separate discussions regarding the relationships between tension, length, and velocity in the activated muscle and stress and strain in the tendon. But our task is more complex when they are merely parts of a larger musculoskeletal model (Figure 2.22). Typically, the modeler is given the instantaneous positions and velocities of the skeletal segments, and the task is to determine the musculotendon forces, apply them to the skeletal segments, and then compute the motions resulting from these muscle actions. Musculotendon lengths (ℓ^{MT}), velocities (v^{MT}), and muscle activations ($a(t)$) are the easy quantities to compute. But when tendon elasticity and length (ℓ^{T}) are included, it becomes more difficult to compute the force that the musculotendon actuator will deliver.

Properties of the tendinous elements come into play during all normal voluntary contractions, but may or may not affect resulting motions of the limb segments significantly. Although seemingly obvious, the contributions made by tendons are often neglected for reasons of simplicity, and muscle tension-length and force-velocity properties alone are used. Neglecting the series tendon elasticity can be justified in cases where tendon stretch is negligible, *e.g.,* when stress is small and/or tendon length is short. But if the tendon stretches an amount approaching the fiber length of a particular muscle, the tension-length characteristics of the actuator will differ significantly from the tension-length behavior of the muscle alone (Figure 2.23). In the figure, the horizontal axis for musculotendon length is normalized to the sum of the relaxed length of the muscle and tendon, $\ell_o^M + \ell_r^T$. This enables plots of different tendon and muscle lengths to be plotted together and easily compared. Increasing tendon length tends to distort the peak of the active muscle tension-length curve to the right. As $\ell_r^T \rightarrow \infty$ the active curve underlying the total tension curve deforms from a somewhat bell-shaped curve toward a right triangular form.

As previously mentioned, muscles crossing the distal joints are particularly susceptible, as the ratios of tendon rest length to optimal muscle fiber length (ℓ_r^T / ℓ_o^M) are greater at the distal extremities and approach a value of 10 in humans. Muscles crossing the human knee exhibit ratios of about 5, while those at the human hip are less than 1, leading to the conclusion that tendon elasticity is important at the ankle, and should be included at the knee, but can probably be neglected at the hip (Hoy *et al.,* 1989; Zajac, 1989). Alexander (1984) reports the extreme case of the camel, which has muscles in the posterior lower legs which are almost entirely composed of tendon. It is thought that these extremely long tendons are an adaptation for energy efficient running.

The inclusion of tendon as an elastic element distinctly separate from the series elastic element of the muscle model has been shown to be important for physiologically realistic models of muscles crossing the knee and ankle (Hoy *et al.,* 1989; Zajac, 1989). At these joints, the general ratios of relaxed tendon to muscle fiber length tend to be about 5 and 10, respectively. This is because at the distal joints, most vertebrates, including humans, are constructed so as to minimize the muscular effort required to hold a limb aloft. Since muscle tissue is heavy, the muscles actuating distal joints are located as close as possible to the trunk. This minimizes the effort required to hold the muscle mass within the limbs against gravity. The forces generated by these muscles are simply transmitted using long tendons. Tendon elasticity is a desirable property to be

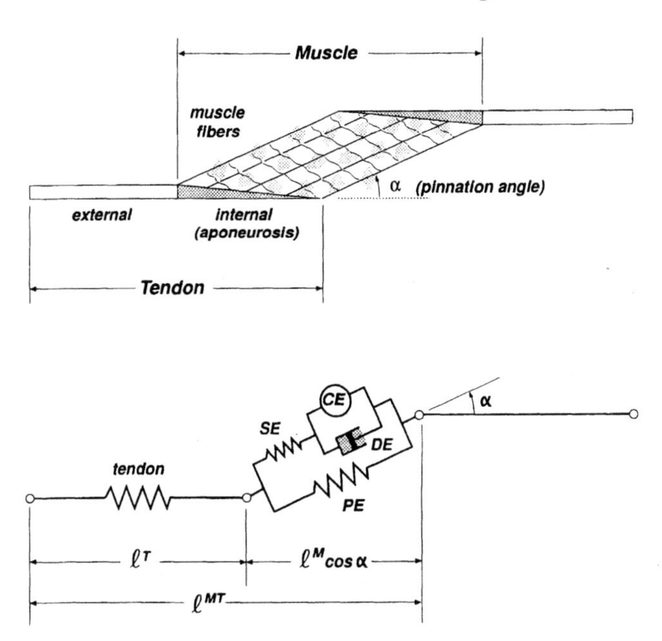

Figure 2.22. Modified Hill type muscle model with added tendon. Simplified structure of the musculotendon actuator (top) and the musculotendon model (bottom). The model is composed of lumped, idealized mechanical components such as springs (tendon, SE, PE), dashpots (DE), and force-generators (CE), that together mimic the macroscopic behavior of biological muscle-tendons.

included within musculoskeletal models. Some researchers choose to include the tendon elasticity as part of the series elastic component of muscle, but there are many advantages of including tendon as a separately modeled entity.

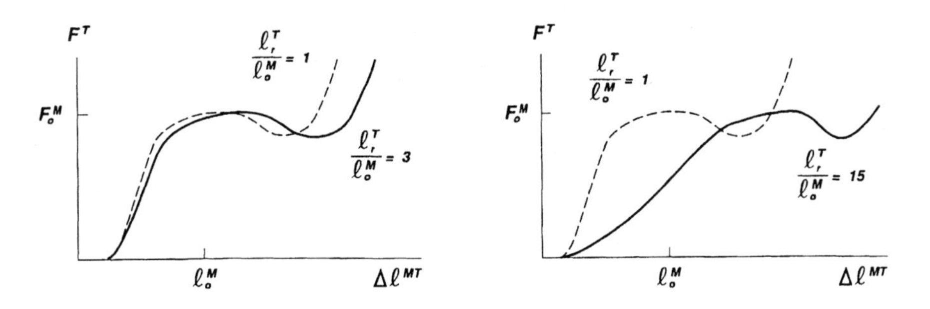

Figure 2.23. The effect of tendon elasticity on the force-length properties of the musculotendon actuator. In the left figure, the ratio of tendon rest length to muscle-fiber length (ℓ_r^T / ℓ_o^M) is 1 in the dashed curve (stiff tendon) and 3 in the solid curve. The right figure contrasts the same dashed curve with one having a relatively long (stretchy) tendon. This ratio is a maximum of about 10 in humans (Zajac, 1989).

A study showing how well-matched muscular contractions might efficiently work together with the elastic parameters of the muscle-tendon compartment during gait was provided by Hof *et al.* (1983). They found that much of the work done on the *triceps surae* complex during early stance phase, as well as work done by the contractile element, was stored in the series elasticity and released during the push-off phase.[8] Ideally, they said, the contractile elements worked in concert with the elastic elements, balancing the contractile force in order to maintain a nearly constant, and efficient, muscle-fiber length. Muscle fibers could then develop the required contractile force at low metabolic cost. Changes in tendon length alone could account for nearly all of the changes in musculotendon length for such "concerted contractions".

While energy dissipation in muscle can be appreciable, the energy dissipated during a quick stretch-relaxation cycle in tendon is reported to be small (Butler *et al.,* 1979). Ker (1981) has also shown that sheep tendons can return 93% of the energy used to stretch the tendons. It is reasonable to conclude that almost all of the energy stored within a tendon can be reused provided it is not dissipated by muscle relaxation. If the muscle maintains a near constant length, the musculotendon actuator can act as a conservative energy storage device, alternately storing and releasing energy during the gait cycle.

Modeling tendon as a separate and distinct element provides the advantage of computing mechanical features (such as energy flows) of muscle and tendon separately. Doing this has some interesting consequences. Considering the muscle fiber length ℓ^M, pinnation angle α, tendon length ℓ^T, and musculotendon length $\ell^{MT} = \ell^M cos\alpha + \ell^T$,

- *When ℓ^M shortens to initiate skeletal movement*, usually ℓ^M will initially shorten, and ℓ^T will initially lengthen.
- *When ℓ^{MT} shortens during a skeletal motion resulting from the musculotendon contraction*, usually ℓ^M and ℓ^T will both shorten, but it is possible for one of the two elements to be lengthening or maintaining the same length.
- *When ℓ^{MT} maintains a constant length*, ℓ^M and ℓ^T can be changing lengths.
- Muscle fiber length ℓ^M can both shorten and lengthen during an increase in muscle tension, and it can shorten and lengthen during a decrease in muscle tension.

This list can be extended a lot further. Hopefully, it has communicated to the reader that muscle does not always shorten when the musculotendon attachments move closer together. The situation is complicated by the interactions of a stretchy tendon and the nonlinear tension-length behavior of muscle.

[8]In this study, tendon was modeled as part of the series elasticity.

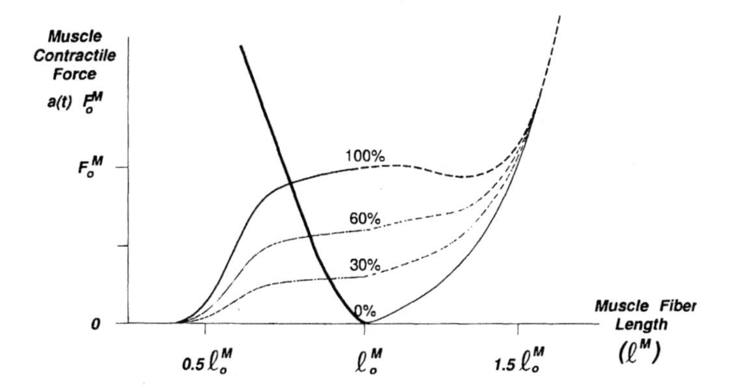

Figure 2.24. Plotting a tendon force-length curve backwards on the muscle tension-length diagram. The total length of the musculotendon actuator is considered to be fixed and equal to the total resting lengths $\ell^{MT} = \ell_o^M + \ell_r^T$ (pinnation ignored). The bottom end of the backwards tendon tension curve must be located on the horizontal axis at length $\ell^M = \ell^{MT} - \ell_r^T = \ell_o^M$. Then the intersection point between the total muscle tension-length diagram at activation $a(t)$ and the backward tendon curve will indicate the equilibrium tension under isometric conditions.

As far as the magnitude of musculotendon force, or equivalently, the magnitude of tendon force (F^T) is concerned, one can envision the interactions between muscle and tendon in a simple way. If a musculotendon actuator is held at some fixed length ℓ^{MT}, the muscle and tendon in series will adjust their lengths until force equilibrium is reached. That is, if the muscle fibers shorten

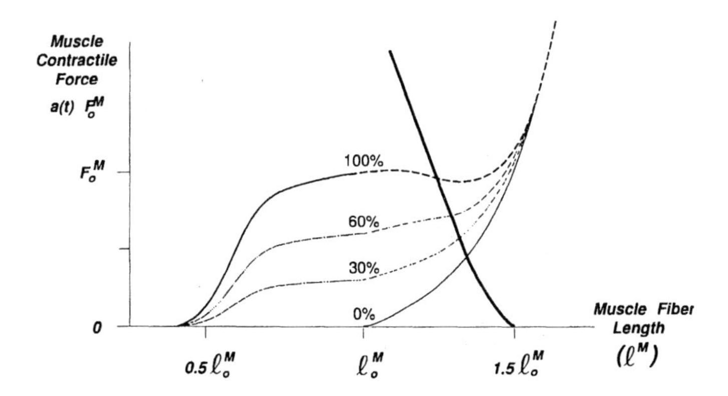

Figure 2.25. Plotting a tendon force-length curve backwards on the muscle tension-length diagram, when the total length of the muscle-tendon is held at length $\ell^{MT} = 1.5\ell_o^M + \ell_r^T$ (pinnation ignored). The bottom end of the tendon curve is moved to the right from its location in Figure 2.24 to new position $\ell^M = \ell^{MT} - \ell_r^T = 1.5\ell_o^M$. Again, the intersection point between the total muscle tension-length diagram at activation $a(t)$ and the backward tendon curve will indicate the equilibrium tension under isometric conditions.

in accordance with their activation, length, and velocity dependent properties, the tendon must lengthen. Since tendon is elastic, it will stretch until its tension equals the contractile force exerted by the muscle along the tendon axes.

Under isometric conditions this can be illustrated by converting the tendon stress-strain or force-strain curve into a force-length curve, and plotting it *backwards* on the isometric muscle force-length curves plotted at various activation levels (see Figures 2.24 and 2.25). The bottom end (the point of zero force) on the backwards tendon force-length curve must coincide with muscle length,

$$\ell^M = \ell^{MT} - \ell_r^T \,. \tag{2.68}$$

Also, if the pinnation angle α is nonzero, the tension-length curves for each activation level $a(t)$ plotted must be reduced by multiplication with $cos\alpha$. After these steps are accomplished, the tendon tension at equilibrium can be graphically determined by simply locating the intersection of the backward tendon curve and the total muscle tension-length curves.

In order to numerically quantify what each element is doing, usually a *musculotendon modeling* computer code is used. While they may appear to be rather complicated in code form, they simply use a computer to iteratively find the equilibrium tension. Such a code includes a *muscle model*, which includes the tension-length-velocity characteristics of a generic muscle in normalized form (Zajac, 1989). Lengths are normalized to ℓ_o^M, velocities are normalized to $v_{max} = 10\ell_o^M \, s^{-1}$, and forces are normalized by dividing by F_o^M. Normalization allows a single subprogram to compute the tension for many muscles of different size and length. Individual muscle parameters (ℓ_o^M, F_o^M, pinnation angle α), and instantaneous values of activation ($a(t)$), muscle fiber length (ℓ^M), and contraction velocity ($v^M = -\dot{\ell}^M$) are passed into the generic subprogram, and the tensions F^M for specific muscles are computed. These values are obtained from original dissection, or from tables (Brand *et al.*, 1982; Fukunaga *et al.*, 1992, 1996; Friederich and Brand, 1990; Hoy *et al.*, 1989; Narici, 1992; Seireg and Arvikar, 1989; Van der Helm and Yamaguchi, 2000; Wickiewicz *et al.*, 1983; Yamaguchi *et al.*, 1990). Carhart (2000) has compiled one such table in Appendix B.

Also contained within a typical musculotendon modeling code is a *tendon model* which computes tendon stretch as a function of the magnitude of tendon force F^T. Under dynamic conditions, a transient buildup of muscle-tendon force is produced by a change in muscle activation, $\dot{F}^T \equiv dF^T/dt$. $\dot{F}^T$ may be obtained using the equations governing muscle-tendon contraction dynamics, which are functions of the instantaneous tendon force F^T, the lengths and velocities of the muscles and tendons, and the current activation (Zajac, 1989),

$$\frac{dF^T}{dt} = f\left(F^T,\, \ell^{MT},\, v^{MT},\, a(t)\right) \tag{2.69}$$

This buildup of force by way of muscle and tendon interactions is referred to as *musculotendon contraction dynamics.*[9]

It should be noted that the time constants for the tendon portion of the model are several orders of magnitude longer than the time constants for muscle contraction. Therefore, one usually simplifies the analysis by considering the muscle contraction and relaxation cycle to be so short as to nullify any creep exhibited by the tendon. Tendon is replaced by a linear or nonlinear spring, and the damping characteristics of tendon are usually ignored for quick movements.

A useful value to compute using a musculotendon modeling code is *power*, which is the rate of performing mechanical work,

$$P^i = \frac{\int \vec{F}^i \cdot d\vec{s}}{dt} \tag{2.70}$$

$$= F^i v^i \tag{2.71}$$

where superscript (^{i}) is used to indicate muscle (^{M}), tendon (^{T}), or musculotendon (^{MT}). An abbreviation for tension $F^i \equiv |\vec{F}^i|$ and contraction velocity $v^i = -\dot{\ell}^i$ are used in the equation.

Power is defined *positively* here when the muscle, tendon, or musculotendon element shortens and produces work on the system. The system can also perform *negative work* on the muscles and tendons, by stretching them when a tensile force is present. When energy is stored in tendon, much of it is available for creating skeletal movements provided it is not dissipated by a decrease in muscle tension. When the muscle fibers are stretched, some energy is stored

[9] (*The following is explained in footnote form as this material may not make sense until after Chapters 6 and 7 are read.*) By including F^T as a *state variable* in the dynamic equations, contraction dynamics can be taken into account during dynamic simulations of musculoskeletal motion. Most of the elements within the *vector of states* are basic variables such as positions and velocities that provide a precise instantaneous description of the system. Additional variables can be added if it is important to compute the dynamics of these variables. Additional time derivatives, for instance $\dot{F}^T$, must also be computed on an on-going basis as the simulation proceeds from initial time $t = 0$ to the end time.

During a simulation, ℓ^{MT} and v^{MT} are computed using the methods of vector kinematics described in Chapter 4. The muscle length and velocity ℓ^M and v^M are determined or estimated, then used to find muscle force F^M in accordance with the muscle force-length-velocity properties at the current activation level $a(t)$. An imbalance in forces $F^M \cos\alpha$ and F^T leads to a rise or fall in F^T, and hence, allows the computation of $\dot{F}^T$.

Activation dynamics, the transient response of muscle activation to a change in "neural" control input, can also be included in dynamic simulations by including muscle activations as elements of the state. It is usually desirable to include *both* activation and contraction dynamics when performing simulations in which final results are to be obtained. One simple way of describing muscle activation dynamics is as a first-order response to stepwise changes in the "neural" controls, which makes it easy to compute $\dot{a}(t)$.

By appending both $a(t)$ and $F^T(t)$ to the state vector for each muscle included in the dynamic simulation model, the derivatives of these variables can be computed at the current time and integrated "for free" along with the positions and velocities of the system. This process allows them to be predicted at future times (as $t \to t + \Delta t$). By constantly updating and incorporating the instantaneous values of $a(t)$ and $F^T(t)$ as time proceeds, the limitations in muscle force buildup and decay, and the interplay of forces between muscle and tendon may be included within a musculoskeletal model.

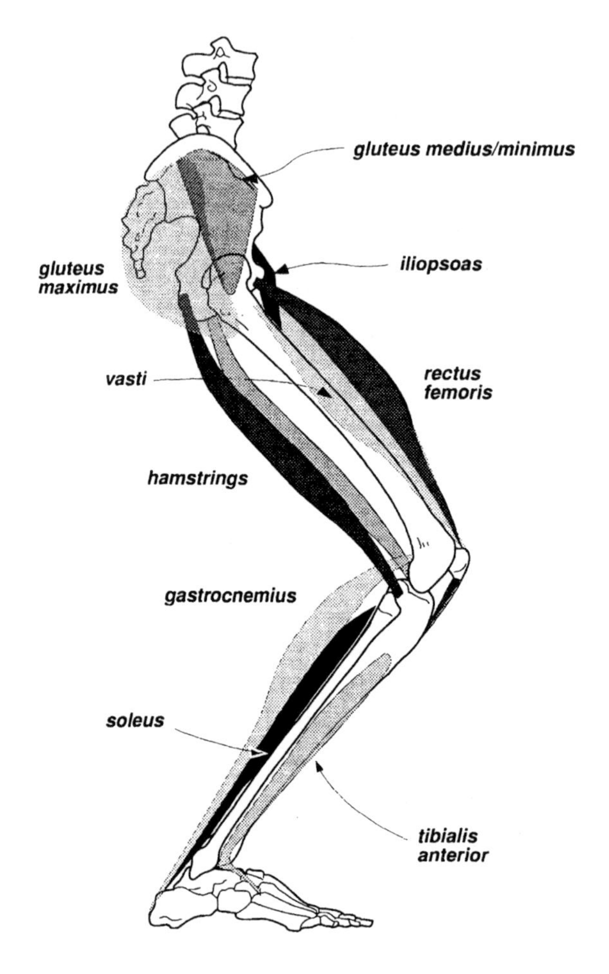

Figure 2.26. Some major muscle groups of the lower extremity.

within the elastic elements of the muscle, but large amounts of energy are mechanically dissipated as heat because of the presence of viscous damping within the muscle. Mechanical energy dissipation should not be confused with the metabolic energy required to create and maintain the tensile force in muscle.

In order to interpret musculotendon power curves, the key facts to remember are relatively simple.

- Tensions F^T and $F^M cos\alpha$ are equal and the lengths ℓ^T and $\ell^M cos\alpha$ are additive.

- *Muscle always generates a tensile force,* but it can shorten only if the tendon force is less than the force the muscle is capable of generating.

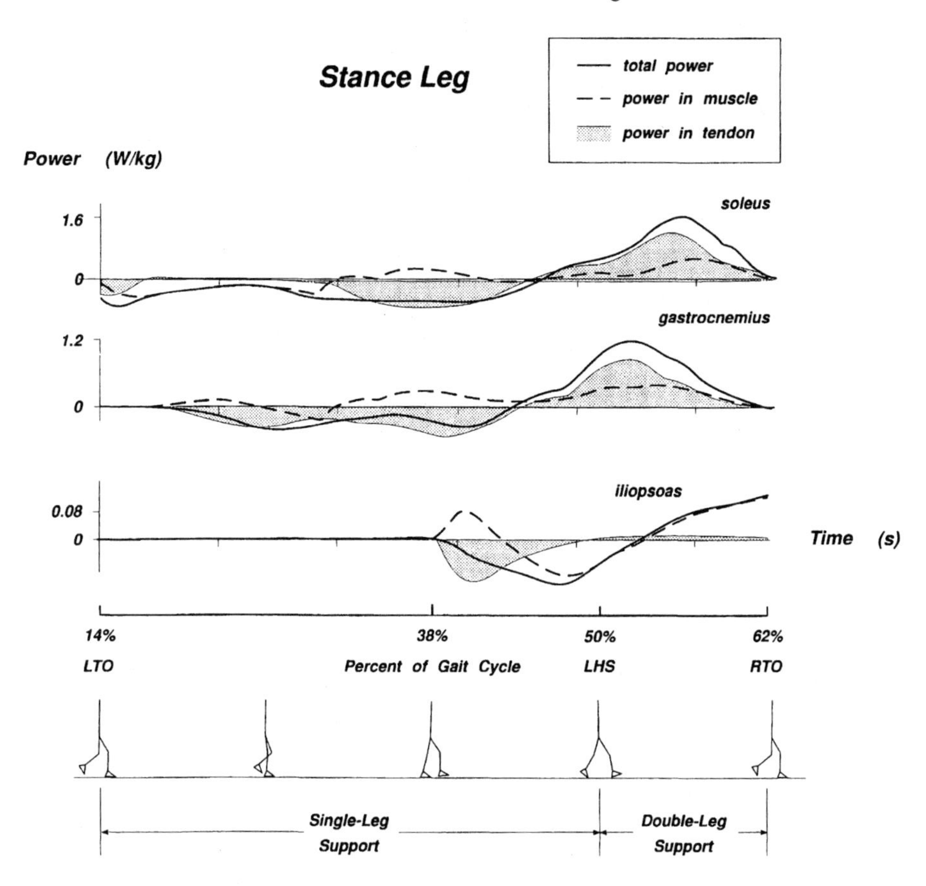

Figure 2.27. Power stored and expended by the major plantarflexors and hip flexor during gait. Power is the rate of performing work, and is the product of tension and velocity. Muscle power, defined as $P^M = F^M v^M$ and plotted as thin solid lines, is positive when the muscle shortens ($v^M > 0$) and performs work on the structures it is attached to. Tendon power is defined in the same way, $P^T = F^T v^T$, and is plotted as the shaded regions. Total musculotendon power is the sum of the muscle power and the tendon power (thick solid lines). In the top two curves, note that the tendon power is stored within the tendon at a low negative rate during the stance phase of gait, and expended during "pushoff" at a higher positive rate. The total negative shaded area corresponds to the energy stored in the tendon, and is equal to the total positive shaded area because the energy stored in the tendon was recovered. In the *iliopsoas* muscle (which is a grouping of two muscles, *iliacus* and *psoas major*), note that the initial phase of muscle contraction simply stretches the tendon because the positive work of the muscle causes negative energy to stored in the tendon.

- ℓ^T *always shortens when force decreases, and lengthens when force increases.* Negative tendon power indicates increasing force levels in both the muscle and tendon because they are connected in series.

- When power is zero in a muscle, tendon, or musculotendon, either the tensile force is zero, its shortening velocity is zero, or both are zero.

By keeping these thoughts in mind, the observer can infer much from the muscle-tendon power curves without requiring additional information regarding the muscle lengths, tendon lengths, and musculotendon lengths.

For instance, the curves provided in Figure 2.27 show the muscle power, tendon power, and musculotendon power for three muscles during normal human gait. Figure 2.26 provides some guidance as to where these muscles are in the lower extremity. In the top two curves of Figure 2.27, energy is absorbed from the contractile elements and from retarding the forward rotation of the lower leg about the ankle during the "stance phase." The energy storage within the tendons appears on the power curves as negative powers. For each muscle, the total area enclosed between its tendon power curve and the horizontal axis indicates the amount of energy stored within the tendon. Later, as body weight is shifted from the stance leg to the contralateral leg, the energy stored as passive stretch is released at a higher rate of energy expenditure during the "push-off phase" of gait. The energy expended from the tendon is the area enclosed on the positive side of the horizontal axis. When the positive areas and the negative areas are nearly the same, very little energy dissipation has occurred.

The third curve of Figure 2.27 is included as an illustration of muscle force buildup, which causes shortening in the muscle and lengthening in the tendon. This is seen in the figure as positive muscle power and negative tendon power. The musculotendon unit does not exhibit as much power as the tendon because it is moving with smaller velocity. As the tendon continues to absorb energy, it must continue lengthening and increasing in tensile force. (*Why?*) It is not until the tendon stops stretching that the musculotendon power begins to match the muscle power. This implies that the tension in the tendon remains relatively constant, and that further movements of muscle create matching movements of the skeletal structure.

2.5 SUMMARY

In summary, the study of muscle, tendon, and the interactions between them can barely be introduced in a single chapter. The emphasis in this chapter was placed on studying how aggregate tensile force – the primary interest for a biodynamicist – is affected by muscle length, velocity, pinnation, activation, and tendon elasticity. The topics of muscle modeling and musculotendon modeling were also introduced using basic models without mass. It is expected that any reader desiring to know the current state of the art will go directly to the literature to compare and contrast the latest models and methods before proceeding to his or her own modeling topic.

That being said, enough background has been given for a reader to become conversant with a muscle physiologist, a muscle modeler, or a musculoskeletal modeler. However, what was presented here only scratches the surface of a broad and deep ocean of knowledge. One could easily devote a lifetime of study in a subset of this area, and many have. Should you be privileged to have this opportunity, you will find a ready audience for your discoveries.

2.6 EXERCISES

1. A muscle is stretched by attaching a weight $W = 2F_o^M/3$. Assuming no tendon stretch, ignoring dynamic effects, and making reasonable assumptions regarding muscle properties,

   (a) What will be the muscle's length at equilibrium if the muscle is at rest (activation $a(t) = 0$)?

   (b) If the muscle is now activated to 50%, what will the new muscle length be?

   (c) What if $a(t) = 1.0$?

   Use the active and passive force-length diagrams to explain your answers in Parts (a) to (c).

2. Find the shortening velocity for which maximum power is developed by a muscle, assuming,

   (a) A linear force-velocity curve with a slope of $-F_o^M/v_{max}$.

   (b) A force-velocity curve obeying Hill's equation. Start by using the alternate form of Hill's equation,

$$v^M = -\dot{\ell}^M = \frac{F_o^M - F^M}{(F^M + a)/b}. \tag{2.72}$$

   Assume that force is constant in the tendon for Parts (a) and (b). Shortening velocity is defined as positive when the muscle, which has length ℓ, is shortening.

3. *Design of an exercise machine.* A muscle is to be exercised by designing a weightlifting mechanism composed of a pulley, cam, weight, and cable arrangement. For simplicity the joints, bones, *etc.* are neglected and the muscle is considered to be wrapped around the pulley and affixed at its end. The pulley and cam are welded rigidly to each other so that they move together as a unit. The pulley of uniform radius r is pinned at its center, C, which also serves as the center of rotation for the cam. Cables wrapped about the pulley and cam are connected to the muscle and weight, respectively. As shown in Figure 2.28, a concentric contraction of the muscle will serve

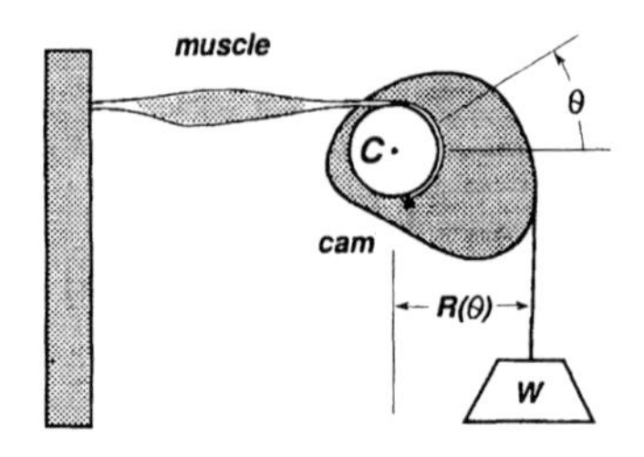

Figure 2.28. Figure for Problem 2.3.

to rotate the pulley-cam structure in a counterclockwise direction, raising the weight. Your task is to determine the optimal displacement $R(\theta)$ of the vertical cable from C, as a function of rotation angle θ, so as to maximally and uniformly exercise the muscle.

(a) Find a simple mathematical expression (*i.e.*, sine, cosine, *etc.*) that approximately describes muscle force F^M as a function of length ℓ^M.

(b) Derive $R(\theta)$ using your answer from Part (a). Assume that $r = \ell_o^M/\pi$ and that $\theta = 0$ when $\ell^M = 1.5\ell_o^M$.

4. Graph the transient behavior of the Maxwell model during a stretch/relaxation cycle at two different loading rates, as shown in Figure 2.29. Case A shows a fast stretch/relaxation, while Case B depicts a slow stretch/relaxation. Begin by writing $F(t)$ as a linear function of time, and solve for elongation $x(t)$ in both cases. Sketch the ascending and descending curves of force $F(t)$ versus deflection $x(t)$, and graphically estimate the energy lost during the stretch/relaxation cycle.

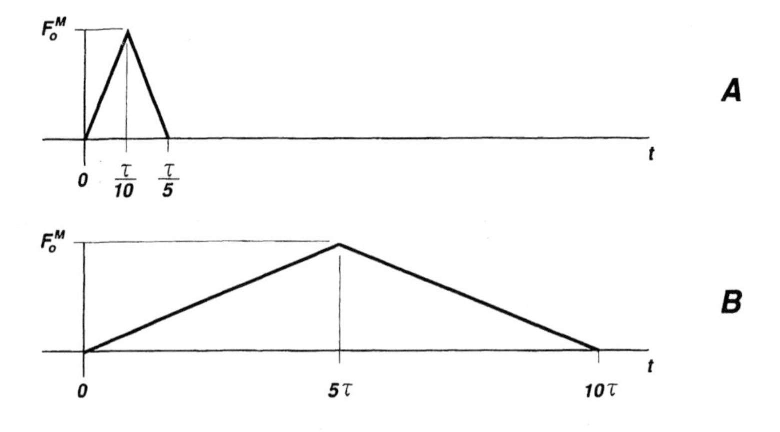

Figure 2.29. Figure for Problem 2.4.

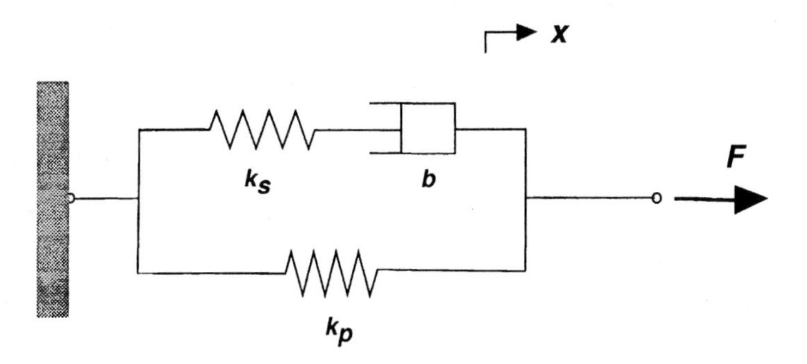

Figure 2.30. Kelvin model for Problem 2.5.

5. *Derivation of differential equation for a Kelvin model.* Referring to Figure 2.30, let x_{k_s} be the extension of the series elasticity, x_b the extension of the dashpot, and x_{k_p} the extension of the parallel elasticity. Similarly, F_{k_s}, F_{k_p}, and F_b are the magnitudes of the tensile forces in the series and parallel springs, and the dashpot, respectively. The following relations are in effect,

$$F_{k_s} = F_b \tag{2.73}$$

$$x_{bk_s} = x_b + x_{k_s} \tag{2.74}$$

$$F = F_{k_s} + F_{k_p} \tag{2.75}$$

$$x = x_{bk_s} = x_{k_p} \,. \tag{2.76}$$

(a) Form an equivalent expression for the quantity $F + (b\dot{F}/k_s)$ using equations for force development in springs and dashpots, and the relations you have. Write this in equation form, with $F + (b\dot{F}/k_s)$ on the left hand side, and your expression on the right hand side. Simplify the equation to obtain a first order differential equation relating F, $\dot{F}$, x, and $\dot{x}$. Put your answer into the following form,

$$F + \tau_\epsilon \dot{F} = k_p \left(x + \tau_\sigma \dot{x}\right) \,, \tag{2.77}$$

where

$$\tau_\epsilon = \frac{b}{k_s} \tag{2.78}$$

$$\tau_\sigma = \frac{b}{k_p}\left(1 + \frac{k_p}{k_s}\right) . \tag{2.79}$$

τ_ϵ is the *relaxation time constant for constant strain* (constant deformation). τ_σ is the *relaxation time constant for constant stress* (constant load).

(b) Solve this differential equation for $x(t)$ when the load is constant ($\dot{F} = 0$). Plot and compare your answer to Figure 2.16.

(c) Solve this differential equation for $F(t)$ under conditions of constant deformation ($\dot{x} = 0$). Plot and compare your answer to Figure 2.16.

6. *Muscle tendon interactions.* A muscle and tendon are attached together in series. The musculotendon actuator is held at fixed length $D = \ell_r^T + 1.25\ell_o^M$, where ℓ_r^T is the resting tendon length. D is defined by two immovable supports. At time t_o, a stimulus $c(t)$ is applied to the muscle which causes it to contract. Assuming first-order activation dynamics, and ignoring dynamical effects (inertia), plot the *approximate* time courses of:

   (a) activation $a(t)$;

   (b) muscle length ℓ^M;

   (c) tendon length ℓ^T;

   (d) tendon force F^T.

   For simplicity, assume that tendon stiffness, k^T, is linear and given by,

$$k^T = 10\left(\frac{F_o^M}{\ell_o^M}\right). \tag{2.80}$$

   Use time constants $\tau_{act} = 12$ ms, and $\tau_{deact} = 24$ ms.

7. *Does a Sharp Dip Exist in the Muscle Force-Length Curve for Muscles Having Long Tendons?* Some curves in the literature show a total (active + passive) isometric muscle force-length curve that has a sharp dip at lengths greater than the optimal fiber length ℓ_o^M. For instance, McMahon (1984) shows the following curves for the *gastrocnemius* and *sartorius* muscles. Note that *gastrocnemius* (left) does not exhibit the sharp dip, whereas *sartorius* does.

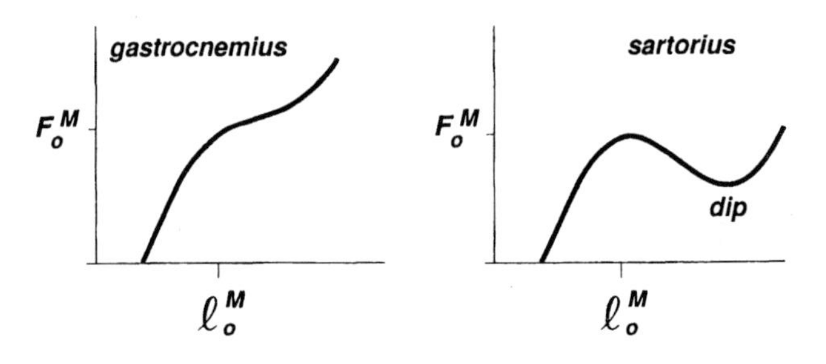

Figure 2.31. Figure for Problem 2.7.

Let's create the following hypothetical situation. Supposing that the *sartorius* muscle was attached in series with a long tendon, and that the entire musculotendon unit, fully activated, was slowly (*i.e.*, quasistatically) stretched from a short length to a long length. Explain what would happen as the muscle fiber length approached and surpassed its optimal fiber length. You can use qualitative graphs of the tendon length versus time and the fiber length versus time to aid you in your explanation.

Note: This situation probably explains why *sartorius* does not have a long tendon. In fact, $\ell_o^M = 57.9$ cm, whereas resting tendon length $\ell_r^T = 4.0$ cm for this biarticular muscle spanning the hip and knee. Thus, the ratio of tendon length to muscle fiber length is only 0.07, whereas in the hamstrings (a grouping of biarticular hip-knee muscles on the posterior side of the femur) it is 3.85.

8. Muscle is always attached to bones via a stretchy tendon. Therefore, the muscle model is actually connected to bone via a series elastic spring, k^T. We can group this elasticity from both sides of the muscle into a single elasticity on one side of the muscle only (see Figure 2.22). For simplicity, consider the tendon and muscle elasticities to be linear.

   Select a muscle from Appendix B, and compute the ratio ℓ_r^T / ℓ_o^M. Given a fixed muscle-tendon length equal to the relaxed lengths of the muscle plus tendon, and using values from the Appendix for tendon stiffness, optimal muscle length, and optimal muscle force, write a computer program or utilize a spreadsheet to find the equilibrium length and force of the muscle and tendon in series when the muscle activation is specified as an input.

   Hint: Start by creating a table that approximates the isometric force-length curve of the muscle at 100% activation. Then, when a different activation level is specified, the table can be scaled accordingly.

9. Muscle, tendon, and musculotendon power curves are provided in Figure 2.32 for the hip abductors, hamstrings, and dorsiflexor muscles of the *swing* leg during normal human walking. Referring to the muscle diagram of Figure 2.26, what observations can be made regarding the interactions of muscle and tendon? Consider factors such as energy storage and release, energy dissipation, tension buildup, tendon to muscle length ratios, muscle and tendon velocities, *etc.*

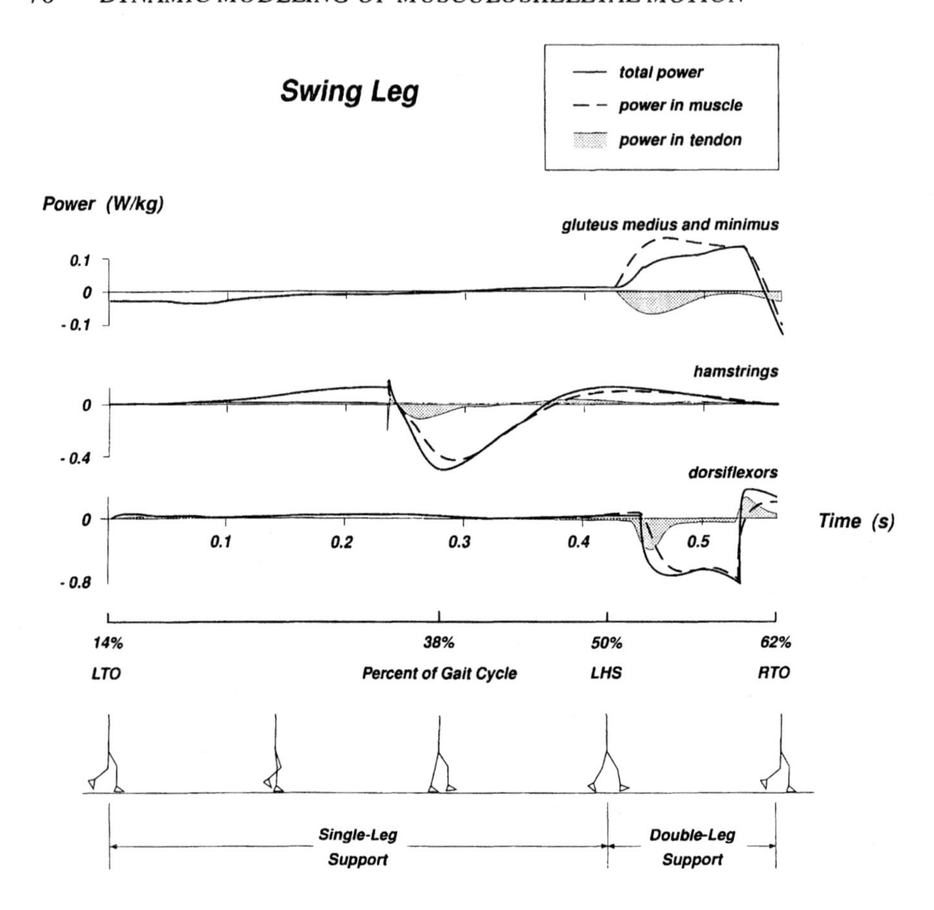

Figure 2.32. Figure for Problem 2.9, showing power produced by the muscles, tendons, and musculotendon actuators of the swing leg during normal human gait. Negative power is produced during eccentric contractions.

References

Alexander, R. M. (1984) "Walking and Running." *American Scientist*, V. 72, no. 4, pp. 348-354.

Brand, R. A., Crowninshield, R. D., Wittstock, C. E., Pedersen, D. R., Clark, C. R., and Van Krieken, F. M. (1982) "A model of lower extremity muscular anatomy." *J. Biomechanical Engineering*, V. 104, pp. 304-310.

Butler, D. L., Grood, E. S., Noyes, F. R., and Zernicke, R. F. (1979) "Biomechanics of Ligaments and Tendons," in *Exercise and Sport Sciences Review.* Franklin Institute Press, Philadelphia, PA, V. 6, pp. 125-183.

Friederich, J. A. and Brand, R. A. (1990) "Muscle fiber architecture in the human lower limb." *Technical Note, Journal of Biomechanics*, V. 23, pp. 91-95.

Fung, Y. C. B. (1967) "Elasticity of soft tissues in simple elongation." *American Journal of Physiology*, V. 213, pp. 1532-1544.

Fukunaga, T., Roy, R. R., Shellock, F. G., Hodgson, J. A., Day, M. K., *et al.* (1992) "Physiological cross-sectional area of human leg muscles based on magnetic resonance imaging." *J. Orthop. Res.*, V. 10, n. 6, pp. 928-934.

Fukunaga, T., Roy, R. R., Shellock, F. G., Hodgson, J. A., and Edgerton, V. R. (1996) "Specific tension of human plantar flexors and dorsiflexors." *J. Appl. Physiol.*, V. 80, n. 1, pp. 158-165.

Gordon, A. M., Huxley, A.F., and Julian, F. J. (1966)"The variation in isometric tension with sarcomere length in vertebrate muscle fibres." *Journal of Physiology*, V. 184, pp. 170-192.

He, Jiping, Levine, William S., and Loeb, Gerald E. (1991) "Feedback gains for correcting small perturbations to standing posture." *IEEE Transactions on Automatic Control*, V. 36, n. 3, pp. 322-332.

Henneman, E., and Olson, C.B. (1965) "Relations between structure and function in the design of skeletal muscles." *Journal of Neurophysiology*, V. 28, pp. 581-598.

Hill, A. V. (1938) "The heat of shortening and the dynamic constants of muscle." *Proc. Roy. Soc. B. (Lond.)*, V. 126, pp. 136-195.

Hill, A. V. (1950) "The series elastic component of muscle." *Proc. Roy. Soc. B. (Lond.)*, V. 137, pp. 273-280.

Hill, A. V. (1953) "The mechanics of active muscle." *Proc. Roy. Soc. B. (Lond.)*, V. 141, pp. 104-117.

Hill, D. K. (1968) "Tension due to interaction between the sliding filaments in resting striated muscle. The effect of stimulation." *Journal of Physiology*, V. 199, pp. 637-684.

Hof, A. L., Geelen, B. A., and Van Den Berg, J. (1983) "Calf muscle moment, work and efficiency in level walking: Role of series elasticity." *Journal of Biomechanics*, V. 16, n. 7, pp. 523-537.

Hoy, M. G., Zajac, F. E., and Gordon, M. R. (1989) "A musculoskeletal model of the human lower extremity: The effect of muscle, tendon, and moment arm on the knee and ankle. *Journal of Biomechanics*, V. 22, pp. 157-169.

Huxley, A. F. (1974) "Review lecture: Muscular contraction." *Journal of Physiology*, V. 243, pp. 1-43.

Ker, R. F. (1981) "Dynamic tensile properties of the plantaris tendon of sheep (*Ovis aries*)." *J. Exp. Biol.*, V. 93, pp. 283-302.

McMahon, Thomas A. (1984) *Muscles, Reflexes, and Locomotion.* Princeton University Press, Princeton, NJ.

Narici, M. V., Landoni, L., and Minetti, A. E. (1992) "Assessment of human knee extensor muscles stress from *in vivo* physiological cross-sectional area and strength measurements." *Eur. J. Appl. Physiol.*, V. 65, n. 5, pp. 438-434.

Pinto, J. G. and Fung, Y. C. (1973) "Mechanical properties of the heart muscle in the passive state." *Journal of Biomechanics*, V. 6, pp. 597-616.

Pollack, Gerald H. (1990) *Muscles and Molecules: Uncovering the Principles of Biological Motion*. Ebner and Sons, Seattle, WA.

Seireg, A. and Arvikar, R. J. (1989) *Biomechanical Analysis of the Musculoskeletal Structure for Medicine and Sports*. Hemisphere Publishing Corporation, New York.

Van der Helm, Frans C. T., and Yamaguchi, G. T. (2000) "Morphological data for the development of musculoskeletal models: An update." Appendix 1 in *Biomechanics and Neural Control of Posture and Movement*,, pp. 717-773.

Wickiewicz, T. L., Roy, R. R., Powell, P. L., and Edgerton, V. R. (1983) "Muscle architecture of the human lower limb." *Clinical Orthopaedics and Related Research*, V. 179, pp. 275-283.

Wilkie, D. R. (1956) "Measurement of the series elastic component at various times during a single muscle twitch." *J. Physiology (Lond.)*, V. 134, pp. 527-530.

Winters, J. M. and Crago, P. E. (2000) *Biomechanics and Neural Control of Posture and Movement*. Springer-Verlag, New York, NY. *Provides a section relating to recent developments in muscle modeling.*

Woo, Savio L-Y., and Buckwalter, Joseph A. (ed.) (1988) *Injury and Repair of the Musculoskeletal Soft Tissues*. American Academy of Orthopaedic Surgeons, 40001, Rosemont, IL.

Yamaguchi, G. T. (1989) *Feasibility and Conceptual Design of Functional Neuromuscular Stimulation Systems for the Restoration of Natural Gait to Paraplegics Based on Dynamic Musculoskeletal Models*. Ph.D. Dissertation, Department of Mechanical Engineering, Stanford University, Stanford, CA.

Yamaguchi, G. T. and Zajac, F. E. (1990) "Restoring unassisted natural gait to paraplegics with functional neuromuscular stimulation: A computer simulation study". *IEEE Transactions on Biomedical Engineering*, V. 37, n. 9, pp. 886-902.

Yamaguchi, G. T., Sawa, A. G. U., Moran, D. W., Fessler, M. J., and Winters, J. M. (1990) "A survey of human musculotendon actuator parameters," Appendix in *Multiple Muscle Systems: Biomechanics and Movement Organization*, J. M. Winters and S. L.-Y. Woo (eds.), Springer-Verlag, New York.

Zajac, F.E. (1989) "Muscle and tendon: Properties, models, scaling and application to biomechanics and motor control." *CRC Critical Reviews in Biomedical Engineering*, V. 17, n. 4, pp. 359-411.

II
DEFINING SKELETAL KINEMATICS

Chapter 3

RIGID BODIES AND REFERENCE FRAMES

Objective – The ability to simultaneously work in multiple reference frames greatly simplifies kinematic and dynamic analyses in three dimensions. In this chapter, the reader is introduced to direction cosines, which provide the key to analyzing models with multiple degrees of freedom.

3.1 GENERALIZED COORDINATES

Systems of moving bodies can be characterized by the number of generalized coordinates n necessary and sufficient to uniquely define the configuration of the system at any particular instant in time. This integer value n is referred to as the number of *degrees of freedom* possessed by the system, and does not include parameters fixed by geometry.

As an example, suppose point P is constrained to follow a line L which is fixed in space and passes through a fixed point O. One parameter, say x, is needed to determine the distance from O to P. Because points are considered to be infinitesimally small, the rotational orientation of P about L is unnecessary, and the system is said to possess one degree of freedom.

If P was a bead of finite size constrained to move along the line, one would have to consider both the linear position of P on L, and its rotational angle about line L. This system would require two degrees of freedom to be specified to uniquely define its instantaneous configuration. Another example of a two degree of freedom system is the two link planar linkage shown in Figure 1.1. Knowing the two rotation angles q_1 and q_2 completely specifies the system configuration, because point A_o is fixed in space. Even though several coordinates would be needed to locate point A_o, these coordinate values are not counted as degrees of freedom because they are fixed parameters of the system.

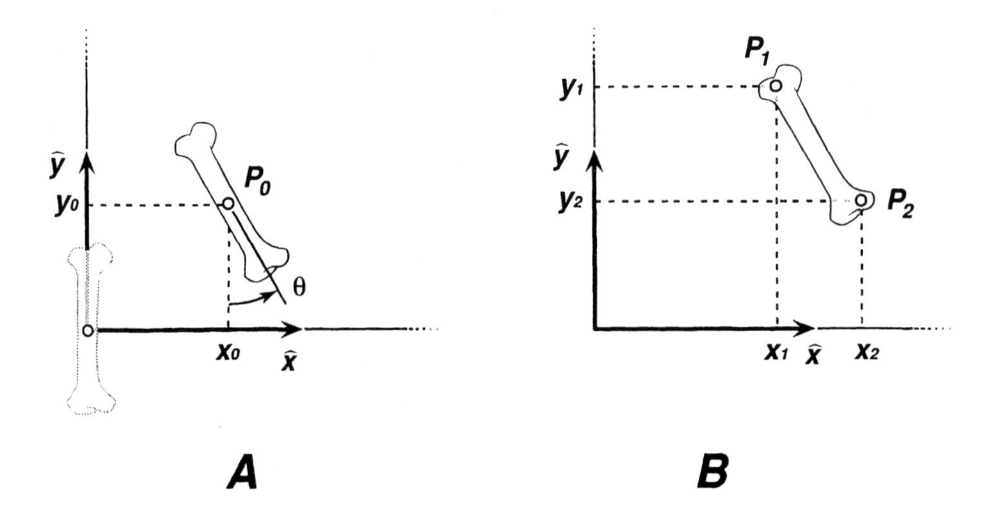

Figure 3.1. A rigid body translating and rotating in a plane. Its instantaneous configuration may be specified by the coordinates of one point and an angle (A, requiring three parameters), or by the coordinates of two points (B, requiring four parameters).

3.1.1 SINGLE RIGID BODIES IN THE PLANE AND IN CARTESIAN SPACE

Three degrees of freedom are necessary and sufficient to uniquely define the position and orientation of an object confined to planar motion. For example, the horizontal ($\hat{x}$) and vertical ($\hat{y}$) locations of a point P_0 on the object and a rotation angle (θ) as shown in Figure 3.1A. A different set of information, for example the point locations of two points $P_1(x_1, y_1)$ and $P_2(x_2, y_2)$ of the rigid body in the plane of motion can also be sufficient to uniquely define the location and orientation of the body (Figure 3.1B). However the latter set must specify four pieces of information while the previous set only needs three known parameters. The *minimum* number of parameter values required to determine the instantaneous configuration is the number of degrees of freedom. Thus, an object moving in planar motion has three, not four, degrees of freedom.

By the same reasoning, a rigid body moving unconstrained in Cartesian space has six degrees of freedom (Figure 3.2). For example, the specification of a point location in 3-D space requires three variables x_p, y_p, and z_p. The location of one point on the rigid body thus requires three degrees of freedom. Three more degrees of freedom are required to define the orientation of the object (three rotation angles, for instance).

Constraints act to reduce the degrees of freedom via the following equation,

$$n = N_{free} - c \tag{3.1}$$

where n is the number of degrees of freedom of the constrained system, N_{free} is the number of degrees of freedom for a "free" or unconstrained system, and c is the number of *constraint equations* that can be written. A "hard" constraint equation defines a relationship among the system variables using an equality ($=$). For instance, if we had three variables x, y, and z, the simple equations $z = 5$ and $x + y = -1$ are examples of constraint equations that can be written. Each constraint equation reduces the number of degrees of freedom possessed by the system by one. Other types of constraint equations exist, but may or may not result in a reduction of the system degrees of freedom.[1]

In terms of real world situations, *configurational constraints* are constraint functions as they relate the system variables using an equality (*e.g.*, $\gamma = \alpha + \beta$ or $z_p = x_p$ for the system depicted in Figure 3.2). Thus, for every configurational constraint, the number of degrees of freedom possessed by the system decreases by one. *Boundary constraints* involve an inequality and/or an equality (*e.g.*, $y_p > 5$ or $\alpha \leq \beta$). These constraints only reduce the number of degrees of freedom during time periods when the equality is in effect.

Motion constraints are a third kind of constraint which exist when the first time derivatives of the system variables can be related via equality equations (*e.g.*, $\dot{\gamma} = 2\dot{\alpha}$, or $\dot{z}_p = -\alpha\dot{x}_p$). Of course, one can always derive a "motion constraint" by differentiating a configurational constraint. That is, if $\gamma = \alpha - \beta$, one might be tempted to call $\dot{\gamma} = \dot{\alpha} - \dot{\beta}$ a "motion constraint".

[1]For instance, see *motion constraints* later in this section.

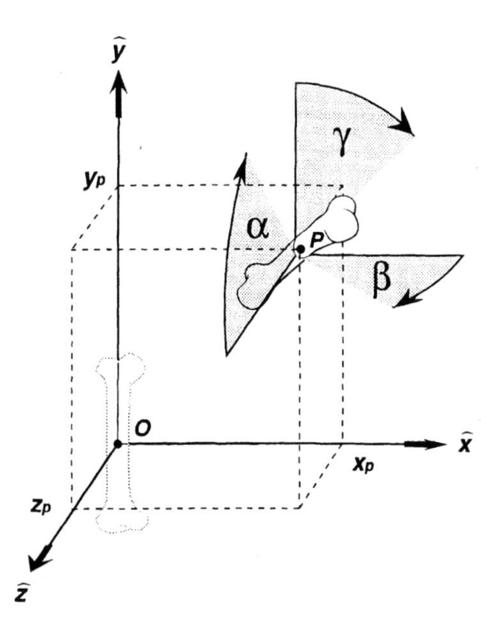

Figure 3.2. A rigid body translating and rotating in three dimensional space from its original location centered at point O. Its instantaneous configuration is specified by a minimum of six quantities, for instance the coordinates of one point and three angles. The instantaneous configuration can also be specified using two points and an angle, or three points, but these require seven and nine parameters, respectively.

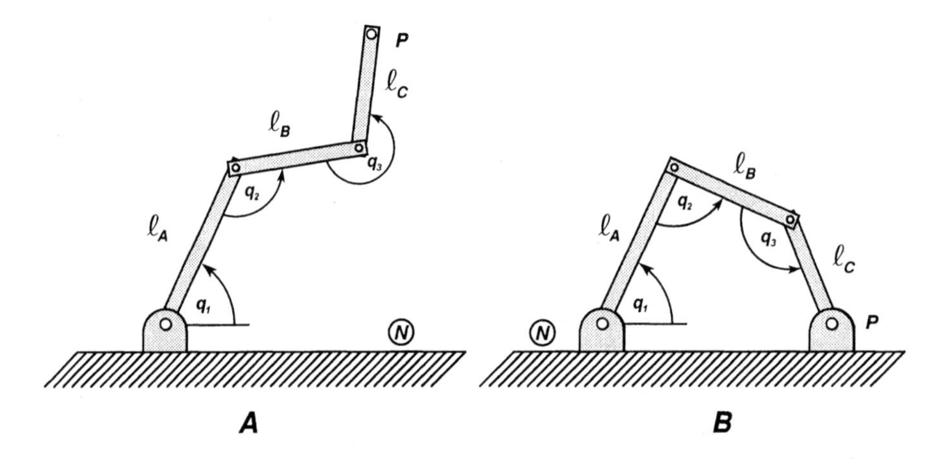

Figure 3.3. *A*. A system of three rigid links. If each joint has one degree of freedom, the system has a total of three degrees of freedom. *B*. When the endpoint P is rigidly pinned to the "ground", the system becomes a four bar linkage having only one degree of freedom.

To resolve this problem, motion constraints are distinguished from configurational constraints by examining the lowest order equation that can be written to physically describe the system. In the example just stated, if the zeroth order equation (*i.e.*, no time derivatives) describes an actual physical relationship between the system variables, then the system is said to be subject to a configurational constraint. It is called a configurational constraint even though the first derivative of that equation is also mathematically correct. On the other hand, if the first order equation describes an actual physical relationship between the velocities, but the corresponding zeroth order equation does not, then the system is said to be subject to a motion constraint. In other words, systems subject to configurational constraints also have constraints to their motions, but these are not called motion constraints. Systems subject to real motion constraints are not necessarily subject to corresponding configurational constraints.

3.1.2 EXAMPLE – THE FOUR-BAR LINKAGE

Three degrees of freedom uniquely define the configuration of the three-link planar system shown in Figure 3.3A when the endpoint P of link C is free. A set of angular measures defining the orientations of links A, B, and C is one set of generalized coordinates that uniquely specify the system configuration.

If, however, the endpoint is pinned to the base frame N as in Figure 3.3B, a fourth link is created (N) and the endpoint P is no longer free to move with respect to N. In this case, point P has lost its ability to move in the $\hat{x}$ (horizontal) and in the $\hat{y}$ (vertical) directions – minus two degrees of freedom in all. Hence the four-bar linkage has only one degree of freedom because two of

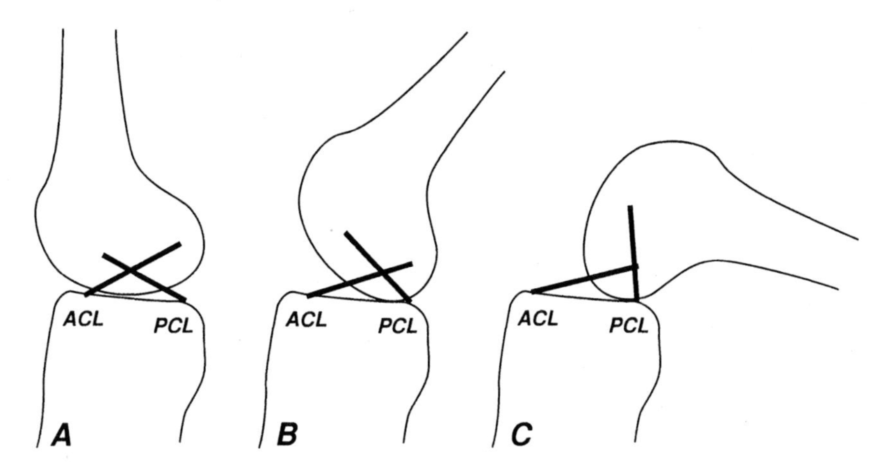

Figure 3.4. Depiction of the four bar linkage model of the knee proposed by Strasser (1917). The anterior cruciate (ACL) and posterior cruciate (PCL) ligaments remain extended and nearly constant in length as the joint is flexed. Because they do not go slack, they are considered to be two of the rigid links. The bony interconnections between the attachments to the tibia and the femur form the other two links.

the three original degrees of freedom were lost by pinning the endpoint. In fact, only one degree of freedom is needed to uniquely define the configuration of the system, such as the angle q_1 shown. The other angles, q_2 and q_3, are functions ($\mathcal{F}$ and $\mathcal{G}$) of the lengths ℓ_A, ℓ_B, and ℓ_C of links A, B, and C, respectively, and the single variable angle q_1,

$$q_2 = \mathcal{F}(q_1) \quad (3.2)$$

$$q_3 = \mathcal{G}(q_1) \; . \quad (3.3)$$

The study of four-bar linkages encompasses enough material to form an entire graduate course on the subject. Strasser (1917) even used a four-bar linkage to describe the mechanical action of the anterior and posterior cruciate ligaments of the knee (Figure 3.4). The anterior cruciate was considered to be one link, and the posterior cruciate ligament another. Even though they were not rigid, it was observed that they remained taut and maintained the same length throughout the range of flexion and extension. The bony spans between the connections on the femur and tibia formed the remaining two links. Though the crossed four-bar model of the knee has been disproven, the Strasser model did establish the primary mechanical interactions during flexion and extension.[2] The model has even been extended to make four-bar polycentric hinges (a hinge with a moving center of rotation) for prosthetic knee joints.

[2]This model does not account for the smaller three dimensional motions that are evident. For instance, during weight bearing, the geometry of the articulating surfaces greatly affects the knee motions.

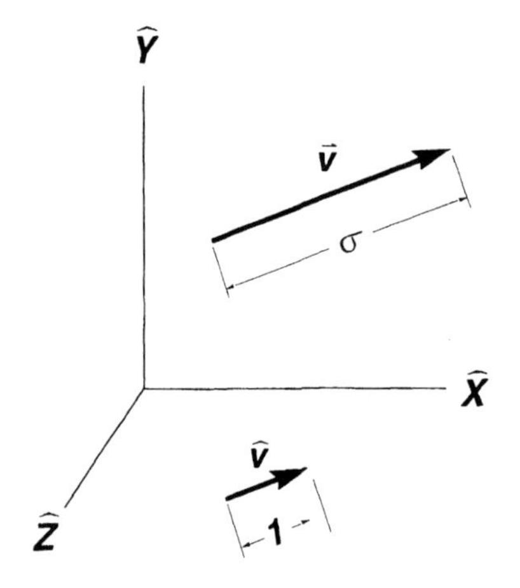

Figure 3.5. A vector $\vec{v}$ with scalar magnitude σ. A unit vector $\hat{v}$ which is parallel to $\vec{v}$ but has a magnitude of 1 is formulated by dividing $\vec{v}$ by its own magnitude, $\hat{v} = \vec{v}/|\vec{v}| = \vec{v}/\sigma$.

3.2 VECTORS AND SCALARS

A *vector* is a mathematical representation communicating both magnitude and direction. Vectors can represent physical quantities such as forces, moments, torques, positions, velocities, accelerations, angular velocities, and angular accelerations. For example, the long arrow in Figure 3.5 represents the vector $\vec{v}$. The length of the vector $\vec{v}$ is its *scalar* magnitude $|\vec{v}|$, and its direction is indicated by the direction of the arrow.

A vector with a magnitude of one is known as a *unit vector*. Here, a unit vector is indicated by a carrot overstrike (ˆ) instead of an arrow written above the name of the vector. A unit vector parallel to any vector $\vec{v}$ can always be found by dividing $\vec{v}$ by its magnitude,

$$\hat{v} = \frac{\vec{v}}{|\vec{v}|} \tag{3.4}$$

When a unit vector $\hat{v}$ is used to express the direction of a vector $\vec{v}$, a *scalar* multiplier σ may be used to conveniently express the magnitude $|\vec{v}|$,

$$\vec{v} = \sigma\,\hat{v} \tag{3.5}$$

The quantity σ is known as the *measure number* of the vector $\vec{v}$, and is a scalar because it only encodes magnitude (a scalar quantity), and not direction.

Vectors are added by placing them head to tail in accordance with general rules for vector addition. Any vector $\vec{u}$, in three-dimensional space can be formulated by algebraically summing three noncoplanar vectors,

$$\vec{u} = \alpha\,\vec{a} + \beta\,\vec{b} + \gamma\,\vec{c} \tag{3.6}$$

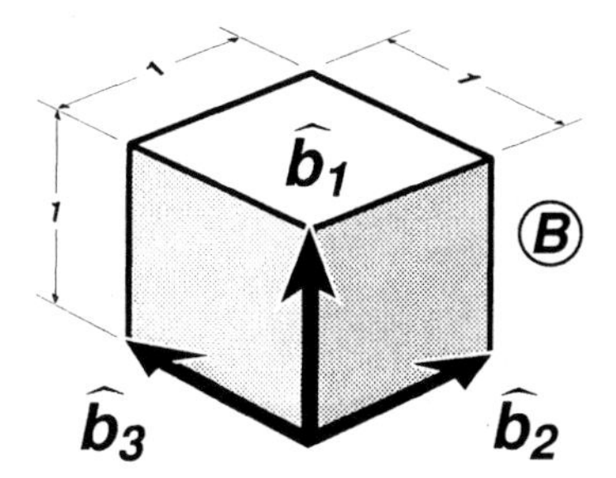

Figure 3.6. Mutually perpendicular unit vectors.

where $\vec{a}$, $\vec{b}$, and $\vec{c}$, are any three noncoplanar vectors, and α, β, and γ are three scalar measure numbers. Similarly, any vector $\vec{u}$ may be decomposed into component vectors in any three noncoplanar directions. The most common decomposition utilizes a *basis* (*i.e.,* a set of three noncoplanar vectors) comprised of *mutually perpendicular unit vectors,*

$$\vec{u} \;=\; \delta\,\hat{\kappa}_1 + \epsilon\,\hat{\kappa}_2 + \phi\,\hat{\kappa}_3\,, \tag{3.7}$$

where $\hat{\kappa}_1$, $\hat{\kappa}_2$, and $\hat{\kappa}_3$ are any three mutually perpendicular unit vectors, and δ, ϵ, and ϕ, are three scalar measure numbers as before.

Mathematically, any three noncoplanar vectors can be used to define a basis in three dimensional space. Hereafter in this text, however, a set of basis vectors will refer to the most commonly used basis, which is a set of mutually perpendicular unit vectors arranged in right-handed fashion (Figure 3.6). For example, if rigid body B has a reference frame affixed to it comprised of basis vectors $\hat{b}_1$, $\hat{b}_2$, and $\hat{b}_3$, the vectors are usually considered to have unit length, are mutually perpendicular, and arranged so the cross product $\hat{b}_1 \times \hat{b}_2$ equals $\hat{b}_3$ according to the right-hand rule (Figure 3.7). If the fingers of the right hand are curled in the same direction as the circular arrow, the extended thumb will yield the direction vector. In the same way, a counterclockwise circular arrow will be used here to indicate a vector pointing out of the drawing, and a clockwise circular arrow will be used to show a vector pointing perpendicularly into the paper (Figure 3.8).

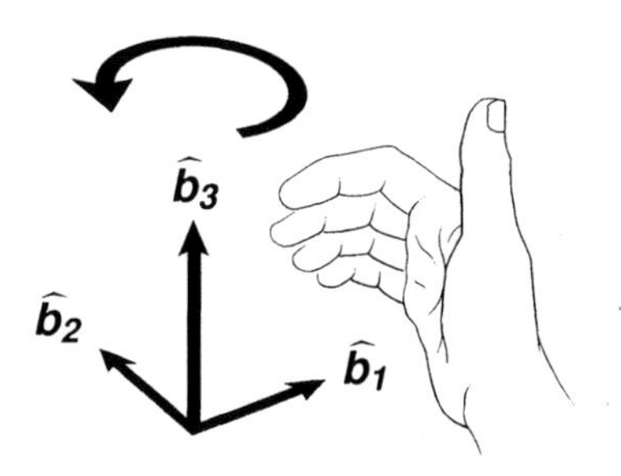

Figure 3.7. Right hand rule for vector cross products. When the fingers of the right hand are initially aligned with vector $\hat{b}_1$, and curled toward $\hat{b}_2$ through the acute angle, then $\hat{b}_3$ in a right handed system points in the same direction as the outstretched right thumb.

Figure 3.8. Right hand rule and sign convention used to draw vectors pointing perpendicularly into or out of the page. A "circular arc" with an arrowhead is shown on the page. When the fingers of the right hand are curled with the fingertips pointing in the same direction as the arrowhead, the direction of the extended right thumb indicates the direction of the vector.

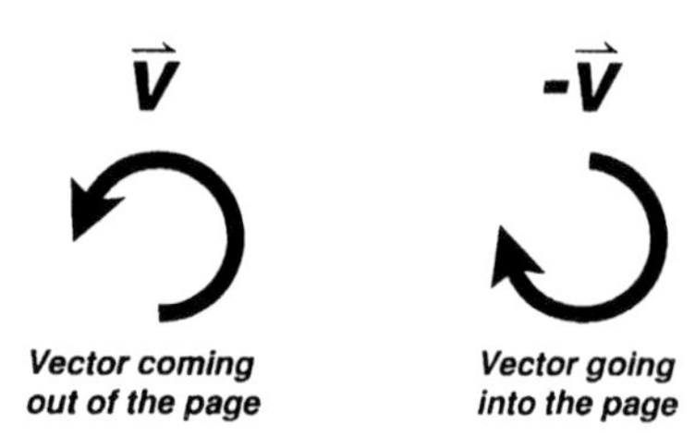

3.2.1 NOTATION

Following the notation used by Kane and Levinson (1985), the expression ${}^{A}\vec{v}^{P}$ will be used to indicate the *velocity of point P relative to reference frame A*. It is especially important to note the *left superscript* of the vector quantity, which is used to indicate the reference frame the vector is defined in. For instance, the velocity of a point P fixed in reference frame A has zero velocity in A,[3]

$$ {}^{A}\vec{v}^{P} = 0\,. \tag{3.8} $$

If reference frame A is moving without rotation and with constant velocity $u\vec{n}_1$ relative to reference frame N, then all the points of A will move at the same velocity $u\vec{n}_1$ relative to N, and therefore,

$$ {}^{N}\vec{v}^{P} = u\vec{n}_1\,. \tag{3.9} $$

Similarly, ${}^{N}\vec{p}^{P}$ will be used to indicate the *position of P in reference frame N*, meaning that ${}^{N}\vec{p}^{P}$ will be a vector originating from the origin of N and ending at its destination point P. It is often more convenient to use an alternative notation for position vectors which uses two right superscripts and no left superscript. For example, the position vector directed from point P to point Q in reference frame B would be represented as $\vec{p}^{PQ}$, with the first letter named as the point of the vector's tail, and the rightmost letter being the point at the vector's head (Figure 3.9). Note that the left superscript B has been dropped, as three superscripts (B on the left, P and Q on the right) would be redundant.

In this text, the letters $\vec{p}$, $\vec{v}$, and $\vec{a}$ will *always* be used to indicate position, velocity, and acceleration, respectively. For these linear quantities, the superscripts will always be *points* on the right-hand side, and reference frames on the left-hand side. In angular quantities, the Greek letter $\vec{\omega}$ will be reserved to express angular velocity, and $\vec{\alpha}$ will be used to express angular acceleration.

[3] Technically, this should read ${}^{A}\vec{v}^{P} = \vec{0}$ (with a vector sign over the zero), but a zero vector is often written without the vector sign. In any event, it still points nowhere!

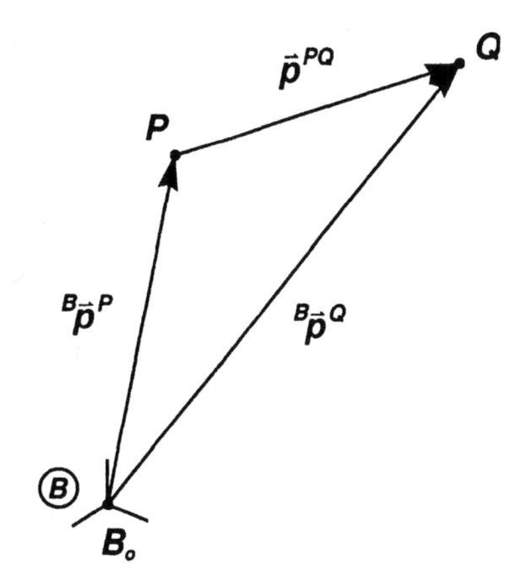

Figure 3.9. Notation for position vectors.

The superscripts for angular quantities will always contain one rigid body reference frame on each side of the vector quantity. For instance, ${}^{S}\vec{\alpha}^{D}$ indicates the angular acceleration of body D in reference frame S, and ${}^{N}\vec{\omega}^{B}$ represents the angular velocity of rigid body B in reference frame N.

3.3 RIGID BODIES, REFERENCE FRAMES, AND BASIS VECTORS

Rigid bodies and reference frames will be used interchangeably in this text even though the terms literally have different meanings. A *rigid body* is considered to be a structure that maintains a constant form despite the application of forces which cause the body to move. Another way to think of a rigid body is that all the particles making up the body have fixed locations relative to each other – and thus the body cannot fracture, expand, distort, or otherwise change any of its macroscopic descriptors (moment of inertia, center of mass location, *etc.*) throughout the time interval encompassed by the analysis. A reference frame can be affixed to each rigid body, which moves and rotates along with the body's movements as if rigidly attached. Because both the reference frame and the body are rigid, and affixed to one another, the motions of the reference frame and the body are equivalent.

Figure 3.10 illustrates these concepts. The tibia is shown hanging from the knee in two positions with respect to a fixed femur. With the tibia considered to behave approximately as a rigid body, reference frames and basis vectors can be affixed to the shank in such a way that the "1" unit basis vector of each reference frame points anteriorly, the "2" basis vector points superiorly, and the "3" basis vector points laterally when standing erectly. Figure 3.11 extends this to the thigh and foot. During erect stance, (Figure 3.11A), the basis vectors are defined in alignment with the N basis vectors. When the segments are moved

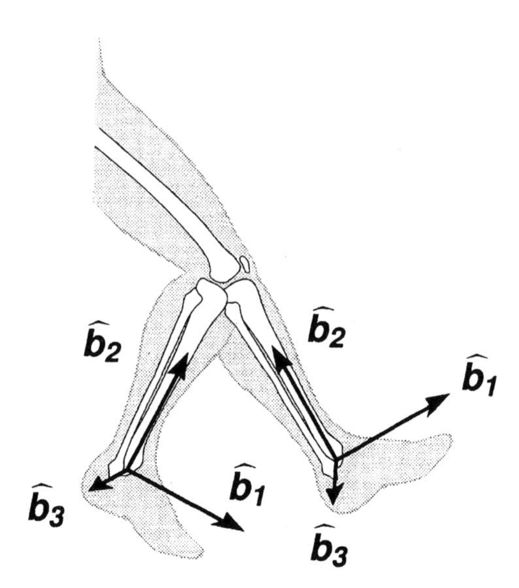

Figure 3.10. Basis vectors affixed to the shank.

(Figure 3.11B), the basis vectors move because the vectors are rigidly affixed to the segments.

It is important to note the distinction between a set of right-handed mutually perpendicular basis vectors and a reference frame. Reference frames have well-defined origins, whereas bases only define the directions of the individual basis vectors. Reference frames are therefore bases with origins defined in Cartesian space.

3.4 DIRECTION COSINES

Direction cosines define angular relationships between bases. Consider a planar movement of the shank (lower leg) with respect to the ground during level walking as depicted in Figure 3.12. The ground (segment N) presents a horizontal surface, with $\hat{n}_1$ pointing in the direction of travel and $\hat{n}_2$ pointing vertically upward. The "plane of motion" is defined by basis vectors $\hat{n}_1$ and $\hat{n}_2$. Reference frame A defines the orientations of the shank with respect to N. Angle q_1 describes the magnitude of angular rotation between A and N in the plane of motion.

The relationships between the N basis vectors and the A basis vectors are derived via vector addition. As shown in Figure 3.12B, the $\hat{a}_1$ vector can be decomposed into composite horizontal and vertical components. Since all of the individual basis vectors have lengths equal to 1, the horizontal component of $\hat{a}_1$ has magnitude $(1)\cos(q_1)$ and points in the $+\hat{n}_1$ direction. The vertical component has magnitude $(1)\sin(q_1)$ and direction $+\hat{n}_2$. Thus,

$$\hat{a}_1 = \cos(q_1)\hat{n}_1 + \sin(q_1)\hat{n}_2 \,. \tag{3.10}$$

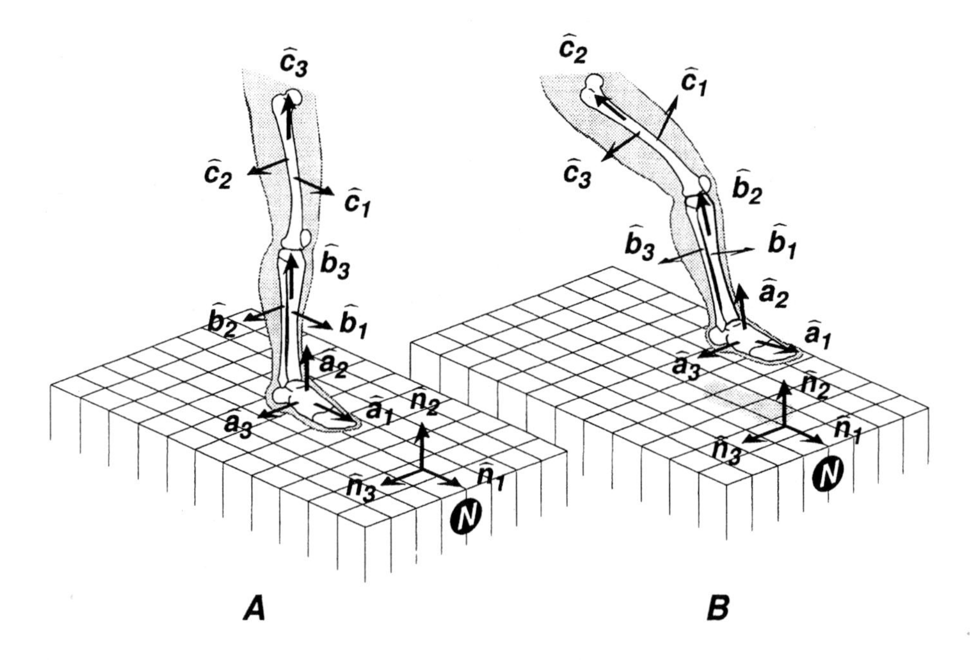

Figure 3.11. Basis vectors affixed to the ground (N), foot, shank, and thigh.

Similarly, $\hat{a}_2$ can be constructed from composite vectors in the $-\hat{n}_1$ and $+\hat{n}_2$ directions (Figure 3.12C):

$$\hat{a}_2 = \sin(q_1)(-\hat{n}_1) + \cos(q_1)\hat{n}_2\,. \tag{3.11}$$

The basis vector $\hat{a}_3$ remains parallel to $\hat{n}_3$, and thus is equivalent,

$$\hat{a}_3 = \hat{n}_3\,. \tag{3.12}$$

Organizing these into a table with the "fixed" (or more proximal) reference frame on the left and the "moving" (or more distal) reference frame above and to the right,

${}^{N}R^{A}$	$\hat{a}_1$	$\hat{a}_2$	$\hat{a}_3$
$\hat{n}_1$	$\cos q_1$	$-\sin q_1$	0
$\hat{n}_2$	$\sin q_1$	$\cos q_1$	0
$\hat{n}_3$	0	0	1

(3.13)

This table, which defines the $\hat{n}_1$, $\hat{n}_2$, and $\hat{n}_3$ components of vectors $\hat{a}_1$, $\hat{a}_2$, and $\hat{a}_3$ (and the $\hat{a}_1$, $\hat{a}_2$, and $\hat{a}_3$ components of vectors $\hat{n}_1$, $\hat{n}_2$, and $\hat{n}_3$), is referred to as a *table of direction cosines.* The direction cosines themselves are the elements of the table – and include the sines and numerals, too! To save

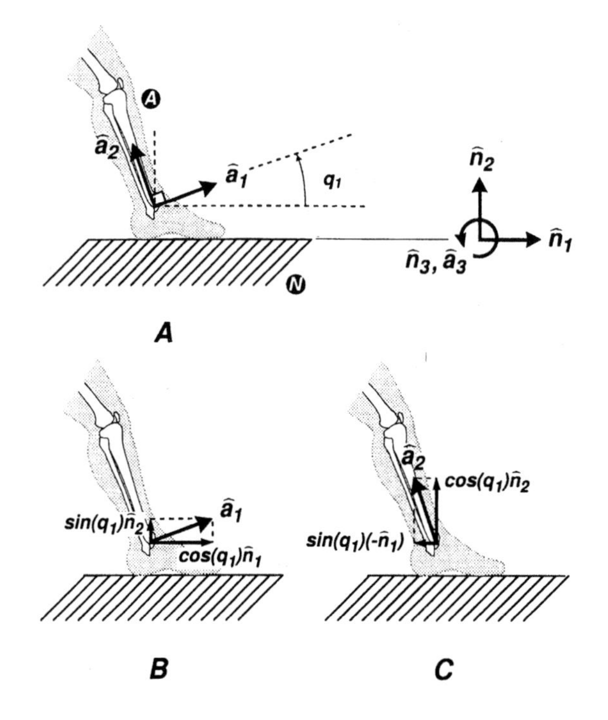

Figure 3.12. A. Basis vectors affixed to the ground (reference frame N), and shank (reference frame A) are used to derive the direction cosines between N and A. B. Unit basis vector $\hat{a}_1$ is broken down into the vector sum of two unit vectors in the N reference frame. C. Vector $\hat{a}_2$ is also broken down into components.

writing, $\cos(q_1)$ and $\sin(q_1)$ are abbreviated as c_1 and s_1,

${}^N R^A$	$\hat{a}_1$	$\hat{a}_2$	$\hat{a}_3$
$\hat{n}_1$	c_1	$-s_1$	0
$\hat{n}_2$	s_1	c_1	0
$\hat{n}_3$	0	0	1

(3.14)

It is evident from this construction there is only one degree of freedom relating the N and A bases (angle q_1). This table is typical of a simple rotation about the common $\hat{a}_3$ and $\hat{n}_3$ axis. A simple rotation is one in which the direction (but not necessarily the location) of the axis of revolution between frames remains fixed. Here, no matter how much A is rotated from N by angle q_1, $\hat{a}_3$ and $\hat{n}_3$ remain fixed in orientation and parallel to each other, and hence the rotation is classified as a simple rotation. Direction cosines from simple rotations about a common basis vector always have two cosines aligned diagonally, one positive and one negative sine aligned along the other diagonal, and one row and one column each with two zeros and a one. The one shows that the axis of the rotation is fixed in direction and parallel to both the $\hat{a}_3$ and $\hat{n}_3$ directions.

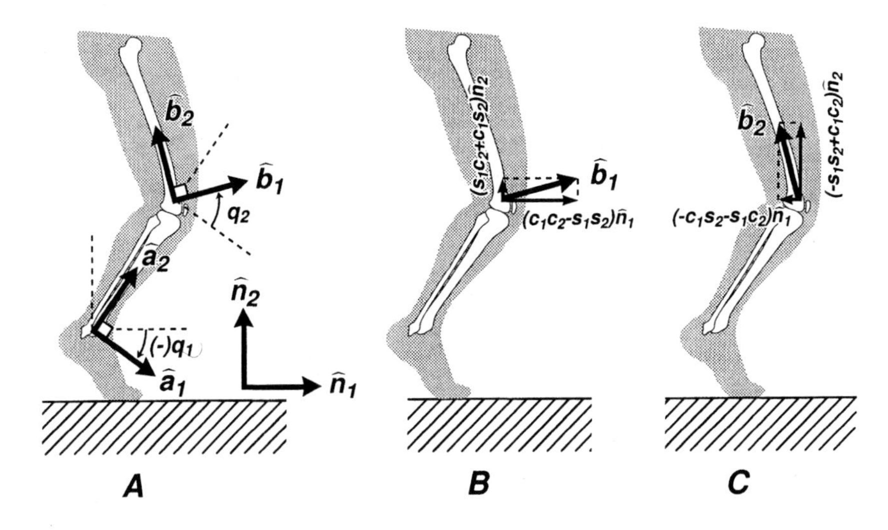

Figure 3.13. Basis vectors affixed to the ground (N), shank (A), and thigh (B) used to derive the direction cosines between N and A, A and B, and N and B.

3.4.1 DIRECTION COSINES IN PLANAR MOTIONS

Now, let's include a rotation at the knee as well as the previous rotation at the ankle. If rigid body B rotates by angle q_2 from rigid body A, and A rotates from N as defined before, then additional direction cosines may be defined. Figure 3.13 depicts such a case when the motions of the femur (B) and the shank (A) are coplanar. The following direction cosine tables now describe the simple rotation from N to A, and the simple rotation from A to B,[4]

${}^N R^A$	$\hat{a}_1$	$\hat{a}_2$	$\hat{a}_3$
$\hat{n}_1$	c_1	$-s_1$	0
$\hat{n}_2$	s_1	c_1	0
$\hat{n}_3$	0	0	1

(3.15)

${}^A R^B$	$\hat{b}_1$	$\hat{b}_2$	$\hat{b}_3$
$\hat{a}_1$	c_2	$-s_2$	0
$\hat{a}_2$	s_2	c_2	0
$\hat{a}_3$	0	0	1

(3.16)

[4]Helpful hint: Note that angle q_1 is shown positively in Figure 3.12, and negatively in Figure 3.13. The direction cosine table is identical for both positive and negative angles, but the numerical values of s_1 and $-s_1$ are different. In order to make sure the direction cosine tables are derived with the correct sign in front of each sine function, the analyst should rotate the basis vectors (mentally or on paper) until the angle between them is *positive and less than 90 degrees.*

A compound rotation from N to B can easily be defined via substitution.

$$\hat{n}_1 = c_1\hat{a}_1 - s_1\hat{a}_2 \tag{3.17}$$
$$= c_1\left(c_2\hat{b}_1 - s_2\hat{b}_2\right) - s_1\left(s_2\hat{b}_1 + c_2\hat{b}_2\right) \tag{3.18}$$
$$= \left(c_1c_2 - s_1s_2\right)\hat{b}_1 - \left(c_1s_2 + s_1c_2\right)\hat{b}_2 \tag{3.19}$$

$$\hat{n}_2 = s_1\hat{a}_1 + c_1\hat{a}_2 \tag{3.20}$$
$$= s_1\left(c_2\hat{b}_1 - s_2\hat{b}_2\right) + c_1\left(s_2\hat{b}_1 + c_2\hat{b}_2\right) \tag{3.21}$$
$$= \left(s_1c_2 + c_1s_2\right)\hat{b}_1 + \left(-s_1s_2 + c_1c_2\right)\hat{b}_2 \tag{3.22}$$

$$\hat{n}_3 = \hat{a}_3 \tag{3.23}$$
$$= \hat{b}_3 \tag{3.24}$$

Therefore, the table defining the direction cosines between the N and the B bases is,

${}^N R^B$	$\hat{b}_1$	$\hat{b}_2$	$\hat{b}_3$
$\hat{n}_1$	$c_1c_2 - s_1s_2$	$-c_1s_2 - s_1c_2$	0
$\hat{n}_2$	$s_1c_2 + c_1s_2$	$-s_1s_2 + c_1c_2$	0
$\hat{n}_3$	0	0	1

(3.25)

This logic can be easily extended to systems having many more planar degrees of freedom.

3.4.2 DIRECTION COSINES IN 3-D

The direction cosines for any sequence of simple rotations can be performed in exactly the same way as the above example. Using the direction cosine tables for each simple rotation, the relationships between basis vectors can be defined for a noncoplanar compound rotation. Let's say, for example, that a third simple rotation by angle q_3 is defined about the $\hat{b}_2$ axis in the example above, bringing into existence a fourth reference frame C (Figure 3.14). Anatomically, angle q_3 corresponds to a rotation of the pelvis about the axis of the femoral shaft.

The direction cosines from B to C are derived by first drawing the two sets of basis vectors in such a way that *the viewing direction is parallel to the rotation axis.* In other words, the drawing of the basis vectors depicts the rotation axis (here, $\hat{b}_2$ and $\hat{c}_2$) as pointing perpendicularly into or out of the drawing (depicted as a circular arc with an arrowhead, Figure 3.14). One should also check if right-handed reference frames and bases are used throughout, and if each drawing made from a different viewpoint correctly reflects the right-handed nature of the coordinate frames (*i.e.,* that the "1" axis crossed by the "2" axis yields the "3" axis). An arc defining the angle of positive rotation is also drawn to complete the diagram.

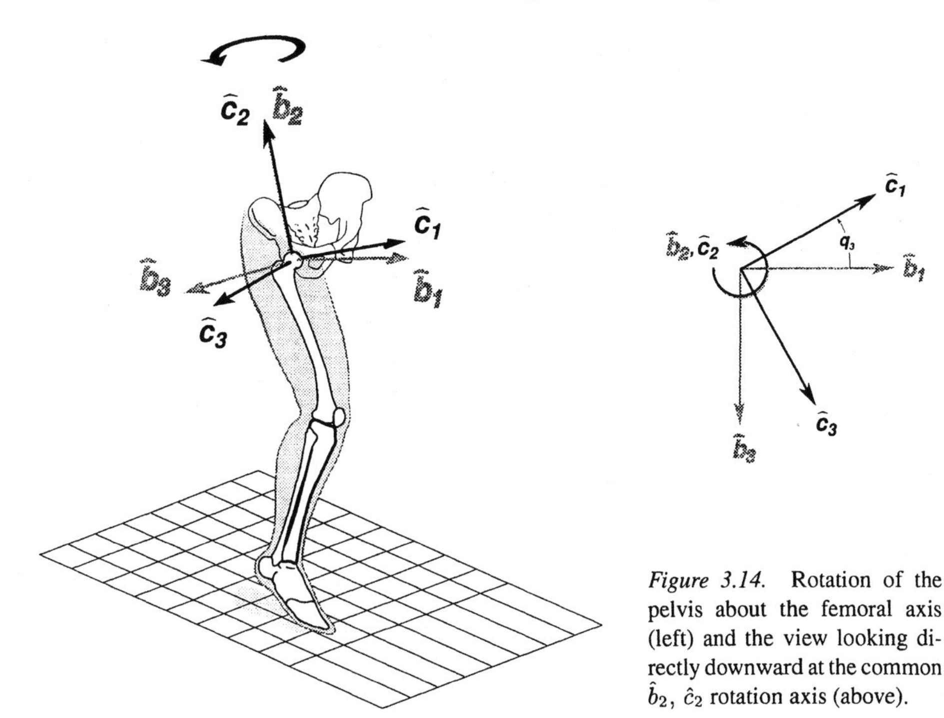

Figure 3.14. Rotation of the pelvis about the femoral axis (left) and the view looking directly downward at the common $\hat{b}_2$, $\hat{c}_2$ rotation axis (above).

Usually, the two bases are defined so they are coincident when the rotation angle is zero. One basis can be thought of as being "fixed" (usually drawn with horizontal and vertical axes), and the other basis as "rotating". For the purpose of defining direction cosines (only), the reader will find it less confusing to *always redraw the rotated basis with the rotation angle positive and less than 90 degrees.*[5]

Next, derive the direction cosines by selecting one of the "rotating" vectors at a time, and decomposing them into a vector sum of "fixed" vectors multiplied by the appropriate sines or cosines. In our example, this yields,

$$\hat{c}_3 = \cos(q_3)\hat{b}_3 + \sin(q_3)\hat{b}_1 \,, \tag{3.26}$$

$$\hat{c}_1 = \cos(q_3)\hat{b}_1 + \sin(q_3)(-\hat{b}_3) \,, \tag{3.27}$$

$$\hat{c}_2 = \hat{b}_2 \,. \tag{3.28}$$

[5]Though direction cosines can be derived via other means, it is extremely important that they be absolutely correct. An error at this stage is both disastrous and insidious because it is difficult to find and will permeate every subsequent equation. Therefore, the reader is urged to develop a consistent and well practiced methodology, resisting the urge to take shortcuts.

Hence, with the "fixed" frame on the left, and the "rotating" frame listed above and to the right, the simple rotation is represented by:

${}^B R^C$	$\hat{c}_1$	$\hat{c}_2$	$\hat{c}_3$
$\hat{b}_1$	c_3	0	s_3
$\hat{b}_2$	0	1	0
$\hat{b}_3$	$-s_3$	0	c_3

(3.29)

If the direction cosines relating the N frame to the C frame are desired, they can be found via substitution using shorthand notation for sine and cosine and the tables listed in Equations 3.25 and 3.29,[6]

$$\hat{c}_1 = c_3\hat{b}_1 - s_3\hat{b}_3 \tag{3.30}$$

$$= c_3\left[(c_1c_2 - s_1s_2)\,\hat{n}_1 + (s_1c_2 + c_1s_2)\,\hat{n}_2\right] - s_3\left[\hat{n}_3\right], \tag{3.31}$$

$$\hat{c}_2 = \hat{b}_2 \tag{3.32}$$

$$= \left[(-c_1s_2 - s_1c_2)\,\hat{n}_1 + (-s_1s_2 + c_1c_2)\,\hat{n}_2\right], \tag{3.33}$$

$$\hat{c}_3 = s_3\hat{b}_1 + c_3\hat{b}_3 \tag{3.34}$$

$$= s_3\left[(c_1c_2 - s_1s_2)\,\hat{n}_1 + (s_1c_2 + c_1s_2)\,\hat{n}_2\right] + c_3\left[\hat{n}_3\right]. \tag{3.35}$$

From these equations, the entries in the table can be made,

${}^N R^C$	$\hat{c}_1$	$\hat{c}_2$	$\hat{c}_3$
$\hat{n}_1$	$c_3(c_1c_2 - s_1s_2)$	$-c_1s_2 - s_1c_2$	$s_3(c_1c_2 - s_1s_2)$
$\hat{n}_2$	$c_3(s_1c_2 + c_1s_2)$	$-s_1s_2 + c_1c_2$	$s_3(s_1c_2 + c_1s_2)$
$\hat{n}_3$	$-s_3$	0	c_3

(3.36)

3.5 ROTATION MATRICES

Matrix multiplication is an alternative to using tedious substitution in finding the table of direction cosines from N to C. First, the nine entries in each table of direction cosines are written in matrix form (*i.e.*, in the same row-column order but without the basis vectors). The table above written in matrix form is,

$${}^N R^C = \begin{bmatrix} c_3(c_1c_2 - s_1s_2) & -c_1s_2 - s_1c_2 & s_3(c_1c_2 - s_1s_2) \\ c_3(s_1c_2 + c_1s_2) & -s_1s_2 + c_1c_2 & s_3(s_1c_2 + c_1s_2) \\ -s_3 & 0 & c_3 \end{bmatrix} \tag{3.37}$$

[6]Note the importance of putting carrots over unit vectors. Without the carrots, it would be difficult to distinguish between, for instance, the unit vectors $\hat{c}_i$ and the shorthand notation for cosines c_i ($i = 1, 2, 3$).

${}^N R^C$ is called "the rotation matrix from N to C", with the superscripts being read from left to right. Note that the *left* superscript corresponds to the basis indicated by the basis vectors written on the *left*-hand side of the direction cosine table, and the *right* superscript corresponds to the basis indicated by the basis vectors written *above and on the right side of* the direction cosine table. As another example, the rotation matrices from N to B and B to C (Equations 3.25 and 3.29) from the previous sections are given below,

$$ {}^N R^B = \begin{bmatrix} c_1c_2 - s_1s_2 & -c_1s_2 - s_1c_2 & 0 \\ s_1c_2 + c_1s_2 & -s_1s_2 + c_1c_2 & 0 \\ 0 & 0 & 1 \end{bmatrix} \tag{3.38} $$

$$ {}^B R^C = \begin{bmatrix} c_3 & 0 & s_3 \\ 0 & 1 & 0 \\ -s_3 & 0 & c_3 \end{bmatrix} \tag{3.39} $$

The reader may wish to verify that the rotation matrix (and hence the table of direction cosines) from N to C given above in Equation 3.37 may be obtained by multiplying the following rotation matrices,

$$ {}^N R^C = \left({}^N R^B\right)\left({}^B R^C\right) \tag{3.40} $$

$$ = \left({}^N R^A\right)\left({}^A R^B\right)\left({}^B R^C\right). \tag{3.41} $$

One can think of the interior superscripts A and B as "cancelling each other" when they appear as adjacent right and left superscripts, leaving only the exterior superscripts N on the left and C on the right. The order of the matrix multiplication is important, and the cancellation method serves as a safeguard against performing a matrix multiplication in the wrong order.

3.5.1 PROPERTIES OF ROTATION MATRICES

Rotation matrices relating one set of basis vectors to another are 3×3 examples of *orthonormal* matrices. Orthonormal matrices have several important properties which can be exploited to simplify kinematic and dynamic biomechanical analyses.

By definition, orthonormal matrices have rows and columns containing the measure numbers of unit vectors. Hence, it is almost trivial to state that each row and column has a vector magnitude of exactly 1. It is less apparent when looking at the product of several matrix multiplications (see, for example, Equation 3.48) that the sums of the squares of each row and column is 1. By definition, the row elements of orthonormal matrices form vectors perpendicular to each other, and the column elements of orthonormal matrices form vectors perpendicular to each other.[7]

Two useful properties are delivered via the transpose. The rotation matrix from B to A, ${}^{B}R^{A}$, is easily determined from the transpose of ${}^{A}R^{B}$,

$$
{}^{B}R^{A} = \left({}^{A}R^{B}\right)^{T}. \tag{3.42}
$$

Furthermore, the inverse of an orthonormal matrix is also its transpose,

$$
\left({}^{A}R^{B}\right)^{-1} = \left({}^{A}R^{B}\right)^{T}. \tag{3.43}
$$

These relations are useful in terms of simplifying the analysis of biomechanical motions.

3.5.2 EULER ANGLES AND OTHER ROTATION SETS

The famous mathematician Euler recognized that a series of three rotations could be used to uniquely define the orientation of a rigid body in three-dimensional space. Though there are many sets of three sequential rotations which lead to a unique orientation of a rigid body, the most commonly used are *Euler rotations.*

To cast this section into a form familiar to biomechanists, consider the rotations of the pelvis relative to a fixed femur. Initially, the pelvis reference frame P is considered to be coincident with the right femoral reference frame F, oriented such that $\hat{f}_3$ points anteriorly, $\hat{f}_1$ points laterally, and $\hat{f}_2$ points inferiorly (Figure 3.15A). This orientation is chosen because the resulting Euler rotations will roughly correspond to physiological rotation angles. The Euler rotations form a $\hat{3} - \hat{1}' - \hat{3}''$ set (a $\hat{z} - \hat{x}' - \hat{z}''$ sequence in Cartesian space), meaning the first rotation occurs about the *"3"* axis, the second rotation occurs about the once rotated *"1"* axis, and the third rotation occurs about the twice rotated *"3"* axis. Each rotation is a simple rotation, and brings the P frame into an intermediate orientation. Figure 3.15B shows the pelvis in its final orientation after the three rotations have been performed. These three rotations are described in sequential order (*i, ii, iii*) below.

i) Rotation about the common $\hat{f}_3$, $\hat{p}'_3$ axis. From its initial orientation coincident with the F frame, the first intermediate orientation P' is achieved via a simple rotation about the common $\hat{f}_3, \hat{p}'_3$ axis (Figure 3.15C). The drawing shows the first two basis vectors of F and P' superimposed. It is understood that the $\hat{f}_3$ and $\hat{p}'_3$ axes are colinear, and point perpendicularly out of the paper.

[7] When working with motion tracking systems, it is wise to check and make sure that the hardware and software actually delivers orthonormal matrices. Otherwise, "rotation matrices" which are not quite orthonormal will introduce scaling errors during matrix vector multiplication.

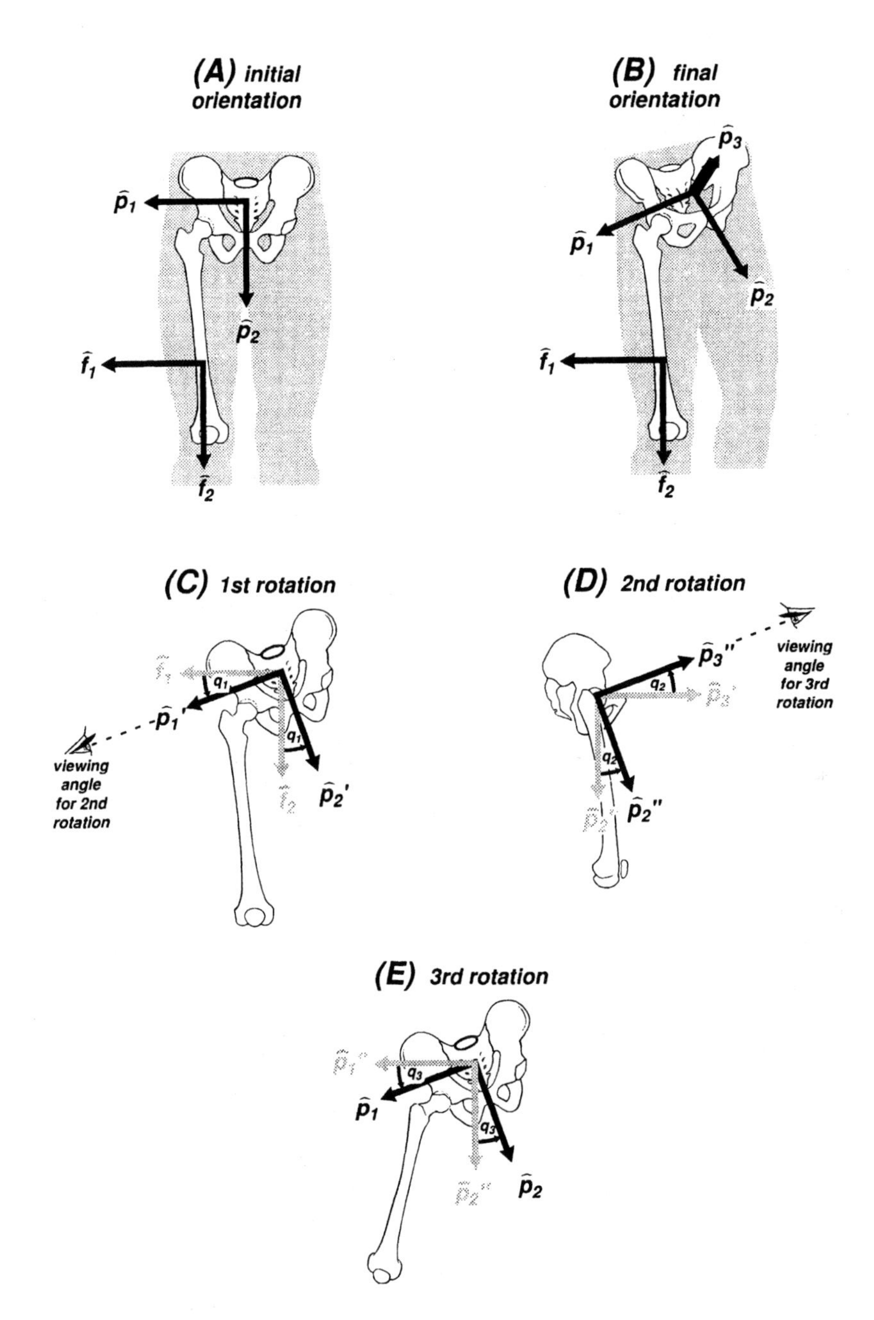

Figure 3.15. Euler rotations of the pelvis (P) relative to the femur (F).

Angle q_1 is defined positively when the rotation of $\hat{p}_1^{'}$ is counterclockwise from

$\hat{f}_1$, consistent with the angular representation for the $\hat{f}_3$, $\hat{p}_3^{'}$ axis. The view is defined with the $\hat{f}_3$, $\hat{p}_3^{'}$ unit vectors pointing out of the paper (directly at the viewer), and with the angle q_1 positive and less than 90 degrees.

${}^{F}R^{P'}$	$\hat{p}_1^{'}$	$\hat{p}_2^{'}$	$\hat{p}_3^{'}$
$\hat{f}_1$	c_1	$-s_1$	0
$\hat{f}_2$	s_1	c_1	0
$\hat{f}_3$	0	0	1

(3.44)

A negative rotation corresponds to pelvic list (drooping of the pelvis to the left in the frontal plane) or hip adduction. Abduction of the hip occurs with a positive rotation angle.

The position of the viewer's eye looking along the next rotation axis ($\hat{p}_1^{'}$) is also shown on the left side of (Figure 3.15C). If the eye were placed in this position looking toward the hip joint, the $\hat{p}_2^{'}$ axis will appear to be pointing directly downward, and the $\hat{p}_3^{'}$ axis will point to the right as shown shaded in grey in Figure 3.15D.

ii) Rotation about the common $\hat{p}_1^{'}$, $\hat{p}_1^{''}$ axis. The next intermediate orientation $P^{''}$ is obtained via a second simple rotation about the common $\hat{p}_1^{'}$, $\hat{p}_1^{''}$ axis (Figure 3.15D). A drawing of the basis vectors looking directly at the common $\hat{p}_1^{'}$, $\hat{p}_1^{''}$ axis is made as an aid to the derivation of the table elements.

${}^{P'}R^{P''}$	$\hat{p}_1^{''}$	$\hat{p}_2^{''}$	$\hat{p}_3^{''}$
$\hat{p}_1^{'}$	1	0	0
$\hat{p}_2^{'}$	0	c_2	$-s_2$
$\hat{p}_3^{'}$	0	s_2	c_2

(3.45)

When q_1 is small, a positive value of the rotation angle q_2 approximates hip extension. In the unusual event that q_1 is large, it would be incorrect to refer to q_2 as the hip extension angle. This is because hip extension is usually defined in the vertical sagittal plane, and if the hip is abducted or adducted the rotation axis $\hat{p}_1^{'}$, which is perpendicular to the $\hat{p}_2^{'}$, $\hat{p}_3^{'}$ plane, will not be perpendicular to the vertical sagittal plane. The position of the viewer's eye looking along the last rotation axis ($\hat{p}_3^{''}$, $\hat{p}_3$) is shown on the upper right side of Figure 3.15D. With the eye placed in this position looking toward the hip joint, the $\hat{p}_2^{''}$ axis will appear to be pointing downward, and the $\hat{p}_1^{''}$ axis will point to the left.

iii) Rotation about the common $\hat{p}''_3$, $\hat{p}_3$ axis. The final rotation (Figure 3.15E) leads to the orientation of the pelvis P.

${}^{P''}\!R^{P}$	$\hat{p}_1$	$\hat{p}_2$	$\hat{p}_3$
$\hat{p}''_1$	c_3	$-s_3$	0
$\hat{p}''_2$	s_3	c_3	0
$\hat{p}''_3$	0	0	1

(3.46)

The table of direction cosines relating the femoral (F) and pelvic (P) reference frames is obtained most simply via matrix multiplication,

$$ {}^{F}\!R^{P} = \left({}^{F}\!R^{P'}\right)\left({}^{P'}\!R^{P''}\right)\left({}^{P''}\!R^{P}\right), \tag{3.47}$$

which yields the rotation matrix ${}^{F}\!R^{P}$ and its corresponding table of direction cosines,

${}^{F}\!R^{P}$	$\hat{p}_1$	$\hat{p}_2$	$\hat{p}_3$
$\hat{f}_1$	$c_1c_3 - s_1c_2s_3$	$-c_1s_3 - s_1c_2c_3$	s_1s_2
$\hat{f}_2$	$s_1c_3 + c_1c_2s_3$	$-s_1s_3 + c_1c_2c_3$	$-c_1s_2$
$\hat{f}_3$	s_2s_3	s_2c_3	c_2

(3.48)

Direction cosines for virtually any compound rotation can be found easily using this methodology. One needs only to define the sequence of simple rotations comprising the compound rotation, and to perform the matrix multiplications correctly.

3.6 TRANSFORMATION OF COORDINATES

The matrix ${}^{F}\!R^{P}$ is called a *rotation matrix* because in a computational sense, the rotation of a rigid body is equivalent to changing the basis with which each point of that body is described. Consider the following two cases that arise often in biomechanics,

- *Case I. A rigid body whose outer surface is described by a collection of points P_i, $(i = 1, 2, 3, \cdots, N)$ is to be rotated and translated.* A computer cannot infer the shape and form of a real object. We must "digitize" the surface of the object into distinct point locations in order for the computer to calculate the new point locations following rotation and translation.

- *Case II. A point location of interest, P_i, fixed on a rigid body segment is to be tracked as the body segment is rotated and translated.* For example, markers affixed to the surface of body segments must be tracked to compute joint angles in experimental analyses. In musculoskeletal modeling, the origin and insertion locations of a musculotendon actuator must be tracked

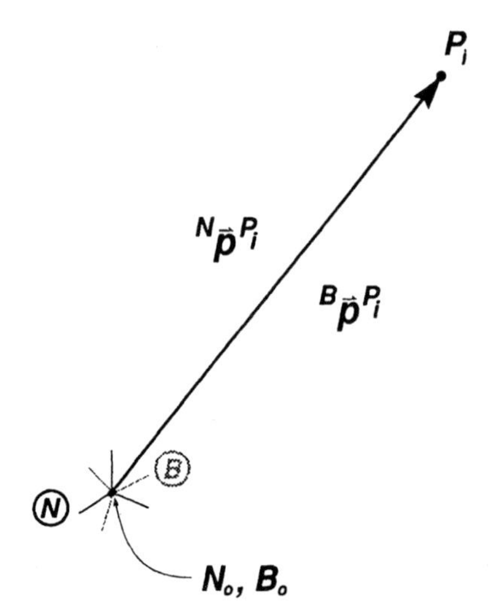

Figure 3.16. A rotated position vector does not change its magnitude or direction. Its measure numbers do change in numerical values because the vector is re-expressed using a rotated basis.

in order to compute the length and velocity of the musculotendon during joint flexions and extensions.

In each case, the rigid body rotation changes the location of point(s) P_i in space relative to an inertial reference frame such as the ground. However, the point location(s) remains fixed in the frame of the rigid body itself.

Assume for the purpose of illustration that a reference frame B affixed to the rigid body is originally coincident with the inertial reference frame N. A rigid body rotation of B about an axis passing through the common origins of frames B and N will bring the B reference frame to a new orientation. If we express the vector $^N\vec{p}^{P_i}$ drawn from the origin to point P_i in terms of the basis vectors of B, the measure numbers of $^N\vec{p}^{P_i}$ will change even though the original vector $^N\vec{p}^{P_i}$ will remain completely unchanged (*i.e.*, it will retain its original direction and magnitude as shown in Figure 3.16),

$$^N\vec{p}^{P_i} = \alpha_1\hat{n}_1 + \beta_1\hat{n}_2 + \gamma_1\hat{n}_3 \tag{3.49}$$

$$= \alpha_2\hat{b}_1 + \beta_2\hat{b}_2 + \gamma_2\hat{b}_3\,. \tag{3.50}$$

In Equation 3.49, $^N\vec{p}^{P_i}$ is "expressed" using the basis vectors of reference frame N, where Equation 3.50 expresses the same vector using the basis vectors of frame B.

It is an easy matter to compute the new measure numbers $(\alpha_2, \beta_2, \gamma_2)$ given the rotation matrix $^BR^N$ and a vector composed of the original measure numbers

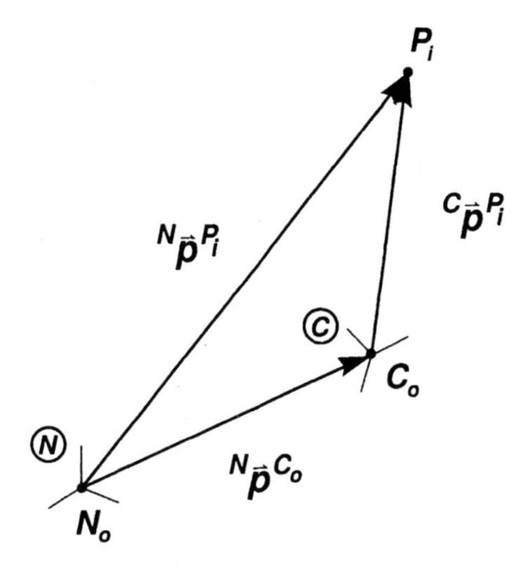

Figure 3.17. Transformation of a position vector involves a translation between origins ($^{N}\vec{p}^{C_o}$) and a rotation.

$(\alpha_1, \beta_1, \gamma_1)$,

$$\begin{bmatrix} \alpha_2 \\ \beta_2 \\ \gamma_2 \end{bmatrix} = {}^{B}R^{N} \begin{bmatrix} \alpha_1 \\ \beta_1 \\ \gamma_1 \end{bmatrix} . \tag{3.51}$$

In symbolic form,

$$^{N}\vec{p}^{P_i} = {}^{B}R^{N} \; {}^{N}\vec{p}^{P_i} \tag{3.52}$$

where the apparent contradiction results because premultiplying the vector $^{N}\vec{p}^{P_i}$ by the matrix $^{B}R^{N}$ *does not change the vector* – it only serves to *re-express* $^{N}\vec{p}^{P_i}$ in terms of a different set of basis vectors. $^{N}\vec{p}^{P_i}$ still points from the origin of the N reference frame to the point P_i. This fact is obvious when $^{B}R^{N}$ is expressed along with the basis vectors (as in Equations 3.49 and 3.50), but it is not readily apparent when only the measure numbers are given (as in Equation 3.51).[8] Because the measure numbers have changed to indicate the position of point P_i using the basis vectors of B, it may appear that the vector has likewise changed. Because a digital computer works only with the measure numbers, we can easily fool it into thinking that the vector has actually rotated, whereas in fact we have simply re-expressed it in terms of the B basis vectors which have been rotated in the *opposite* direction.

A full *transformation of coordinates* includes a translation along with the rotation. For example, transforming the coordinates of a point P_i measured

[8]In this case, because the origins of the B and N reference frames are coincident, $^{N}\vec{p}^{P_i}$ is actually equal to $^{B}\vec{p}^{P_i}$ (the position vector of P_i in reference frame B). That is, these two vectors have the same magnitude and point in the same direction even though they have different measure numbers. When the origins of N and B are in different locations, $^{N}\vec{p}^{P_i} \neq {}^{B}\vec{p}^{P_i}$.

in one reference frame into coordinates measured in another reference frame is shown in Figure 3.17. The reference frame N has origin N_o and a third reference frame C has its origin C_o some distance from N_o. The vector ${}^N\vec{p}^{C_o}$ (expressed in the basis vectors of reference frame N) defines the position of C_o in N. The position vector ${}^C\vec{p}^{P_i}$ from the origin of C to point P_i (expressed in the basis vectors of C) is related to ${}^N\vec{p}^{P_i}$ by either of the following equations,

$$ {}^N\vec{p}^{P_i} = {}^N R^C \, {}^C\vec{p}^{P_i} + {}^N\vec{p}^{C_o} \quad (3.53) $$

$$ {}^C\vec{p}^{P_i} = {}^C R^N \left({}^N\vec{p}^{P_i} - {}^N\vec{p}^{C_o} \right) \quad (3.54) $$

where ${}^N\vec{p}^{P_i} = \alpha_3\hat{n}_1 + \beta_3\hat{n}_2 + \gamma_3\hat{n}_3$, ${}^C\vec{p}^{P_i} = \alpha_1\hat{c}_1 + \beta_1\hat{c}_2 + \gamma_1\hat{c}_3$, and ${}^N\vec{p}^{C_o} = (dx)\,\hat{n}_1 + (dy)\,\hat{n}_2 + (dz)\,\hat{n}_3$.

It is understood from the discussion following Equation 3.52 that the rotation-matrix-to-vector multiplication only serves to re-express vectors from one reference frame to another, changing the measure numbers but not changing the original vector's magnitude or direction. Parentheses are included to ensure that vectors are expressed in the proper bases before they are added together. Equation 3.53 is the more commonly used of the two expressions. Equation 3.54 can be useful when it is more convenient to work in the basis vectors of C.

Numerically, the operation in Equation 3.53 is expressed as:

$$ \begin{bmatrix} \alpha_3 \\ \beta_3 \\ \gamma_3 \end{bmatrix} = \left({}^N R^C \begin{bmatrix} \alpha_1 \\ \beta_1 \\ \gamma_1 \end{bmatrix} \right) + \begin{bmatrix} dx \\ dy \\ dz \end{bmatrix} \quad (3.55) $$

Note that the matrix premultiplication by ${}^N R^C$ serves to change the measure numbers to a common basis *prior to adding the translation.* This explains why the translation by adding $[dx \quad dy \quad dz]^T$ in Equation 3.53 is performed *after* the matrix-vector multiplication. A common computational trick adds a trivial fourth equation $1 = 1$ to the three equations expressed in matrix form in Equation 3.55. This allows the rotation by change of basis and translation to be accomplished in a single matrix-vector multiplication,

$$ \begin{bmatrix} \alpha_3 \\ \beta_3 \\ \gamma_3 \\ 1 \end{bmatrix} = {}^N T^C \begin{bmatrix} \alpha_1 \\ \beta_1 \\ \gamma_1 \\ 1 \end{bmatrix} \quad (3.56) $$

where the 3×3 rotation matrix ${}^N R^C$ has been imbedded into the 4×4 transformation matrix, ${}^N T^C$,

$$ {}^N T^C = \left[\begin{array}{ccc|c} & & & dx \\ & {}^N R^C & & dy \\ & & & dz \\ \hline 0 & 0 & 0 & 1 \end{array} \right] \quad (3.57) $$

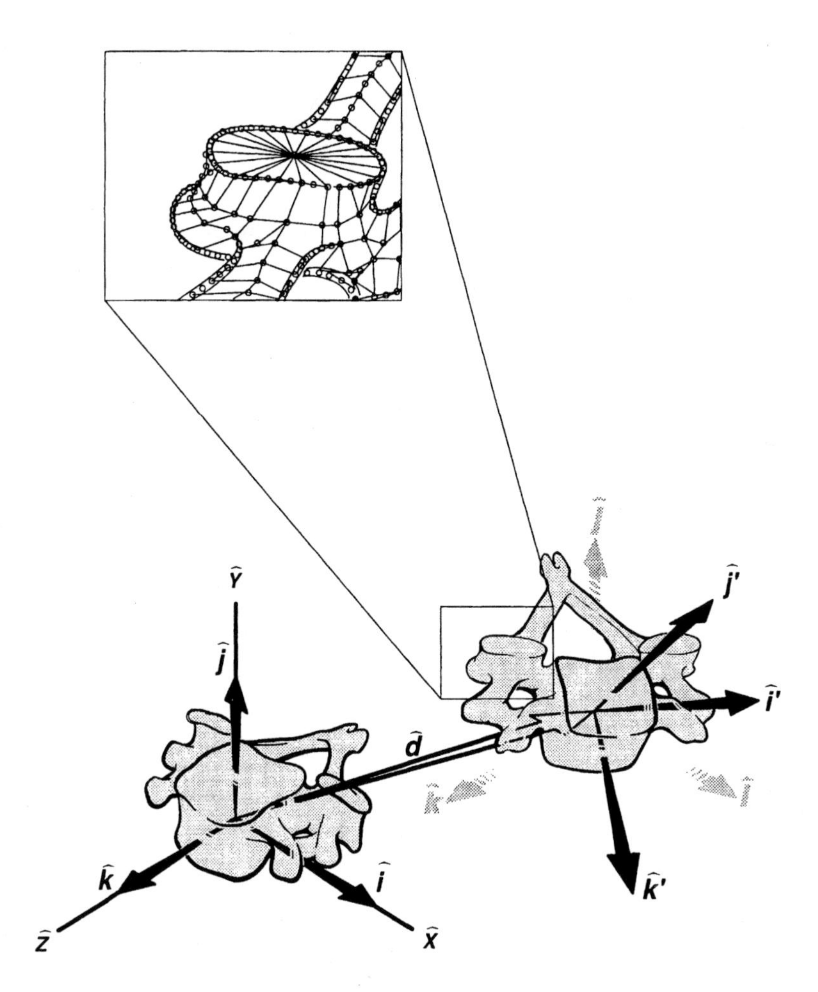

Figure 3.18. A computer graphic representation of a rigid body can be both translated and rotated using vector transformations. The position vectors of each point of the surface of the rigid body are transformed by matrix-vector multiplication. Using the transformed measure numbers in the original (not the rotated) basis creates the illusion of translation and rotation.

This trick is called a *homogeneous transformation*, and is a useful trick commonly employed in computer graphics (Mortenson, 1985). Suppose you wanted to represent the surface of a bone (rigid body C, for instance) by 3000 points in Cartesian space (Figure 3.18). These points are most conveniently located by a series of vectors from the origin defined in reference frame C to each point in turn, ${}^{C}\vec{p}^{P_i} \quad (i = 1, 2, \ldots, 3000)$. When rotation and translation of the entire body occur, a new view of the bone must be generated from the new locations of each point describing the bony surface in view. Thus, 3000 individual transformations must take place before the computer-graphic workstation can redraw the bone in its new location and orientation. Once the rotation angles and translations are defined, the transformation matrix (Equation 3.56) is computed and used to multiply each of the vectors ${}^{C}\vec{p}^{P_i} \quad (i = 1, 2, \ldots, 3000)$ in turn. This arduous process was enabled in real time with the advent of the Silicon Graphics IRIS computers in the 1980s. A special chip was designed and produced (called the "Graphics Engine") by founder Dr. James Clark to perform the 4×4 matrix-vector multiplications in silicon hardware rather than via software programs stored in memory.

3.7 EXERCISES

1. The human pelvis and femur are joined together with a "ball-in-socket" joint that keeps the centroid of the femoral head (the "ball") fixed with respect to the center of the acetabulum (the "socket"). How many degrees of freedom does it take to specify the position and orientations of the femur and pelvis when constrained by such a joint? Your answer will depend upon whether you consider the joint constraints to be "soft" or "hard."

2. The human elbow forms a junction of three bones that come together in close proximity to the elbow. Figures 4.13 and 4.14 may help to visualize these. The humerus and ulna are joined by a cylindrical joint which maintains a nearly fixed axis with respect to these bones (a one degree of freedom joint). The proximal end of the radius is cupped, and maintains contact with the lateral condyle of the distal humerus. The distal end of the radius is attached with ligaments to the distal end of the ulna. Ligaments do not stretch much, but they do not create a "hard" constraint, either.

   Pronation of the forearm can be done in such a way as to maintain the centroid of the wrist (the junction between the distal radius and ulna) in a fixed location in three-dimensional space. Considering the wrist centroid to be fixed in space as a "hard" endpoint constraint, how many degrees of freedom are needed to specify the positions and orientations of the ulna and radius? Assume that the humerus is fixed in space.

3. Link N is connected to A, A to B, and B to C. Find direction cosines from N to A, N to B, and N to C for the following systems. In all cases, the reference frames are coincident when the rotation angle is 0.

   (a) Rotation of A from N by angle θ_1 about the common $\hat{n}_2$, $\hat{a}_2$ axis; rotation of B from A by angle θ_2 about the common $\hat{a}_2$, $\hat{b}_2$ axis; and rotation of C from B by angle θ_3 about the common $\hat{b}_3$, $\hat{c}_3$ axis.

(b) Rotation of A from N by angle θ_1 about the common $\hat{n}_1$, $\hat{a}_1$ axis; rotation of B from A by angle θ_2 about the common $\hat{a}_2$, $\hat{b}_2$ axis; and rotation of C from B by angle θ_3 about the common $\hat{b}_3$, $\hat{c}_3$ axis.

4. The Euler rotations were defined in Section 3.5.2 as a sequence of rotations, $\hat{z} - \hat{x}' - \hat{z}''$ or $\hat{3} - \hat{1}' - \hat{3}''$. It is also common to define these as $\hat{z} - \hat{y}' - \hat{z}''$. Find the direction cosines for this alternate sequence of Euler rotations.

5. Find the Euler angles commonly referred to as α, β, γ where $\alpha = q_1$, $\beta = q_2$, $\gamma = q_3$ using the definitions of Section 3.5.2, given the following instantaneous values of the rotation matrix:

$$^{F}R^{P} = \begin{bmatrix} 0.826358 & 0.563079 & 0.008593 \\ 0.394664 & -0.568180 & -0.722090 \\ -0.401710 & 0.600097 & -0.691750 \end{bmatrix} \tag{3.58}$$

Either the matrix from Problem 4 or Equation 3.48 may be used to find the answers.

6. Sometimes there is a need to separately digitize the front and back views of an object or a set of markers. For instance, a laser scanner can only see one side of an object at a time to digitize a solid object, or a set of cameras might only be able to see the markers placed on one side of an opaque body.

Devise a procedure (transformation) that will convert the "back" set of marker coordinates into the reference frame of the "front" set. Assume a minimum of three markers are visible in both the "front" and "back" sets, and that these markers have different coordinates in the two views.

Note: This problem requires some foreknowledge regarding the numerical solution of n equations and n unknowns when the equations are expressed in linear system form, $A\vec{x} = \vec{b}$. Students unfamiliar with this may gain some insights by reading Section 7.1.1.

References

Kane, T. R., and Levinson, D. A. (1985) *Dynamics: Theory and Applications.* McGraw-Hill, New York, NY.

Kapandji, I. A. (1970) *The Physiology of the Joints – Volume 2, Lower Limb.* L. H. Honoré (transl.), Churchill Livingstone, Edinburgh, UK.

Mortenson, Michael E. (1985) *Geometric Modeling.* John Wiley & Sons, New York.

Strasser, H. (1917) *Lehrbuch der Muskel-und Gelenkmechanik.* Springer, Berlin.

Chapter 4

VECTOR BASED KINEMATICS

Objective – To find the positions and velocities of every point at which a force acts, the angular velocities of every body on which a torque acts, and the accelerations of every mass center.

4.1 VECTOR COMPUTATIONS

4.1.1 THE VECTOR DOT PRODUCT

Given any two vectors defined in a common reference frame, $\vec{v}$ and $\vec{w}$,

$$\vec{v} = v_1\hat{a}_1 + v_2\hat{a}_2 + v_3\hat{a}_3 \tag{4.1}$$
$$\vec{w} = w_1\hat{a}_1 + w_2\hat{a}_2 + w_3\hat{a}_3\,, \tag{4.2}$$

the dot, or "scalar" product of $\vec{v}$ and $\vec{w}$ is computed as follows,

$$\begin{aligned}
\vec{v}\cdot\vec{w} &= (v_1\hat{a}_1)\cdot(w_1\hat{a}_1 + w_2\hat{a}_2 + w_3\hat{a}_3) \\
&\quad + (v_2\hat{a}_2)\cdot(w_1\hat{a}_1 + w_2\hat{a}_2 + w_3\hat{a}_3) \\
&\quad + (v_3\hat{a}_3)\cdot(w_1\hat{a}_1 + w_2\hat{a}_2 + w_3\hat{a}_3) && (4.3)\\
&= v_1w_1\,(\hat{a}_1\cdot\hat{a}_1) + v_1w_2\,(\hat{a}_1\cdot\hat{a}_2) + v_1w_3\,(\hat{a}_1\cdot\hat{a}_3) \\
&\quad + v_2w_1\,(\hat{a}_2\cdot\hat{a}_1) + v_2w_2\,(\hat{a}_2\cdot\hat{a}_2) + v_2w_3\,(\hat{a}_2\cdot\hat{a}_3) \\
&\quad + v_3w_1\,(\hat{a}_3\cdot\hat{a}_1) + v_3w_2\,(\hat{a}_3\cdot\hat{a}_2) + v_3w_3\,(\hat{a}_3\cdot\hat{a}_3) && (4.4)\\
&= v_1w_1\,(\hat{a}_1\cdot\hat{a}_1) + v_2w_2\,(\hat{a}_2\cdot\hat{a}_2) + v_3w_3\,(\hat{a}_3\cdot\hat{a}_3) && (4.5)\\
&= v_1w_1 + v_2w_2 + v_3w_3\,. && (4.6)
\end{aligned}$$

Performing the dot products in an orthonormal basis is simple because each basis vector is dotted with vectors of the same basis so that the vector pairs

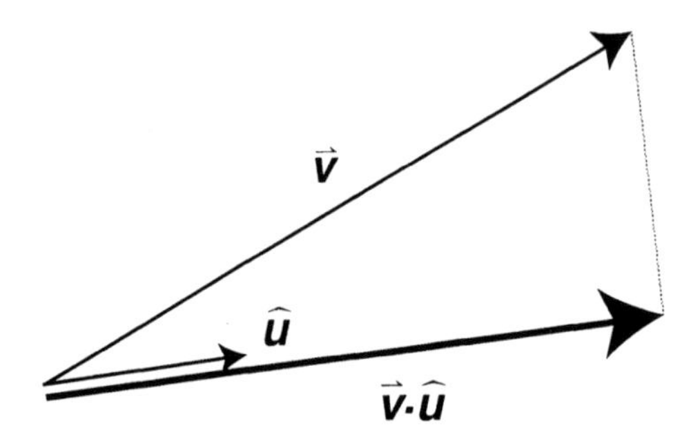

Figure 4.1. Dot product of vectors $\vec{v}$ and a unit vector $\hat{u}$.

are either parallel or perpendicular. Any unit vector dotted with itself yields a scalar value of 1 (*i.e.*, $\hat{a}_i \cdot \hat{a}_i = 1$), and any vector dotted with a vector that is perpendicular to itself yields a scalar value of 0, (*i.e.*, $\hat{a}_i \cdot \hat{a}_j = 0$, where $i \neq j$).

When dot products are desired in *mixed* bases, the operation is simplified through the use of the direction cosine tables. Suppose basis vector $\hat{a}_i$ is to be dotted with the basis vector $\hat{b}_j$. Let's assume we have the table of direction cosines relating reference frames A and B, written so that the basis vectors $\hat{a}_1$, $\hat{a}_2$, and $\hat{a}_3$ are listed on the left hand side of the table while the basis vectors $\hat{b}_1$, $\hat{b}_2$, and $\hat{b}_3$ are listed on top of the table,

${}^AR^B$	$\hat{b}_1$	$\hat{b}_2$	$\hat{b}_3$
$\hat{a}_1$	r_{11}	r_{12}	r_{13}
$\hat{a}_2$	r_{21}	r_{22}	r_{23}
$\hat{a}_3$	r_{31}	r_{32}	r_{33}

(4.7)

The dot product of $\hat{a}_i$ with $\hat{b}_j$ may then be simply read off the table as the element r_{ij} listed in the ith row and the jth column of the table.

Proof. Since the vector $\hat{a}_i$ can be expressed using the basis vectors $\hat{b}_1$, $\hat{b}_2$, and $\hat{b}_3$, we can form the dot product of $\hat{a}_i$ with $\hat{b}_j$ in a common basis,

$$\hat{a}_i = r_{i1}\hat{b}_1 + r_{i2}\hat{b}_2 + r_{i3}\hat{b}_3\,, \tag{4.8}$$

and thus,

$$\hat{a}_i \cdot \hat{b}_j = \left(r_{i1}\hat{b}_1 + r_{i2}\hat{b}_2 + r_{i3}\hat{b}_3\right) \cdot \hat{b}_j \tag{4.9}$$

$$= r_{i1}\left(\hat{b}_1 \cdot \hat{b}_j\right) + r_{i2}\left(\hat{b}_2 \cdot \hat{b}_j\right) + r_{i3}\left(\hat{b}_3 \cdot \hat{b}_j\right)\,, \tag{4.10}$$

and for the usual case where the basis vectors $\hat{b}_1$, $\hat{b}_2$, and $\hat{b}_3$ are mutually perpendicular,

$$\hat{b}_i \cdot \hat{b}_j = 0 \qquad (i \neq j) \tag{4.11}$$

$$\hat{b}_i \cdot \hat{b}_j = 1 \qquad (i = j)\,. \tag{4.12}$$

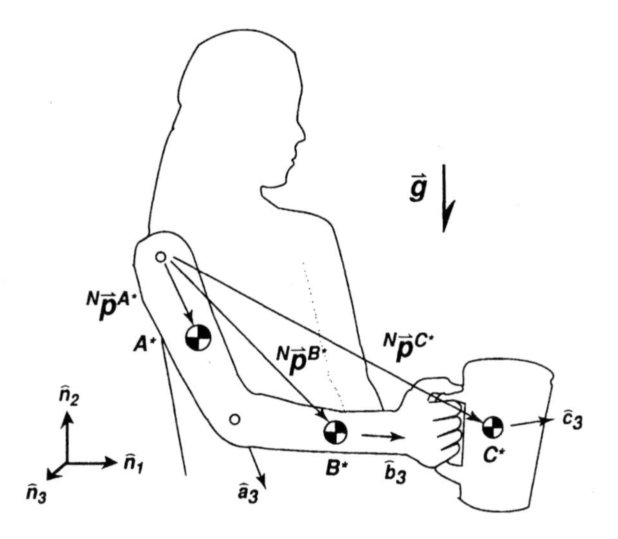

Figure 4.2. A system having three mass centers A^*, B^*, and C^*.

Therefore, for $j = 1, 2$, or 3,

$$\hat{a}_i \cdot \hat{b}_j = r_{ij} \,. \tag{4.13}$$

One can also write this as the dot product of two vectors expressed in different bases,

$$\vec{v} \cdot \vec{w} \;=\; \vec{v}^T \; {}^A\!R^B \; \vec{w} \tag{4.14}$$

where $\vec{v}$ is expressed in the basis vectors of A, and $\vec{w}$ is expressed in the basis vectors of B. Working from right to left in Equation 4.14, pre-multiplication of $\vec{w}$ by ${}^AR^B$ re-expresses $\vec{w}$ from B basis vectors into A basis vectors. Next, the dot product is completed by pre-multiplication with $\vec{v}^T$.

A common usage of the dot product is to find the component $\vec{w}$ of a vector in a particular direction (Figure 4.1). The magnitude of the component of vector $\vec{v}$ in the direction defined by vector can be found by dotting $\vec{v}$ with a unit vector $\hat{u}$ parallel to $\vec{u}$. For instance, if the magnitude of the vertical component of $\vec{v}$ is desired, one can simply take the dot product of $\vec{v}$ with the unit vector pointing in the vertical direction. The following example will illustrate this method.

4.1.1.1 EXAMPLE – POTENTIAL ENERGY OF A SYSTEM HAVING THREE LINKS

A quantity commonly computed in biomechanical linkages is the *potential energy of the system.* Consider the system shown in Figure 4.2, consisting of links A, B, and C moving in an inertial reference frame N. The mass centers A^*, B^*, and C^* of these bodies are located at distances of ρ_A, ρ_B, and ρ_C from their proximal ends and along the longitudinal axes of the bodies, which are oriented in the $\hat{a}_3$, $\hat{b}_3$, and $\hat{c}_3$ directions, respectively. Using m_A, m_B, and m_C as the masses, the potential energy (PE) of the system relative to the origin of N is computed by finding the elevations of the mass centers. These elevations are easily derived as the dot products of $\hat{n}_2$ with ${}^N\vec{p}^{A^*}$, ${}^N\vec{p}^{B^*}$, and ${}^N\vec{p}^{C^*}$,

$$PE = \hat{n}_2 \cdot g \left[m_A \, {}^N\vec{p}^{A^*} + m_B \, {}^N\vec{p}^{B^*} + m_C \, {}^N\vec{p}^{C^*} \right] \tag{4.15}$$

$$\begin{aligned} &= \hat{n}_2 \cdot g \, m_A \left(\rho_A \hat{a}_3\right) \\ &\quad + \hat{n}_2 \cdot g \, m_B \left(\ell_A \hat{a}_3 + \rho_B \hat{b}_3\right) \\ &\quad + \hat{n}_2 \cdot g \, m_C \left(\ell_A \hat{a}_3 + \ell_B \hat{b}_3 + \rho_C \hat{c}_3\right) \end{aligned} \tag{4.16}$$

$$\begin{aligned} &= g \left(m_A \rho_A + m_B \ell_A + m_C \ell_A\right) \left(\hat{n}_2 \cdot \hat{a}_3\right) \\ &\quad + g \left(m_B \rho_B + m_C \ell_B\right) \left(\hat{n}_2 \cdot \hat{b}_3\right) \\ &\quad + g \, m_C \rho_C \left(\hat{n}_2 \cdot \hat{c}_3\right) . \end{aligned} \tag{4.17}$$

Finally, if the rotation matrices ${}^NR^A$, ${}^NR^B$, and ${}^NR^C$ are known, the dot products $\hat{n}_2 \cdot \hat{a}_3$, $\hat{n}_2 \cdot \hat{b}_3$, and $\hat{n}_2 \cdot \hat{c}_3$, can be read directly from the tables. That is, $\hat{n}_2 \cdot \hat{a}_3$ is the scalar element of ${}^NR^A$ residing in the second ($\hat{n}_2$) row, third ($\hat{a}_3$) column, $\hat{n}_2 \cdot \hat{b}_3$ is the scalar element of ${}^NR^B$ residing in the second row, third column, and $\hat{n}_2 \cdot \hat{c}_3$ is the scalar element of ${}^NR^C$ residing in the second row, third column.

4.1.2 THE VECTOR CROSS PRODUCT

While the dot product may be defined in mixed bases, *the vector cross product must be performed using a common basis* (Figure 4.3). Using two arbitrary vectors $\vec{v}$ and $\vec{w}$ defined in the basis vectors of A,

$$\vec{v} = v_1 \hat{a}_1 + v_2 \hat{a}_2 + v_3 \hat{a}_3 \tag{4.18}$$

$$\vec{w} = w_1 \hat{a}_1 + w_2 \hat{a}_2 + w_3 \hat{a}_3 , \tag{4.19}$$

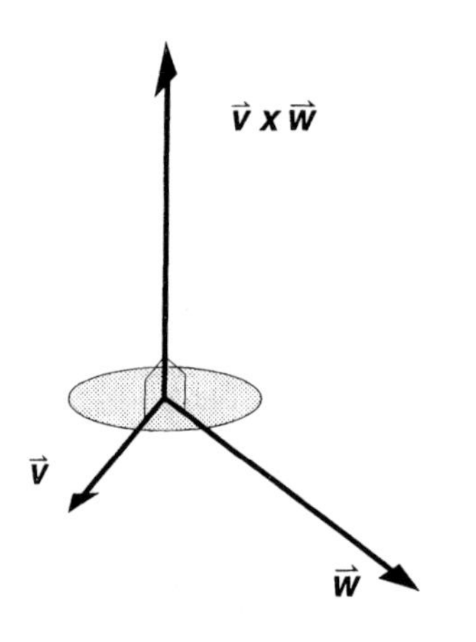

Figure 4.3. Cross product of vectors $\vec{v}$ and $\vec{w}$. The direction of the cross product is determined using the right hand rule (Figure 3.7).

the cross product is the following,

$$\vec{v}\times\vec{w} \;=\; (v_2w_3 - v_3w_2)\,\hat{a}_1-(v_1w_3 - v_3w_1)\,\hat{a}_2+(v_1w_2 - v_2w_1)\,\hat{a}_3\,, \quad (4.20)$$

where the measure numbers of $\hat{a}_i$ are formed from the determinants of the $\hat{a}_j$ and $\hat{a}_k$ $(i \neq j;\; i \neq k)$ measure numbers, respectively, of $\vec{v}$ and $\vec{w}$. The reader is cautioned to note the sign change in the $\hat{a}_2$ component.

A good way to perform this operation is to write the first vector of the cross product directly above the second, as in the equations for $\vec{v}$ and $\vec{w}$ above. The components $\vec{a}_i$ should be ordered sequentially ($i = 1, 2, 3$) and aligned vertically so that the ith components are written directly above and below one another. A finger is placed over each component in turn, and the measure numbers are formed from the determinants of the other measure numbers. For instance, the $\hat{a}_1$ component of the cross product is found by placing the finger over the terms $v_1\hat{a}_1$ and $w_1\hat{a}_1$ in Equations 4.18 and 4.19. The determinant takes the product of the upper left and lower right measure numbers minus the product of the upper right and lower left measure numbers, $(v_2w_3 - v_3w_2)$. The finger is then moved over the $\hat{a}_2$ component, and the respective determinant $v_1w_3 - v_3w_1$ is formed. A minus sign is written in front of the $\hat{a}_2$ component. Finally, the finger is moved over the $\hat{a}_3$ component, and the cross product is completed with the final determinant, $(v_1w_2 - v_2w_1)$.

4.1.3 MOMENT OF A VECTOR

The moment $\vec{M}$ of any vector $\vec{v}$ computed about any point P is defined as,

$$\vec{M} = \vec{p}^{\,PQ} \times \vec{v}\,, \quad (4.21)$$

where Q is *any* point along the line of action of $\vec{v}$ and $\vec{p}^{\,PQ}$ is the position vector drawn from point P to point Q (Figure 4.4).

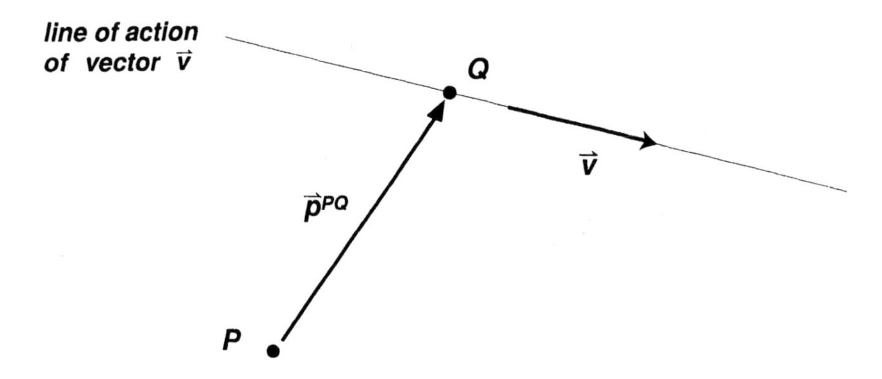

Figure 4.4. Moment of a vector $\vec{v}$ computed about point P. Point Q is *any* point along the line of action of $\vec{v}$. If $\vec{v}$ is replaced by a force $\vec{F}$, the moment is called a *moment of force*.

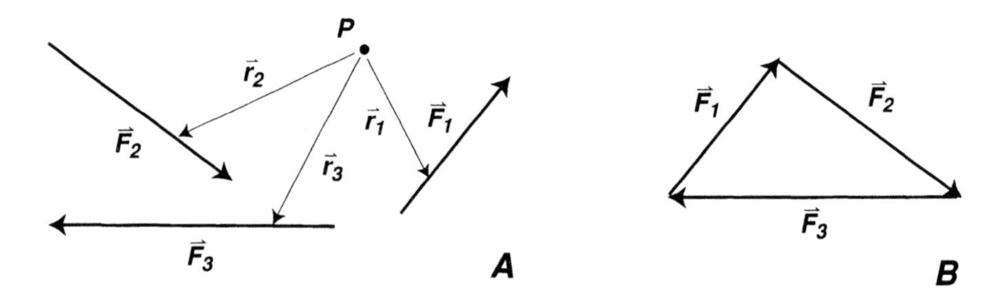

Figure 4.5. A. A couple composed of three forces $\vec{F}_1$, $\vec{F}_2$, and $\vec{F}_3$. When placed head to tail, the vectors create a closed path, which implies that the forces sum to zero. B. The moment of a couple is called a torque, and is the same no matter which point P it is computed about.

4.1.4 MOMENT OF FORCE, TORQUE

Equation 4.21 is most commonly used to compute the *moment of a force* about a particular point. Since force is a vector quantity, we can find the moment $\vec{M}_f$ of force $\vec{F}$ about any point P using the definition above,

$$\vec{M}_f = \vec{p}^{PQ} \times \vec{F}\,, \tag{4.22}$$

where Q is *any point along the line of action of the force* $\vec{F}$ (Figure 4.4) . Note that the computation of a moment first requires the designation of a point P about which the moment is to be computed. That is to say, the moment of force is *point specific* and changes depending upon the location of point P.

Torque is a special type of moment, defined as *the moment of a couple,* where a couple is defined as a set of n forces having no net resultant, *i.e.,*

$$\vec{F}_1 + \vec{F}_2 + \vec{F}_3 + \cdots + \vec{F}_n = 0\,. \tag{4.23}$$

Torque, then, is the algebraic sum of the moments of forces $\vec{F}_1$ through $\vec{F}_n$ about a common point P (see Figure 4.5). An interesting property of torque is that *torque is invariant regardless of which point P is chosen.* Since any point can be chosen, a judicious choice of point P will yield the same result and can save the analyst a lot of work (see the Example following).

Torque and moment of force are therefore different, although in common usage the terms “torque” and “moment” are often interchanged incorrectly. In Kane’s words, “torque is a moment, but a moment is not a torque” – the torque vector is point independent but the moment vector is intimately associated with a specific point (Kane and Levinson, 1985).

Figure 4.6. Free body diagram of the forearm and hand acted upon by gravity ($m_B\vec{g}$). The biceps muscle inserts at point I, and exerts a tensile force $\vec{F_b}$ directed toward the origin of the muscle. Joint reaction force $\vec{R}$ is the summation of all joint contact forces exerted by the distal humerus on the proximal ulna. $\vec{R}$ is applied on the ulna at point E. The inertial force $-m_B\,{}^N\vec{a}^{B^*}$ opposes the acceleration of point B^* in reference frame N. Position vector $\vec{r}_b$ is defined from E to I. The moment of the biceps muscle about E is simply $\vec{r}_b \times \vec{F_b}$. However, the torque exerted on the forearm and hand is the sum of the moments of all of the forces about a single point.

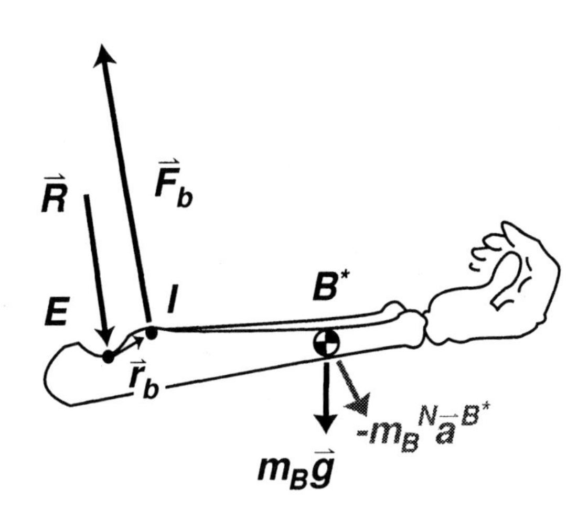

4.1.4.1 EXAMPLE – TORQUE EXERTED ON A RIGID BODY BY A MUSCLE CONTRACTION

Referring to Figure 4.6, it is easy to define the *moment* of the *biceps* muscle about the elbow, E,

$$\vec{M}_b = \vec{r}_b \times \vec{F}_b\,. \tag{4.24}$$

In fact, computing the moment about any specific point requires only that a vector be defined from that point to a point along the line of action of the force.

The *torque* created on the forearm and hand segment (B) is the sum of the moments of *all* the forces acting about a common point. The moments of all the forces are required, because every force shown, including the inertial force $-m_B\,{}^N\vec{a}^{B^*}$, is needed to create a couple,

$$\vec{F}_b + \vec{R} + m_B\vec{g} - m_B\,{}^N\vec{a}^{B^*} = 0\,, \tag{4.25}$$

and a couple is needed to compute the torque. By D'Alembert's principle, the inertial force is treated as if it were any other external force or body force applied directly to the system.

When computing torques, the process is simplified because the *torque will be the same no matter what point it is computed about*. Therefore, since any point can be chosen to compute the individual moments about, a point that lies along the line of action of an *unknown* force is usually chosen. This causes the moment of the unknown force to be zero, and allows the torque to be computed without requiring the force to be determined beforehand.

Following this procedure, the best point to sum moments about in anatomical joints is often the contact point, because the torque can be computed without having to first solve for the joint reaction force $\vec{R}$. In this case, point E is the most convenient point, because it represents the point through which the elbow joint reaction force acts. The torque would be,

$$\begin{aligned}\vec{\tau} &= \left(\vec{r}_b \times \vec{F}_b\right) + \left(\vec{0} \times \vec{R}\right) \\ &\quad + \left(\vec{p}^{EB^*} \times m_B \left(\vec{g} - {}^N\vec{a}^{B^*}\right)\right) . \end{aligned} \tag{4.26}$$

Picking a point *off* of the line of action of $\vec{R}$ would require that $\vec{R}$ be known, and its contribution to the total moment computed.[2]

It is usually not advantageous to sum moments about the centers of mass (in this case, point B^*). One might think that the moments of *two* forces could be eliminated by doing so. But unfortunately, one still has to compute the joint reaction force and determine its moment. Because

$$\vec{R} = -\vec{F}_b - m_B \left(\vec{g} + {}^N\vec{a}^{B^*}\right) = 0 , \tag{4.27}$$

the acceleration ${}^N\vec{a}^{B^*}$ and gravitational force $-m_B\,\vec{g}$ are still incorporated within $\vec{R}$, and remain very much a part of the moment about B^*.

4.1.4.2 EXAMPLE – MUSCLE MOMENT REQUIRED TO HOLD THE PELVIS STATIC IN A GRAVITATIONAL FIELD

In this analysis, the pelvis (rigid body P) rotates about the femoral head (point C) via the Euler rotations of Section 3.5.2, as shown in Figure 4.7. The direction cosines are defined as in Equation 3.48. Furthermore, assume that the femur (body F) is fixed with respect to an inertial reference frame N, and that gravity works downward in the $-\hat{n}_2$ direction on the center of mass, P^*. The gravitational moment about point C is given by

$$\vec{M}_{grav} = \vec{p}^{CP^*} \times Mg\left(-\hat{n}_2\right) , \tag{4.28}$$

where M is the mass of the trunk and pelvis combined. A single muscle, the *gluteus medius*, works to provide most of the abduction needed to counteract the tendency of the pelvis to "list" to the side.[3] The centroids of the muscle origin

[2] A common error is to sum moments about the *instantaneous center of joint rotation (ICR)*, neglecting the contributions to torque provided by a joint reaction force that does not pass through the ICR. In this particular example, point E is both the *focal point* of the joint reaction force and the ICR. This subject is discussed further in Section 5.4.1.

[3] In order to balance all three rotational degrees of freedom, a minimum of *three* muscles would be needed. To keep matters simple here, only the rotation of the pelvis in the frontal plane is considered.

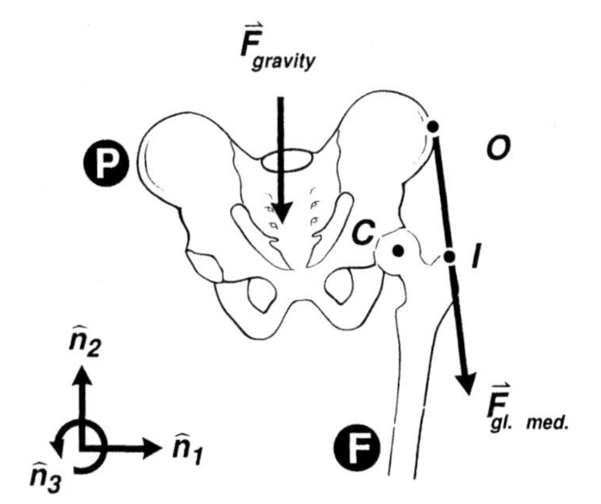

Figure 4.7. Holding the pelvis static in a gravitational field requires the gravitational moment to be balanced by the muscle moment.

and insertion areas are indicated by points O and I, respectively. Measured from point C, the position vectors locating O and I are,

$$\vec{p}^{CI} = i_1\hat{f}_1 + i_2\hat{f}_2 + i_3\hat{f}_3 \tag{4.29}$$

$$\vec{p}^{CO} = o_1\hat{p}_1 + o_2\hat{p}_2 + o_3\hat{p}_3 \,. \tag{4.30}$$

What is the force required in the *gluteus medius* muscle needed to statically counterbalance the gravitational moment about point C in the $\hat{p}_3$ direction?

The moment of the *gluteus medius* muscle about point C is,

$$\vec{M}_{gmed} = \vec{p}^{CO} \times \vec{F} \,, \tag{4.31}$$

where the force $\vec{F}$ is of unknown magnitude. The magnitude of $\vec{F}$ can be solved for if an expression can be written for its moment about C. But since the magnitude of $\vec{F}$ is not known yet, the procedure to compute the unknown force is not immediately obvious.

However, we can formulate an expression for $\vec{F}$ that incorporates an unknown magnitude F and a *known* direction $\hat{k}$, where $\hat{k}$ is a unit vector pointing along the line of action from O to I,

$$\hat{k} = \frac{\vec{p}^{OI}}{|\vec{p}^{OI}|} \,. \tag{4.32}$$

The unknown force $\vec{F}$ is then,

$$\vec{F} = F\hat{k} \,, \tag{4.33}$$

where a straight line of action is assumed. Figure 4.7 shows the vector diagram that relates the pertinent vectors of the problem to each other. Referring to the diagram, and starting from point C, one can get to point I either directly or via point O,

$$\vec{p}^{CO} + \vec{p}^{OI} = \vec{p}^{CI} \tag{4.34}$$

or,

$$\vec{p}^{OI} = \vec{p}^{CI} - \vec{p}^{CO} \tag{4.35}$$
$$= \left(i_1\hat{f}_1 + i_2\hat{f}_2 + i_3\hat{f}_3\right) - \left(o_1\hat{p}_1 + o_2\hat{p}_2 + o_3\hat{p}_3\right) . \tag{4.36}$$

Below, $\vec{p}^{CI}$ is re-expressed in the basis vectors of P. This is because $\vec{p}^{CO}$ and $\hat{k}$ are to be crossed according to Equation 4.31, and therefore must be expressed in the same basis.[4] Using Equation 3.48,

$$\begin{aligned}\vec{p}^{CI} = \; & i_1\left[(c_1c_3 - s_1c_2s_3)\hat{p}_1 + (-c_1s_3 - s_1c_2c_3)\hat{p}_2 + (s_1s_2)\hat{p}_3\right] \\ & + i_2\left[(s_1c_3 + c_1c_2s_3)\hat{p}_1 + (-s_1s_3 + c_1c_2c_3)\hat{p}_2 + (-c_1s_2)\hat{p}_3\right] \\ & + i_3\left[(s_2s_3)\hat{p}_1 + (s_2c_3)\hat{p}_2 + (c_2)\hat{p}_3\right] . \end{aligned} \tag{4.37}$$

Therefore, the vector from O to I is,

$$\vec{p}^{OI} = \beta_1\hat{p}_1 + \beta_2\hat{p}_2 + \beta_3\hat{p}_3 , \tag{4.38}$$

where

$$\begin{aligned}\beta_1 = \; & i_1(c_1c_3 - s_1c_2s_3) + i_2(s_1c_3 + c_1c_2s_3) \\ & + i_3(s_2s_3) - o_1 \end{aligned} \tag{4.39}$$
$$\begin{aligned}\beta_2 = \; & i_1(-c_1s_3 - s_1c_2c_3) + i_2(-s_1s_3 + c_1c_2c_3) \\ & + i_3(s_2c_3) - o_2 \end{aligned} \tag{4.40}$$
$$\beta_3 = i_1(s_1s_2) + i_2(-c_1s_2) + i_3(c_2) - o_3 . \tag{4.41}$$

Using these expressions, $\hat{k}$ is,

$$\hat{k} = \frac{\beta_1\hat{p}_1 + \beta_2\hat{p}_2 + \beta_3\hat{p}_3}{\sqrt{\beta_1^2 + \beta_2^2 + \beta_3^2}} \tag{4.42}$$
$$= \frac{1}{L}\left(\beta_1\hat{p}_1 + \beta_2\hat{p}_2 + \beta_3\hat{p}_3\right) , \tag{4.43}$$

where the length $\left|\vec{p}^{OI}\right| = \sqrt{\beta_1^2 + \beta_2^2 + \beta_3^2}$ has been replaced by the symbol L. The cross product of Equation 4.31 is,

$$\begin{aligned}\vec{M}_{gmed} = \; & \left[o_1\hat{p}_1 + o_2\hat{p}_2 + o_3\hat{p}_3\right] \\ & \times \left[\beta_1\hat{p}_1 + \beta_2\hat{p}_2 + \beta_3\hat{p}_3\right]\frac{F}{L} \end{aligned} \tag{4.44}$$

[4]Because a straight line of muscle action was assumed, one could have alternatively used the cross product $\vec{M}_{gmed} = \vec{p}^{CI^*} \times \vec{F}$ and expressed $\vec{F}$ in the same basis as vector $\vec{p}^{CI^*}$. The workload is the same either way.

$$= \frac{F}{L}\left\{(o_2\beta_3 - o_3\beta_2)\,\hat{p}_1 + (o_3\beta_1 - o_1\beta_3)\,\hat{p}_2 \right.$$

$$\left. + (o_1\beta_2 - o_2\beta_1)\,\hat{p}_3\right\}. \tag{4.45}$$

At this point, there is "only one thing" left to do. The gravitational moment in Equation 4.28 must be computed, expressed in the P basis vectors, and set equal to the moment just computed in Equation 4.45. To solve for the tension F needed to counterbalance the pelvic "list" about the $\hat{p}_3$ axis, the $\hat{p}_3$ components of the gravitational moment of force and muscle moment of force must be equated.[5] The resulting scalar equation is then easily solved for F.

4.1.5 DIRECT TIME DIFFERENTIATION OF A VECTOR

Given a vector $\vec{u}$ expressed in the basis vectors of A, and scalar measure numbers $\mathcal{F}_i$, $(i = 1,\ 2,\ 3)$ that are functions of the generalized coordinates q_i,

$$\vec{u} = \mathcal{F}_1\hat{a}_1 + \mathcal{F}_2\hat{a}_2 + \mathcal{F}_3\hat{a}_3 \tag{4.46}$$

the derivative of $\vec{u}$ with respect to reference frame A is defined as,

$$\frac{{}^A d\vec{u}}{dt} = \frac{d\mathcal{F}_1}{dt}\hat{a}_1 + \frac{d\mathcal{F}_2}{dt}\hat{a}_2 + \frac{d\mathcal{F}_3}{dt}\hat{a}_3 \tag{4.47}$$

$$= \dot{\mathcal{F}}_1\hat{a}_1 + \dot{\mathcal{F}}_2\hat{a}_2 + \dot{\mathcal{F}}_3\hat{a}_3\,, \tag{4.48}$$

where the dot $(\dot{\ })$ is a shorthand form for the first time derivative. The vector $\vec{u}$ must be expressed in the basis vectors of A prior to direct differentiation with respect to time in reference frame A.

The reader is reminded that differentiation of the direction cosines requires the chain rule for differentiation,

$$\frac{d}{dt}(c_i) = -s_i\dot{q}_i \tag{4.49}$$

$$\frac{d}{dt}(s_i) = c_i\dot{q}_i\,. \tag{4.50}$$

4.1.5.1 EXAMPLE – DIRECT TIME DIFFERENTIATION OF A VECTOR.

Let $\vec{v}$ be a vector fixed in reference frame B, for instance,

$$\vec{v} = 1\hat{b}_2 + 3\hat{b}_3\,. \tag{4.51}$$

[5] In general, a single muscle force can only balance the moments about one axis of rotation. In this example, only the major unbalancing component created by the gravitational force is considered.

It is desired to differentiate $\vec{v}$ twice with respect to time in reference frame A. Frame B is rotated from an orientation coincident with reference frame A by successive rotations: angle q_1 about the common $\hat{a}_1$, $\hat{a}_1'$ axis, and angle q_2 about the common $\hat{a}_2'$, $\hat{b}_2$ axis, which leads to the following table of direction cosines,

$$\begin{array}{c|ccc} {}^AR^B & \hat{b}_1 & \hat{b}_2 & \hat{b}_3 \\ \hline \hat{a}_1 & c_2 & 0 & s_2 \\ \hat{a}_2 & s_1 s_2 & c_1 & -s_1 c_2 \\ \hat{a}_3 & -c_1 s_2 & s_1 & c_1 c_2 \end{array} \tag{4.52}$$

Using this table, $\vec{v}$ can be expressed in the basis vectors of A,

$$\begin{aligned} \vec{v} &= 1\,(c_1\hat{a}_2 + s_1\hat{a}_3) + 3\,(s_2\hat{a}_1 - s_1c_2\hat{a}_2 + c_1c_2\hat{a}_3) && (4.53)\\ &= (3s_2)\,\hat{a_1} + (c_1 - 3s_1c_2)\,\hat{a_2} + (s_1 + 3c_1c_2)\,\hat{a_3}\,. && (4.54) \end{aligned}$$

The first time derivative of $\vec{v}$ in reference frame A is,

$$\begin{aligned} \frac{{}^Ad\vec{v}}{dt} &= 3\,[c_2\dot{q}_2]\,\hat{a}_1 + [-s_1\dot{q}_1 - 3\,(s_1(-s_2\dot{q}_2) + (c_1\dot{q}_1)c_2)]\,\hat{a}_2 \\ &\quad + [c_1\dot{q}_1 + 3\,(c_1(-s_2\dot{q}_2) + (-s_1\dot{q}_1)c_2)]\,\hat{a}_3 && (4.55)\\ &= [3c_2\dot{q}_2]\,\hat{a}_1 + [-s_1\dot{q}_1 - 3c_1c_2\dot{q}_1 + 3s_1s_2\dot{q}_2]\,\hat{a}_2 \\ &\quad + [c_1\dot{q}_1 - 3s_1c_2\dot{q}_1 - 3c_1s_2\dot{q}_2]\,\hat{a}_3\,. && (4.56) \end{aligned}$$

Continuing further, the second time derivative of $\vec{v}$ in reference frame A is,

$$\begin{aligned} \frac{{}^Ad^2\vec{v}}{dt^2} &= 3\left[c_2\ddot{q}_2 - s_2\dot{q}_2^2\right]\hat{a}_1 \\ &\quad +\Big[\left(-s_1\ddot{q}_1 - c_1\dot{q}_1^2\right) - 3\Big(c_1(c_2\ddot{q}_1 - s_2\dot{q}_2\dot{q}_1) - s_1\dot{q}_1(c_2\dot{q}_1)\Big) \\ &\quad +3\left(s_1(s_2\ddot{q}_2 + c_2\dot{q}_2^2) + c_1\dot{q}_1(s_2\dot{q}_2)\right)\Big]\hat{a}_2 \\ &\quad +\Big[\left(c_1\ddot{q}_1 - s_1\dot{q}_1^2\right) - 3\Big(s_1(c_2\ddot{q}_1 - s_2\dot{q}_2\dot{q}_1) + c_1\dot{q}_1(c_2\dot{q}_1)\Big) \\ &\quad -3\Big(c_1(s_2\ddot{q}_2 + c_2\dot{q}_2^2) - s_1\dot{q}_1(s_2\dot{q}_2)\Big)\Big]\hat{a}_3 && (4.57)\\ &= \left[3c_2\ddot{q}_2 - 3s_2\dot{q}_2^2\right]\hat{a}_1 \\ &\quad +\Big[(-s_1 - 3c_1c_2)\ddot{q}_1 + (3s_1s_2)\ddot{q}_2 + (-c_1 + 3s_1c_2)\dot{q}_1^2 \\ &\quad +(3s_1c_2)\dot{q}_2^2 + (6c_1s_2)\dot{q}_1\dot{q}_2\Big]\hat{a}_2 \\ &\quad +\Big[(c_1 - 3s_1c_2)\ddot{q}_1 + (-3c_1s_2)\ddot{q}_2 + (-s_1 - 3c_1c_2)\dot{q}_1^2 \\ &\quad +(-3c_1c_2)\dot{q}_2^2 + (6s_1s_2)\dot{q}_1\dot{q}_2\Big]\hat{a}_3\,. && (4.58) \end{aligned}$$

4.1.6 VELOCITY AND ACCELERATION BY TIME DIFFERENTIATION

The location of point P is specified with respect to reference frame A by the following position vector,

$$ {}^{A}\vec{p}^{P} = \mathcal{F}_1 \hat{a}_1 + \mathcal{F}_2 \hat{a}_2 + \mathcal{F}_3 \hat{a}_3 \,. \tag{4.59}$$

Measure numbers $\mathcal{F}_i$ ($i = 1, 2, 3$) are time varying scalar functions of the generalized coordinates q_i. Then, the velocity and acceleration of P in reference frame A may be computed directly using time differentiation,

$$ {}^{A}\vec{v}^{P} = \frac{{}^{A}d\left({}^{A}\vec{p}^{P}\right)}{dt} = \dot{\mathcal{F}}_1 \hat{a}_1 + \dot{\mathcal{F}}_2 \hat{a}_2 + \dot{\mathcal{F}}_3 \hat{a}_3 \tag{4.60}$$

$$ {}^{A}\vec{a}^{P} = \frac{{}^{A}d\left({}^{A}\vec{v}^{P}\right)}{dt} = \ddot{\mathcal{F}}_1 \hat{a}_1 + \ddot{\mathcal{F}}_2 \hat{a}_2 + \ddot{\mathcal{F}}_3 \hat{a}_3 \,. \tag{4.61}$$

The dot ($\dot{\ }$) and double-dot ($\ddot{\ }$) are shorthand forms for the first and second time derivatives,

$$ \dot{\mathcal{F}}_i \equiv \frac{d\mathcal{F}_i}{dt} \tag{4.62}$$

$$ \ddot{\mathcal{F}}_i \equiv \frac{d^2\mathcal{F}_i}{dt^2} \,. \tag{4.63}$$

4.2 ANGULAR VELOCITY AND ANGULAR ACCELERATION

4.2.1 ANGULAR VELOCITY

Consider two rigid body reference frames, A and B. One useful descriptor of the angular motion of B relative to A is the *angular velocity of B in A*, denoted ${}^{A}\vec{\omega}^{B}$, which is defined here as,[6]

$$ {}^{A}\vec{\omega}^{B} \equiv \dot{q}\, \hat{k} \,. \tag{4.65}$$

[6] The following expression (Equation 4.64) is the definition given by Kane and Levinson (1985). It is rigorous but rarely used due to its abstract nature, and thus a simpler "working" definition is introduced here.

Let basis vectors $\hat{b}_1$, $\hat{b}_2$, and $\hat{b}_3$ be mutually perpendicular and arranged so as to form a right-handed coordinate system. Affix $\hat{b}_1$, $\hat{b}_2$, and $\hat{b}_3$ to rigid body B which is rotating with respect to reference frame A. The angular velocity of B in A is defined by Kane and Levinson (1985) as,

$$ {}^{A}\vec{\omega}^{B} \equiv \hat{b}_1 \frac{{}^{A}d\hat{b}_2}{dt} \cdot \hat{b}_3 + \hat{b}_2 \frac{{}^{A}d\hat{b}_3}{dt} \cdot \hat{b}_1 + \hat{b}_3 \frac{{}^{A}d\hat{b}_1}{dt} \cdot \hat{b}_2 \,. \tag{4.64}$$

Each term ${}^{A}d\hat{b}_i/dt$ ($i = 1, 2, 3$) is the velocity of a point P_i at the tip of unit vector $\hat{b}_i$ according to Equation 4.60. The dot products extract scalar components of these linear velocities, which can be called *speeds*. Because the speeds were computed for unit vectors, these linear speeds are equivalent to angular rotation speeds about the three rotation axes.

In the equation, $\hat{k}$ is a unit vector parallel to the *instantaneous rotation axis of B in A,* which is defined as the locus of points having zero velocity in both A and B resulting from the rotational components of the motion. The scalar quantity $\dot{q}$ is called the *angular speed of B in A,* which is defined as the instantaneous time rate of change of q, where q is the angular orientation between corresponding reference lines fixed in A and B on a common *plane of motion* perpendicular to $\hat{k}$.

The angular velocity of body A in reference frame B is related to the angular velocity of body B in A by an almost trivial relationship,

$$ {}^{B}\vec{\omega}^{A} = -{}^{A}\vec{\omega}^{B} \,. \tag{4.66} $$

When reference frame B rotates about reference frame A in such a way that the axis of revolution remains fixed in its direction relative to both A and B throughout a *finite time interval,* then the angular velocity of B in A is considered to be a *simple angular velocity.* The next section explains how simple angular velocities are vectorially added to create kinematic models of physiological joints having multiple degrees of freedom.

4.2.2 ADDITION THEOREM FOR ANGULAR VELOCITIES

The rotation of body B relative to reference frame A is not, in general, well described by a simple angular velocity. However, the rotation of body B relative to A can be expressed as a summation of simple (and not-so-simple) angular velocities. If the angular velocities of intermediate reference frames B' to $B^{(n)}$ are known, then a compound angular velocity may always be decomposed into a summation of simple angular velocities,

$$ {}^{A}\vec{\omega}^{B} = {}^{A}\vec{\omega}^{B'} + {}^{B'}\vec{\omega}^{B''} + {}^{B''}\vec{\omega}^{B'''} + \ldots + {}^{B^{(n)}}\vec{\omega}^{B} \,. \tag{4.67} $$

When the right superscript of one $\vec{\omega}$ is identical to the left superscript of the next $\vec{\omega}$, the reader can think of these interior superscripts as "cancelling" each other.

Example – Compound Angular Velocity following Euler Rotations. The sequence of Euler rotations defined in Section 3.5.2 uniquely specifies the 3-D orientation of the pelvis P with respect to the femoral reference frame F. The

---

For instance, in the third term of the definition, ${}^{A}d\hat{b}_1/dt$ represents the velocity of a point P_1 defined at the tip of the $\hat{b}_1$ vector. Using this definition, ${}^{A}d\hat{b}_1/dt$ can be written as ${}^{A}\vec{v}^{P_1}$. Dot product ${}^{A}\vec{v}^{P_1} \cdot \hat{b}_2 = s$ extracts a particular component of the linear *speed* of point P_1. The component of interest is the component s which lies in the plane perpendicular to the $\hat{b}_3$ rotation axis. Because $\dot{q}_3 = s/|\hat{b}_3| = s$ (the reader might be more familiar with $\omega = v/r$), the quantity ${}^{A}d\hat{b}_1/dt \cdot \hat{b}_2$ simply represents the angular rotation speed about the $\hat{b}_3$ axis.

time derivatives of these rotation angles may be used to define the angular velocity of the pelvis with respect to the femur as well. Each Euler rotation gives rise to a simple angular velocity which may be summed vectorially with the others to yield a compound angular velocity. Specifically,

$$^{F}\vec{\omega}^{P} \;=\; ^{F}\vec{\omega}^{P'} + ^{P'}\vec{\omega}^{P''} + ^{P''}\vec{\omega}^{P}. \tag{4.68}$$

Substituting the angular velocities with their corresponding angular speeds and rotation axes,

$$\begin{aligned} ^{F}\vec{\omega}^{P} &= \dot{q}_1\hat{f}_3 + \dot{q}_2\hat{p}_1^{'} + \dot{q}_3\hat{p}_3^{''} & (4.69)\\ &= \dot{q}_1\left(s_2s_3\hat{p}_1 + s_2c_3\hat{p}_2 + c_2\hat{p}_3\right) \\ &\quad + \dot{q}_2\left(c_3\hat{p}_1 - s_3\hat{p}_2\right) + \dot{q}_3\left(\hat{p}_3\right) & (4.70)\\ &= \left(\dot{q}_1s_2s_3 + \dot{q}_2c_3\right)\hat{p}_1 + \left(\dot{q}_1s_2c_3 - \dot{q}_2s_3\right)\hat{p}_2 \\ &\quad + \left(\dot{q}_1c_2 + \dot{q}_3\right)\hat{p}_3\,. & (4.71) \end{aligned}$$

Note that the angular velocity of body P was expressed in the basis vectors of P. While it is perfectly acceptable to leave an angular velocity vector in mixed bases as in Equation 4.69, doing the little bit of "extra" work to express the angular velocity in the basis vectors of the rotating body almost always saves a huge amount of work later.

4.2.3 ANGULAR ACCELERATION

Suppose the angular velocity $^{A}\vec{\omega}^{B}$ of body B with respect to reference frame A is expressed in the basis vectors of A,

$$^{A}\vec{\omega}^{B} = \dot{\mathcal{F}}_1\hat{a}_1 + \dot{\mathcal{F}}_2\hat{a}_2 + \dot{\mathcal{F}}_3\hat{a}_3\,. \tag{4.72}$$

Measure numbers $\dot{\mathcal{F}}_i$ $(i = 1,\,2,\,3)$ are time derivatives of functions $\mathcal{F}_i$, which are time varying scalar functions of the generalized coordinates q_i. $\dot{\mathcal{F}}_i$, and its derivative $\ddot{\mathcal{F}}_i$, are shorthand forms for the first and second time derivatives of $\mathcal{F}_i$,

$$\begin{aligned} \dot{\mathcal{F}}_i &\equiv \frac{d\mathcal{F}_i}{dt} & (4.73)\\ \ddot{\mathcal{F}}_i &\equiv \frac{d^2\mathcal{F}_i}{dt^2}\,. & (4.74) \end{aligned}$$

Then, the *angular acceleration* of B in reference frame A may be computed directly using time differentiation,

$$^{A}\vec{\alpha}^{B} = \frac{^{A}d\left(^{A}\vec{\omega}^{B}\right)}{dt} = \ddot{\mathcal{F}}_1\hat{a}_1 + \ddot{\mathcal{F}}_2\hat{a}_2 + \ddot{\mathcal{F}}_3\hat{a}_3\,. \tag{4.75}$$

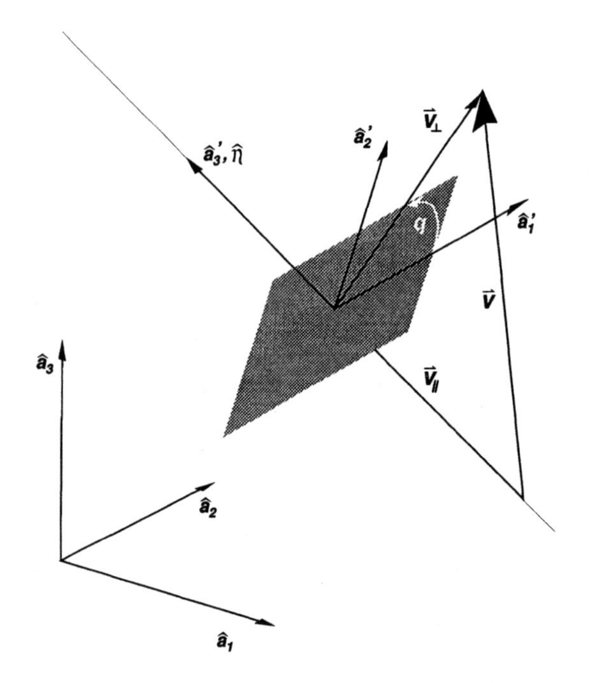

Figure 4.8. Figure for the proof of Equation 4.77. Vector $\vec{v}$ is fixed in reference frame B, which rotates with angular velocity ${}^A\vec{\omega}^B$ in reference frame A. An auxiliary reference frame A' composed of mutually perpendicular basis vectors $\hat{a}_1'$, $\hat{a}_2'$, $\hat{a}_3'$ and fixed in A is aligned so that $\hat{a}_3'$ is coincident with instantaneous rotation axis $\hat{\eta}$. Vector $\vec{v}$ is broken down into two components. $\vec{v}_{||}$ is the component of $\vec{v}$ parallel to $\hat{\eta}$, and $\vec{v}_{\perp}$ is the component perpendicular to $\hat{\eta}$.

As with angular velocity vectors, the angular acceleration of body A in reference frame B is related to the angular acceleration of body B in A,

$$ {}^B\vec{\alpha}^A = -{}^A\vec{\alpha}^B \,. \tag{4.76}$$

4.3 VECTOR CALCULUS VIA CROSS PRODUCTS

4.3.1 DIFFERENTIATION OF A VECTOR USING ANGULAR VELOCITY VECTORS

Any vector $\vec{v}$ *fixed* in reference frame B can be differentiated with respect to time in reference frame A using the angular velocity of B in A rather than performing direct differentiation as defined in Equation 4.48,

$$ \frac{{}^A d\vec{v}}{dt} = {}^A\vec{\omega}^B \times \vec{v} \,. \tag{4.77}$$

This relationship is extremely useful, as it provides an algebraic alternative to the normal process of direct time differentiation using calculus.

Proof. An expression for the left-hand side of Equation 4.77 is found first. Again, $\vec{v}$ is a vector fixed in reference frame B which is rotating with respect to reference frame A. Since ${}^A\vec{\omega}^B$ is a vector quantity, it can be expressed in terms of a simple rotation over an infinitesimally small time interval Δt,

$$ {}^A\vec{\omega}^B = \dot{q}\hat{\eta} \,, \tag{4.78}$$

where $\hat{\eta}$ is a unit vector parallel to ${}^A\vec{\omega}^B$ and fixed in A (see Figure 4.8). Then, $\vec{v}$ can be decomposed into the sum of two vectors, one parallel to $\hat{\eta}$ and the other perpendicular to $\hat{\eta}$,

$$\vec{v}_{||} \equiv \vec{v} \cdot \hat{\eta}\,\hat{\eta} \tag{4.79}$$

$$\vec{v}_{\perp} \equiv \vec{v} - \vec{v}_{||}\,, \tag{4.80}$$

such that

$$\vec{v} = \vec{v}_{||} + \vec{v}_{\perp}\,. \tag{4.81}$$

Let us also define an auxiliary set of mutually orthogonal unit basis vectors $\hat{a}_1^{'}$, $\hat{a}_2^{'}$, $\hat{a}_3^{'}$ fixed in A and oriented such that $\hat{a}_3^{'}$ is coincident with $\hat{\eta}$, $\hat{a}_1^{'}$ is perpendicular to $\hat{\eta}$ and $\hat{a}_2^{'} = \hat{a}_3^{'} \times \hat{a}_1^{'}$. Additionally, $\hat{a}_1^{'}$ forms a reference line fixed in A from which the rotation angle q to $\vec{v}_{\perp}$ is measured. Using these definitions,

$$\vec{v}_{\perp} = L\cos q\hat{a}_1^{'} + L\sin q\hat{a}_2^{'}\,, \tag{4.82}$$

where $L \equiv |\vec{v}_{\perp}|$. Then,

$$\frac{{}^A d\vec{v}}{dt} = \frac{{}^A d}{dt}\left(\vec{v}_{\perp} + \vec{v}_{||}\right) = \frac{{}^A d\vec{v}_{\perp}}{dt} + \frac{{}^A d\vec{v}_{||}}{dt}\,. \tag{4.83}$$

Since $\vec{v}_{||}$, the parallel component of $\vec{v}$ does not change when B rotates with respect to A with angular velocity ${}^A\vec{\omega}^B$, then ${}^A d\vec{v}_{||}/dt = 0$ and

$$\frac{{}^A d\vec{v}}{dt} = \frac{{}^A d\vec{v}_{\perp}}{dt}\,. \tag{4.84}$$

Also, since the basis vectors $\hat{a}_1^{'}$, $\hat{a}_2^{'}$, $\hat{a}_3^{'}$ comprise a reference frame $A^{'}$ that is fixed in A, taking the time derivative of $\vec{v}_{\perp}$ in A is equivalent to taking it in $A^{'}$, or

$$\frac{{}^A d\vec{v}_{\perp}}{dt} = \frac{{}^{A'} d\vec{v}_{\perp}}{dt}\,. \tag{4.85}$$

Thus, to create the desired expression for the left-hand side of Equation 4.77 we need only differentiate Equation 4.82 directly,

$$\frac{{}^A d\vec{v}}{dt} = -L\dot{q}\sin q\hat{a}_1^{'} + L\dot{q}\cos q\hat{a}_2^{'}\,. \tag{4.86}$$

Next, the derivation of an expression for the right-hand side of Equation 4.77 follows.

$${}^A\vec{\omega}^B \times \vec{v} = \dot{q}\hat{\eta} \times \left(\vec{v}_{\perp} + \vec{v}_{||}\right) \tag{4.87}$$

$$= \left(\dot{q}\hat{\eta} \times \vec{v}_{\perp}\right) + \left(\dot{q}\hat{\eta} \times \vec{v}_{||}\right) \tag{4.88}$$

$$= \dot{q}\hat{a}_3' \times \left(L\cos q\hat{a}_1' + L\sin q\hat{a}_2'\right) + (0) \qquad (4.89)$$

$$= \dot{q}L\left(\cos q(\hat{a}_3' \times \hat{a}_1') + \sin q(\hat{a}_3' \times \hat{a}_2')\right) \qquad (4.90)$$

$$= \dot{q}L\left(\cos q\hat{a}_2' + \sin q(-\hat{a}_1')\right). \qquad (4.91)$$

This matches the expression we previously obtained in Equation 4.86.

To obtain the second time derivative of $\vec{v}$ in reference frame A, note that ${}^A\vec{\omega}^B$ cannot simply be crossed again with the expression resulting from performing the cross product on the right hand side of Equation 4.77. The second time derivative must be obtained via *two* cross products,

$$\frac{{}^Ad}{dt}\left(\frac{{}^Ad\vec{v}}{dt}\right) = \frac{{}^Ad}{dt}\left({}^A\vec{\omega}^B \times \vec{v}\right) \qquad (4.92)$$

$$= \frac{{}^Ad}{dt}\left({}^A\vec{\omega}^B\right) \times \vec{v} + {}^A\vec{\omega}^B \times \frac{{}^Ad\vec{v}}{dt} \qquad (4.93)$$

$$= {}^A\vec{\alpha}^B \times \vec{v} + {}^A\vec{\omega}^B \times \frac{{}^Ad\vec{v}}{dt}. \qquad (4.94)$$

4.3.1.1 EXAMPLE – DIFFERENTIATION OF A VECTOR FIXED IN *B* BY A VECTOR CROSS PRODUCT

Following the example in Section 4.1.5, the vector $\vec{v} = 1\hat{b}_2 + 3\hat{b}_3$ can be differentiated using a cross product if ${}^A\vec{\omega}^B$ is expressed in the basis vectors of B,

$${}^A\vec{\omega}^B = \dot{q}_1\hat{a}_1 + \dot{q}_2\hat{b}_2 \qquad (4.95)$$

$$= \dot{q}_1\left(c_2\hat{b}_1 + s_2\hat{b}_3\right) + \dot{q}_2\hat{b}_2 \qquad (4.96)$$

$$= \dot{q}_1c_2\hat{b}_1 + \dot{q}_2\hat{b}_2 + \dot{q}_1s_2\hat{b}_3. \qquad (4.97)$$

The first time derivative of $\vec{v}$ in reference frame A can be found by a vector cross product,

$$\frac{{}^Ad\vec{v}}{dt} = {}^A\vec{\omega}^B \times \vec{v} \qquad (4.98)$$

$$= (3\dot{q}_2 - \dot{q}_1s_2)\,\hat{b}_1 + (-3\dot{q}_1c_2)\,\hat{b}_2 + (\dot{q}_1c_2)\,\hat{b}_3. \qquad (4.99)$$

There is no compelling reason to convert this result into the basis vectors of any other reference frame. It is done here only for the sake of matching this answer with the prior answer (Equation 4.56) obtained via direct time differentiation. Using the direction cosine matrix of Equation 4.52, this becomes,

$$\frac{{}^Ad\vec{v}}{dt} = (3\dot{q}_2 - \dot{q}_1s_2)\left(c_2\hat{a}_1 + s_1s_2\hat{a}_2 - c_1s_2\hat{a}_3\right)$$

$$
\begin{aligned}
&+ (-3\dot{q}_1 c_2)(c_1 \hat{a}_2 + s_1 \hat{a}_3) \\
&+ (\dot{q}_1 c_2)(s_2 \hat{a}_1 - s_1 c_2 \hat{a}_2 + c_1 c_2 \hat{a}_3) \qquad (4.100) \\
= \;& [3c_2 \dot{q}_2 - s_2 c_2 \dot{q}_1 + c_2 s_2 \dot{q}_1] \hat{a}_1 \\
&+ \left[3 s_1 s_2 \dot{q}_2 - s_1 s_2^2 \dot{q}_1 - 3 c_1 c_2 \dot{q}_1 - s_1 c_2^2 \dot{q}_1\right] \hat{a}_2 \\
&+ \left[-3 c_1 s_2 \dot{q}_2 + c_1 s_2^2 \dot{q}_1 - 3 s_1 c_2 \dot{q}_1 + c_1 c_2^2 \dot{q}_1\right] \hat{a}_3 \qquad (4.101) \\
= \;& [3 c_2 \dot{q}_2] \hat{a}_1 \\
&+ [3 s_1 s_2 \dot{q}_2 - s_1 \dot{q}_1 - 3 c_1 c_2 \dot{q}_1] \hat{a}_2 \\
&+ [-3 c_1 s_2 \dot{q}_2 - 3 s_1 c_2 \dot{q}_1 + c_1 \dot{q}_1] \hat{a}_3 \,. \qquad (4.102)
\end{aligned}
$$

4.3.2 DIFFERENTIATION IN TWO REFERENCE FRAMES

Often, the time differentiation of a vector with respect to a particular reference frame can be more easily performed with respect to a different reference frame. Any vector $\vec{v}$ defined in reference frame A and in any other reference frame (say, B) can be differentiated in either reference frame A or B. The first time derivatives of $\vec{v}$ with respect to A and B are related by the following expression involving the angular velocity of B in A,

$$
\frac{{}^A d\vec{v}}{dt} = \frac{{}^B d\vec{v}}{dt} + {}^A\vec{\omega}^B \times \vec{v} \,. \qquad (4.103)
$$

This formula is easy to memorize, as *the same vector* $\vec{v}$ appears in all three terms. However, *it is vitally important that the angular velocity be specified correctly between the two reference frames.* An easy way to remember this is by examining the superscripts. If an "A" superscript appears on the derivative to the left of the equals sign, then an "A" superscript must appear on the left side of the angular velocity. Likewise, a "B" superscript appearing on the derivative to the right side of the equals sign requires a "B" superscript on the right side of the angular velocity. For a further reference on transforming the time derivatives of vectors, see Meriam and Kraige (1997). The proof is from Kane and Levinson (1985).

Proof. Let $\vec{v} = v_1 \hat{b}_1 + v_2 \hat{b}_2 + v_3 \hat{b}_3$. Then,

$$
\begin{aligned}
\frac{{}^A d\vec{v}}{dt} = \;& \left(\frac{dv_1}{dt}\,\hat{b}_1 + \frac{dv_2}{dt}\,\hat{b}_2 + \frac{dv_3}{dt}\,\hat{b}_3\right) \\
&+ \left[v_1 \frac{{}^A d\hat{b}_1}{dt} + v_2 \frac{{}^A d\hat{b}_2}{dt} + v_3 \frac{{}^A d\hat{b}_3}{dt}\right] \qquad (4.104) \\
= \;& \frac{{}^B d\vec{v}}{dt} + \left[v_1 \left({}^A\vec{\omega}^B \times \hat{b}_1\right)\right.
\end{aligned}
$$

$$+ v_2 \left({}^A\vec{\omega}^B \times \hat{b}_2 \right) + v_3 \left({}^A\vec{\omega}^B \times \hat{b}_3 \right) \Big] , \qquad (4.105)$$

where the time derivatives of unit vectors $\hat{b}_1$, $\hat{b}_2$, $\hat{b}_3$ in reference frame A are replaced by cross products following Equation 4.77. Continuing,

$$\frac{{}^A d\vec{v}}{dt} = \frac{{}^B d\vec{v}}{dt} + \Big[\left({}^A\vec{\omega}^B \times v_1\hat{b}_1 \right) + \left({}^A\vec{\omega}^B \times v_2\hat{b}_2 \right) + \left({}^A\vec{\omega}^B \times v_3\hat{b}_3 \right) \Big] \qquad (4.106)$$

$$= \frac{{}^B d\vec{v}}{dt} + \left[{}^A\vec{\omega}^B \times \left(v_1\hat{b}_1 + v_2\hat{b}_2 + v_3\hat{b}_3 \right) \right] \qquad (4.107)$$

$$= \frac{{}^B d\vec{v}}{dt} + \left[{}^A\vec{\omega}^B \times \vec{v} \right] . \qquad (4.108)$$

4.3.2.1 EXAMPLE – DIFFERENTIATING A VECTOR THE EASY WAY

Following the same example problem discussed in Section 4.1.5 and 4.3.1.1, we now seek the first time derivative of the vector $\vec{v} = 1\hat{b}_2 + 3\hat{b}_3$ in the A reference frame. One can avoid expressing $\vec{v}$ in the basis vectors of A, and performing direct differentiation of the direction cosines (which can be quite arduous) in the A reference frame as was done in Section 4.3.1.1. Because $\vec{v}$ is fixed in B, doing the differentiation in B would be much easier, as

$$\frac{{}^B d\vec{v}}{dt} = 0 . \qquad (4.109)$$

This allows us to transform the process of ordinary differentiation of a vector into a cross product, because

$$\frac{{}^A d\vec{v}}{dt} = \frac{{}^B d\vec{v}}{dt} + {}^A\vec{\omega}^B \times \vec{v} \qquad (4.110)$$

$$= 0 + {}^A\vec{\omega}^B \times \vec{v} . \qquad (4.111)$$

Since $\vec{v}$ is defined in the basis vectors of B, ${}^A\vec{\omega}^B$ must be expressed in the same basis. From the earlier examples,

$${}^A\vec{\omega}^B = \dot{q}_1 c_2 \hat{b}_1 + \dot{q}_2 \hat{b}_2 + \dot{q}_1 s_2 \hat{b}_3 . \qquad (4.112)$$

The first time derivative of $\vec{v}$ in reference frame A can be found by a vector cross product in the B basis,

$$\frac{{}^A d\vec{v}}{dt} = \left(\dot{q}_1 c_2 \hat{b}_1 + \dot{q}_2 \hat{b}_2 + \dot{q}_1 s_2 \hat{b}_3 \right) \times \left(1\hat{b}_2 + 3\hat{b}_3 \right) \qquad (4.113)$$

$$= \left(3\dot{q}_2 - \dot{q}_1 s_2 \right) \hat{b}_1 + \left(-3\dot{q}_1 c_2 \right) \hat{b}_2 + \left(\dot{q}_1 c_2 \right) \hat{b}_3 . \qquad (4.114)$$

Again, there is no need to express the differentiated vector in another basis (*e.g.*, the A basis) unless there is a good reason to do so. For instance, if a cross product needed to be performed in reference frame A, then Equation 4.114 can be converted using the table of direction cosines from A to B.

Most of the time, the vector to be differentiated is fixed in some reference frame. This is the reference frame in which the differentiation is usually easiest because the time derivative in Equation 4.103 is zero. The same reference frame also serves as the basis in which the derivative is most easily expressed, as it is the least "polluted" by direction cosines.

4.3.2.2 EXAMPLE – DIFFERENTIATION OF ANGULAR VELOCITY VECTORS

The angular acceleration of any two reference frames A and C can be found by differentiating the angular velocity of C in A in either of the two reference frames A or C. Substituting ${}^{A}\vec{\omega}^{C}$ for $\vec{u}$ (the substituted quantities are enclosed in parentheses),

$$ {}^{A}\vec{\alpha}^{C} = \frac{{}^{A}d\left({}^{A}\vec{\omega}^{C}\right)}{dt} \tag{4.115}$$

$$ = \frac{{}^{C}d\left({}^{A}\vec{\omega}^{C}\right)}{dt} + {}^{A}\vec{\omega}^{C} \times \left({}^{A}\vec{\omega}^{C}\right), \tag{4.116}$$

or, since ${}^{A}\vec{\omega}^{C} \times {}^{A}\vec{\omega}^{C} = 0$,

$$ \frac{{}^{A}d\left({}^{A}\vec{\omega}^{C}\right)}{dt} = \frac{{}^{C}d\left({}^{A}\vec{\omega}^{C}\right)}{dt}. \tag{4.117}$$

This states that the angular acceleration vector of body C rotating with respect to reference frame A may be found by performing time differentiation in *either* the A or C reference frames. This fact is limited to angular velocity vectors, and can always be exploited to perform the differentiation in the frame that has the simplest (*i.e.*, most easily differentiated) expression for the angular velocity.

4.3.2.3 EXAMPLE – VELOCITY OF A MUSCLE

This example is applicable to any muscle having a straight line of action between its origin and insertion. The vector definitions and reference frames for the *gluteus medius* muscle have already been established in Section 4.1.4.2. Therefore, it is convenient to illustrate the methodology using the prior definitions and reference frames of the femur (F) and pelvis (P). As before, the position vectors are defined from the center of the acetabulum (point C) to the centroids of the origination (point O) and insertion areas (point I). Because the hip joint has a fixed center of rotation with respect to both the femur and pelvis, position vector $\vec{p}^{CO}$ is fixed in reference frame P, and vector $\vec{p}^{CI}$ is fixed in reference frame F.

The velocity of the muscle is the time derivative of the position vector from origin to insertion. Because it does not matter which reference frame the derivative is taken with respect to, the derivative will be taken in the F reference frame.

$$\vec{v}^{OI} = \frac{{}^{F}d}{dt}\left(\vec{p}^{OI}\right) \tag{4.118}$$

$$= \frac{{}^{F}d}{dt}\left(\vec{p}^{CI} - \vec{p}^{CO}\right) \tag{4.119}$$

$$= \frac{{}^{F}d}{dt}\left(\vec{p}^{CI}\right) - \frac{{}^{F}d}{dt}\left(\vec{p}^{CO}\right) \tag{4.120}$$

$$= \frac{{}^{F}d}{dt}\left(\vec{p}^{CI}\right) - \left(\frac{{}^{P}d}{dt}\left(\vec{p}^{CO}\right) + {}^{F}\vec{\omega}^{P} \times \vec{p}^{CO}\right) \tag{4.121}$$

Note that $\frac{{}^{F}d}{dt}\left(\vec{p}^{CO}\right)$ in Equation 4.120 was converted to a differentiation in the P reference frame, precisely because $\vec{p}^{CO}$ is fixed in P. Hence,

$$\frac{{}^{F}d}{dt}\left(\vec{p}^{CI}\right) = 0 \tag{4.122}$$

$$\frac{{}^{P}d}{dt}\left(\vec{p}^{CO}\right) = 0, \tag{4.123}$$

and the differentiation of $\vec{p}^{OI}$ in the F reference frame is reduced to a simple cross product,

$$\vec{v}^{OI} = {}^{F}\vec{\omega}^{P} \times \vec{p}^{CO}. \tag{4.124}$$

4.3.3 VELOCITY AND ACCELERATION OF TWO POINTS ON A RIGID BODY

Two extremely useful formulas follow from Equation 4.103 (differentiation of a vector in two reference frames). These formulas relate the velocities and accelerations of two points fixed on a single rigid body,

$${}^{N}\vec{v}^{P} = {}^{N}\vec{v}^{Q} + {}^{N}\vec{\omega}^{B} \times \vec{p}^{QP} \tag{4.125}$$

$${}^{N}\vec{a}^{P} = {}^{N}\vec{a}^{Q} + {}^{N}\vec{\omega}^{B} \times \left({}^{N}\vec{\omega}^{B} \times \vec{p}^{QP}\right) + {}^{N}\vec{\alpha}^{B} \times \vec{p}^{QP}. \tag{4.126}$$

In these equations, P and Q are points fixed on rigid body B, which moves with angular velocity ${}^{N}\vec{\omega}^{B}$ and angular acceleration ${}^{N}\vec{\alpha}^{B}$ with respect to reference

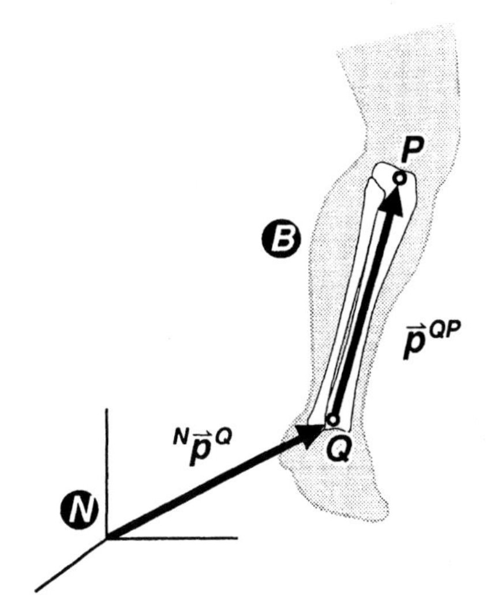

Figure 4.9. Two points fixed on a rigid body B. When the angular velocity $^{N}\vec{\omega}^{B}$ and position vector $\vec{p}^{QP}$ are known, the velocities and accelerations of points P and Q are related.

frame N (Figure 4.9). $\vec{p}^{QP}$ is the position vector pointing *from* point Q *to* point P. $\vec{p}^{QP}$ may be expressed in any basis. Usually, $\vec{p}^{QP}$ is most easily defined and expressed in the body basis, *i.e.,* the basis vectors of the rigid body reference frame (B). Hence, it is most convenient to likewise express the angular velocity and acceleration vectors $^{N}\vec{\omega}^{B}$ and $^{N}\vec{\alpha}^{B}$, respectively, in the basis vectors of the body B prior to performing the vector cross products of Equations 4.125 and 4.126.

Though the formulas look complex at first, the ideas expressed are relatively simple. If the velocity of point Q is known, and the angular velocity of the rigid body B is known, then Equation 4.125 simply states that the velocity of point P can be found by adding a vector cross product to the known velocity $^{N}\vec{v}^{Q}$. Similarly, if the acceleration of point Q is known, and the angular acceleration of B is known, then Equation 4.126 states that the acceleration of point P can be found by adding two vector cross products to the known acceleration $^{N}\vec{a}^{Q}$. Importantly, this means there is an easier, algebraic alternative to finding the velocities and accelerations of moving points without having to perform differential calculus!

These equations are profoundly powerful tools for analyzing biomechanical models or any kinematic chain having well defined "joints" between adjacent links. Beginning with a single point where the velocity is known, the velocities of every point of interest throughout the system can be quickly determined as long as the relative positions and angular velocities are known. Likewise, beginning with a single point at which the acceleration is known, the accelerations of every point of interest on every link of the system can be formulated as long as the the relative positions, angular velocities, and angular accelerations are known. The proof is from Kane and Levinson (1985).

Proof. The proof is direct, and begins by dissociating the derivative of the position vector ${}^N\vec{p}^P$,

$$
\begin{aligned}
{}^N\vec{v}^P &= \frac{{}^Nd\left({}^N\vec{p}^P\right)}{dt} && (4.127)\\
&= \frac{{}^Nd\left({}^N\vec{p}^Q + \vec{p}^{QP}\right)}{dt} && (4.128)\\
&= \frac{{}^Nd\left({}^N\vec{p}^Q\right)}{dt} + \frac{{}^Nd\left(\vec{p}^{QP}\right)}{dt}. && (4.129)
\end{aligned}
$$

Equation 4.103 converts the time differentiation of $\vec{p}^{QP}$ into a trivial differentiation and a cross product,

$$
{}^N\vec{v}^P = {}^N\vec{v}^Q + \left(\frac{{}^Bd\left(\vec{p}^{QP}\right)}{dt} + {}^N\vec{\omega}^B \times \vec{p}^{QP}\right). \qquad (4.130)
$$

Note that $\vec{p}^{QP}$ is fixed in body B, and thus its time derivative in B is zero. This leads to,

$$
\begin{aligned}
{}^N\vec{v}^P &= {}^N\vec{v}^Q + \left(0 + {}^N\vec{\omega}^B \times \vec{p}^{QP}\right) && (4.131)\\
&= {}^N\vec{v}^Q + {}^N\vec{\omega}^B \times \vec{p}^{QP}. && (4.132)
\end{aligned}
$$

A similar procedure is followed to prove Equation 4.126.

$$
\begin{aligned}
{}^N\vec{a}^P &= \frac{{}^Nd\left({}^N\vec{v}^P\right)}{dt} && (4.133)\\
&= \frac{{}^Nd}{dt}\left({}^N\vec{v}^Q + {}^N\vec{\omega}^B \times \vec{p}^{QP}\right) && (4.134)\\
&= \frac{{}^Nd\left({}^N\vec{v}^Q\right)}{dt} + \frac{{}^Nd\left({}^N\vec{\omega}^B\right)}{dt} \times \vec{p}^{QP}\\
&\quad + {}^N\vec{\omega}^B \times \frac{{}^Nd\left(\vec{p}^{QP}\right)}{dt} && (4.135)\\
&= {}^N\vec{a}^Q + {}^N\vec{\alpha}^B \times \vec{p}^{QP}\\
&\quad + {}^N\vec{\omega}^B \times \left(\frac{{}^Bd\left(\vec{p}^{QP}\right)}{dt} + {}^N\vec{\omega}^B \times \vec{p}^{QP}\right) && (4.136)\\
&= {}^N\vec{a}^Q + {}^N\vec{\alpha}^B \times \vec{p}^{QP}\\
&\quad + {}^N\vec{\omega}^B \times \left(0 + {}^N\vec{\omega}^B \times \vec{p}^{QP}\right) && (4.137)
\end{aligned}
$$

$$= {}^{N}\vec{a}^{Q} + {}^{N}\vec{\alpha}^{B} \times \vec{p}^{QP} + {}^{N}\vec{\omega}^{B} \times \left({}^{N}\vec{\omega}^{B} \times \vec{p}^{QP}\right). \quad (4.138)$$

4.3.3.1 EXAMPLE – A η-LINK PLANAR KINEMATIC CHAIN

A planar chain of rigid links is pinned together such that all of the rotation axes are parallel to the $\hat{b}_3$ direction, as shown in Figure 4.10. Each rigid link $B^{(i)}$ $(i = 1, 2, 3, ..., \eta)$ has a set of mutually perpendicular basis vectors $\hat{b}_1^{(i)}$, $\hat{b}_2^{(i)}$, $\hat{b}_3^{(i)}$, arranged so that the $\hat{b}_1^{(i)}$ vector points from the joint closest to the ground (point P_{i-1}) to the joint further away from the ground (point P_i). Angles q_i are measured between the adjacent $\hat{b}_1^{(i)}$ and $\hat{b}_1^{(i-1)}$ axes, and the distances between adjacent joints are given as ℓ_i. To find the velocities and accelerations of each joint, the following procedure is followed using the formulas for the velocity and acceleration of two points on a rigid body.

Position vectors:

$$\vec{p}_1 = \ell_1 \hat{b}_1^{'} \quad (4.139)$$
$$\vec{p}_2 = \ell_2 \hat{b}_1^{''} \quad (4.140)$$
$$\vec{p}_3 = \ell_3 \hat{b}_1^{'''} \quad (4.141)$$
$$\vdots = \vdots$$
$$\vec{p}_\eta = \ell_\eta \hat{b}_1^{(\eta)} \quad (4.142)$$

Angular velocities:

$$^{B}\vec{\omega}^{B'} = \dot{q}_1 \hat{b}_3^{'} \quad (4.143)$$
$$^{B}\vec{\omega}^{B''} = {}^{B}\vec{\omega}^{B'} + {}^{B'}\vec{\omega}^{B''} \quad (4.144)$$
$$= \dot{q}_1 \hat{b}_3^{'} + \dot{q}_2 \hat{b}_3^{''} \quad (4.145)$$
$$= (\dot{q}_1 + \dot{q}_2)\, \hat{b}_3^{''} \quad (4.146)$$
$$^{B}\vec{\omega}^{B'''} = {}^{B}\vec{\omega}^{B''} + {}^{B''}\vec{\omega}^{B'''} \quad (4.147)$$
$$= (\dot{q}_1 + \dot{q}_2 + \dot{q}_3)\, \hat{b}_3^{'''} \quad (4.148)$$
$$\vdots = \vdots$$
$$^{B}\vec{\omega}^{B^{(\eta)}} = {}^{B}\vec{\omega}^{B^{(\eta-1)}} + {}^{B^{(\eta-1)}}\vec{\omega}^{B^{(\eta)}} \quad (4.149)$$
$$= \left(\sum_{i=1}^{\eta} \dot{q}_i\right) \hat{b}_3^{(\eta)} \quad (4.150)$$

Velocities:

$$^{B}\vec{v}^{P_0} = 0 \quad (4.151)$$

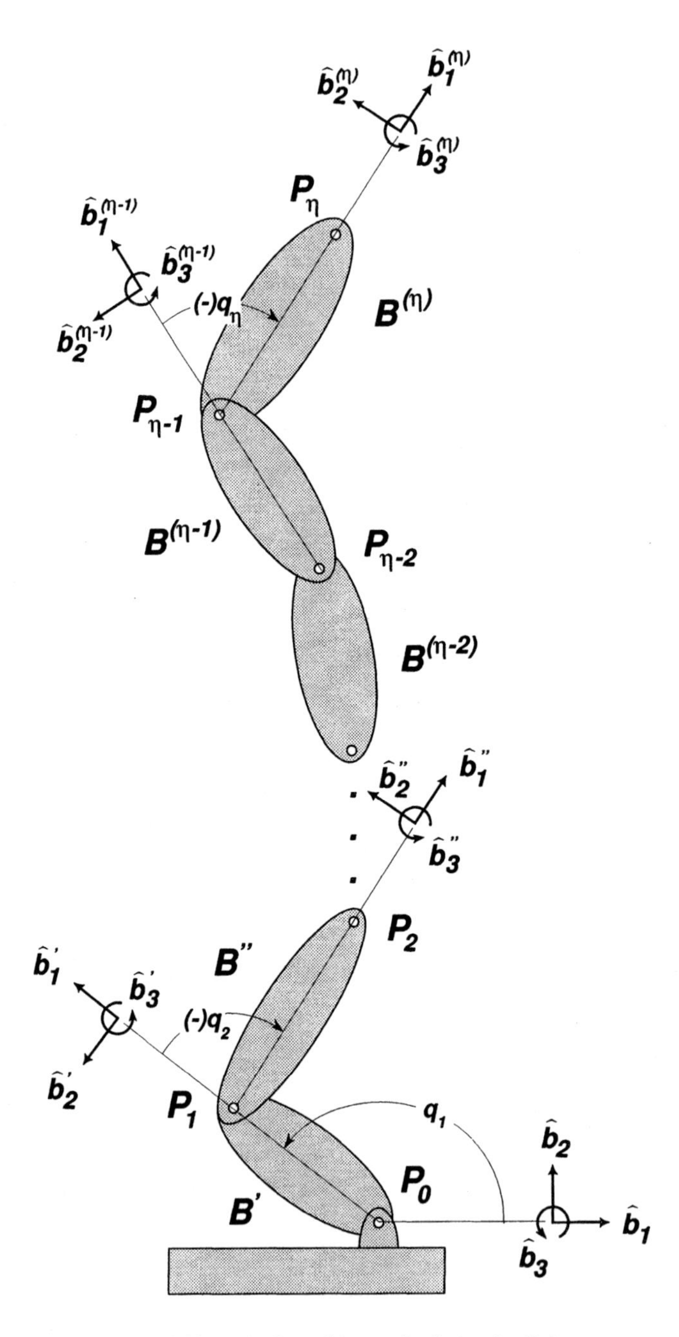

Figure 4.10. A planar kinematic chain of η links.

$$ {}^{B}\vec{v}^{P_1} = {}^{B}\vec{v}^{P_0} + {}^{B}\vec{\omega}^{B'} \times \vec{p}^{P_0 P_1} \tag{4.152}$$
$$ {}^{B}\vec{v}^{P_2} = {}^{B}\vec{v}^{P_1} + {}^{B}\vec{\omega}^{B''} \times \vec{p}^{P_1 P_2} \tag{4.153}$$
$$ \vdots = \vdots \tag{4.154}$$
$$ {}^{B}\vec{v}^{P_\eta} = {}^{B}\vec{v}^{P_{\eta-1}} + {}^{B}\vec{\omega}^{B^{(\eta)}} \times \vec{p}^{P_{\eta-1} P_\eta} \tag{4.155}$$

Accelerations:

$$ {}^{B}\vec{a}^{P_0} = 0 \tag{4.156}$$
$$ {}^{B}\vec{a}^{P_1} = {}^{B}\vec{a}^{P_0} + {}^{B}\vec{\omega}^{B'} \times \left({}^{B}\vec{\omega}^{B'} \times \vec{p}^{P_0 P_1}\right) + {}^{B}\vec{\alpha}^{B'} \times \vec{p}^{P_0 P_1} \tag{4.157}$$
$$ {}^{B}\vec{a}^{P_2} = {}^{B}\vec{a}^{P_1} + {}^{B}\vec{\omega}^{B''} \times \left({}^{B}\vec{\omega}^{B''} \times \vec{p}^{P_1 P_2}\right) + {}^{B}\vec{\alpha}^{B''} \times \vec{p}^{P_1 P_2} \tag{4.158}$$
$$ \vdots = \vdots \tag{4.159}$$
$$ {}^{B}\vec{a}^{P_\eta} = {}^{B}\vec{a}^{P_{\eta-1}} + {}^{B}\vec{\omega}^{B^{(\eta)}} \times \left({}^{B}\vec{\omega}^{B^{(\eta)}} \times \vec{p}^{P_{\eta-1} P_\eta}\right) + {}^{B}\vec{\alpha}^{B^{(\eta)}} \times \vec{p}^{P_{\eta-1} P_\eta} \tag{4.160}$$

4.3.3.2 EXAMPLE – VELOCITY AND ACCELERATION OF THE FOOT DURING A KICK

A chain of rigid links representing the thigh (A), shank (B), and foot (C) is actively swinging toward a ball in a kicking motion. The velocity of the hip joint H is known from experimental measurements and is assumed to be fairly constant from kick to kick. If the angular velocities of bodies A, B, and C are varied, what will the velocity of the contact point F^* of the foot be with respect to the inertial reference frame N?

The velocities are computed by propagating distally from the point of known velocity, which is the hip joint in this example.

$$ {}^{N}\vec{v}^{K} = {}^{N}\vec{v}^{H} + {}^{N}\vec{\omega}^{A} \times \vec{p}^{HK} \tag{4.161}$$
$$ {}^{N}\vec{v}^{A} = {}^{N}\vec{v}^{K} + {}^{N}\vec{\omega}^{B} \times \vec{p}^{KA} \tag{4.162}$$
$$ {}^{N}\vec{v}^{F^*} = {}^{N}\vec{v}^{A} + {}^{N}\vec{\omega}^{C} \times \vec{p}^{AF^*} \tag{4.163}$$

The accelerations may also be found in the same manner.

$$ {}^{N}\vec{a}^{K} = {}^{N}\vec{a}^{H} + {}^{N}\vec{\omega}^{A} \times \left({}^{N}\vec{\omega}^{A} \times \vec{p}^{HK}\right) + {}^{N}\vec{\alpha}^{A} \times \vec{p}^{HK} \tag{4.164}$$

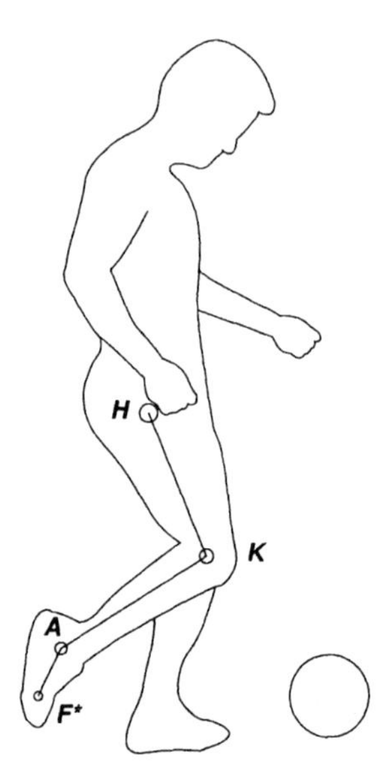

Figure 4.11. A kicking motion.

$$ {}^{N}\vec{a}^{A} = {}^{N}\vec{a}^{K} + {}^{N}\vec{\omega}^{B} \times \left({}^{N}\vec{\omega}^{B} \times \vec{p}^{KA} \right) + {}^{N}\vec{\alpha}^{B} \times \vec{p}^{KA} \quad (4.165) $$

$$ {}^{N}\vec{a}^{F^*} = {}^{N}\vec{a}^{A} + {}^{N}\vec{\omega}^{C} \times \left({}^{N}\vec{\omega}^{C} \times \vec{p}^{AF^*} \right) + {}^{N}\vec{\alpha}^{C} \times \vec{p}^{AF^*} \quad (4.166) $$

Some work in computing the double cross products is saved by recognizing that the terms enclosed in parentheses have already been computed as the second terms in Equations 4.161 through 4.163. In the event that the velocity of the hip joint was not known, one can usually define a point of zero velocity somewhere on the body. For instance, the point of foot to floor contact can be defined as a point with zero velocity. The velocities and accelerations at all other points of interest can then be propagated from that point.

4.3.4 VELOCITY AND ACCELERATION OF ONE POINT MOVING ON A RIGID BODY

Section 4.3.3 explained the relationship between two points *fixed* on the same rigid body. However, situations sometimes arise where the velocity and acceleration of a point moving on a rigid body is desired. In biomechanics, finding the velocity of a marker attached to the skin would be one example. The marker "jiggles" because it has mass and the "rigid" bone segment is covered by nonrigid flesh. In such a case, if the velocity of the point P relative to the rigid body B is known, then the velocity of point P in reference frame N is,

$$ {}^{N}\vec{v}^{P} = {}^{N}\vec{v}^{\bar{B}} + {}^{B}\vec{v}^{P} . \quad (4.167) $$

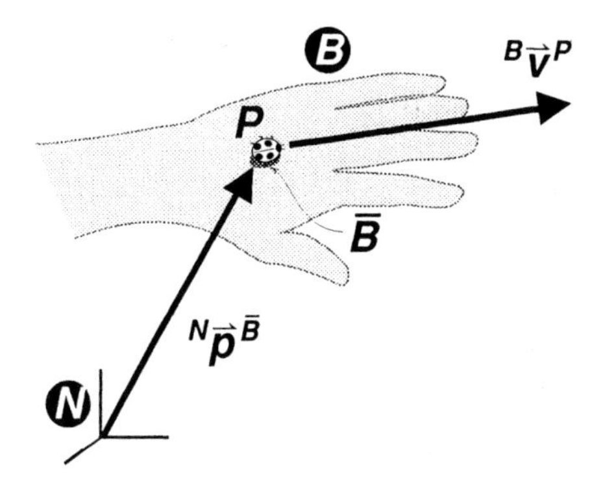

Figure 4.12. Figure and definitions to illustrate one point moving on a rigid body (Equation 4.167). The moving point, P, is depicted as a ladybug crawling upon the hand (body B). Point $\bar{B}$ is the point of the hand which is coincident with P. $\bar{B}$ is shown as a shadow underneath the bug, but in actuality is considered to be stationary with respect to B and coincident with the bug. P has velocity $^{B}\vec{v}^{P}$.

$\bar{B}$ is the point of body B that the point P happens to coincide with at the instant under consideration, and $^{N}\vec{v}^{\bar{B}}$ is the velocity of $\bar{B}$ in N (see Figure 4.12). Similarly, the acceleration of P in N is,

$$^{N}\vec{a}^{P} \;=\; {}^{N}\vec{a}^{\bar{B}} + {}^{B}\vec{a}^{P} + 2\,{}^{N}\vec{\omega}^{B} \times {}^{B}\vec{v}^{P}\,. \tag{4.168}$$

The last term ($2\,{}^{N}\vec{\omega}^{B} \times {}^{B}\vec{v}^{P}$) represents the *coriolis acceleration* due to the angular velocity of B in N and the velocity of point P in B. The *coriolis force* exerted on a mass m at P is equal to $-2m({}^{N}\vec{\omega}^{B} \times {}^{B}\vec{v}^{P})$. There is neither a coriolis acceleration nor a coriolis force if the angular velocity of the body B in N or the linear velocity of the point P on body B are zero. The proof of Equation 4.168 is from Kane and Levinson (1985).

Proof. Let B be a rigid body having angular velocity $^{N}\vec{\omega}^{B}$ in reference frame N. Two points belonging to body B are defined in Figure 4.12. $\bar{B}$ is the point instantaneously coincident with the moving point P, and $\tilde{B}$ which is another point of B. Position vectors from origin O fixed in reference frame N are also defined in the figure. Since $\bar{B}$ and P happen to be coincident at the instant of time under consideration, the velocity of P in N is,

$$^{N}\vec{v}^{P} \;=\; \frac{^{N}d}{dt}\left({}^{N}\vec{p}^{\tilde{B}} + \vec{p}^{\tilde{B}\bar{B}}\right) \tag{4.169}$$

$$\;=\; {}^{N}\vec{v}^{\tilde{B}} + \frac{^{B}d}{dt}\left(\vec{p}^{\tilde{B}\bar{B}}\right) + {}^{N}\vec{\omega}^{B} \times \vec{p}^{\tilde{B}\bar{B}}\,. \tag{4.170}$$

The time rate of change of the position vector from $\tilde{B}$ to $\bar{B}$ is nonzero, because $\bar{B}$ is continually being redefined as point P moves on body B,

$$\frac{^{B}d}{dt}\left(\vec{p}^{\tilde{B}\bar{B}}\right) \;=\; {}^{B}\vec{v}^{P}\,, \tag{4.171}$$

and thus,

$$^{N}\vec{v}^{P} \;=\; {}^{N}\vec{v}^{\tilde{B}} + {}^{B}\vec{v}^{P} + {}^{N}\vec{\omega}^{B} \times \vec{p}^{\tilde{B}\bar{B}}\,. \tag{4.172}$$

To prove Equation 4.167, one needs only note that $\tilde{B}$ can be chosen to be coincident with $\bar{B}$, since $\tilde{B}$ is an arbitrary point of B. This makes $\vec{p}^{\tilde{B}\bar{B}} = 0$, and,

$$ {}^{N}\vec{v}^{P} = {}^{N}\vec{v}^{\tilde{B}} + {}^{B}\vec{v}^{P} . \tag{4.173} $$

The proof of Equation 4.168 begins by differentiating Equation 4.172,

$$ \begin{aligned} {}^{N}\vec{a}^{P} &= \frac{{}^{N}d}{dt}\left({}^{N}\vec{v}^{\tilde{B}}\right) + \left[\frac{{}^{N}d}{dt}\left({}^{B}\vec{v}^{P}\right)\right] \\ &+ \frac{{}^{N}d}{dt}\left({}^{N}\vec{\omega}^{B}\right) \times \vec{p}^{\tilde{B}\bar{B}} + {}^{N}\vec{\omega}^{B} \times \left[\frac{{}^{N}d}{dt}\left(\vec{p}^{\tilde{B}\bar{B}}\right)\right] \end{aligned} \tag{4.174} $$

Equation 4.103 (formula relating the time derivatives in two reference frames), is used to express the bracketed quantities in terms of time derivatives in the B reference frame,

$$ \begin{aligned} {}^{N}\vec{a}^{P} &= {}^{N}\vec{a}^{\tilde{B}} + \left[\frac{{}^{B}d}{dt}\left({}^{B}\vec{v}^{P}\right) + {}^{N}\vec{\omega}^{B} \times {}^{B}\vec{v}^{P}\right] + {}^{N}\vec{\alpha}^{B} \times \vec{p}^{\tilde{B}\bar{B}} \\ &+ {}^{N}\vec{\omega}^{B} \times \left[\frac{{}^{B}d}{dt}\left(\vec{p}^{\tilde{B}\bar{B}}\right) + {}^{N}\vec{\omega}^{B} \times \vec{p}^{\tilde{B}\bar{B}}\right] . \end{aligned} \tag{4.175} $$

As before, $\tilde{B}$ can be chosen to be coincident with $\bar{B}$, eliminating $\vec{p}^{\tilde{B}\bar{B}}$ and allowing $\tilde{B}$ to be replaced by $\bar{B}$,

$$ \begin{aligned} {}^{N}\vec{a}^{P} &= {}^{N}\vec{a}^{\bar{B}} + \left[{}^{B}\vec{a}^{P} + {}^{N}\vec{\omega}^{B} \times {}^{B}\vec{v}^{P}\right] + 0 \\ &+ {}^{N}\vec{\omega}^{B} \times \left[\frac{{}^{B}d}{dt}\left(\vec{p}^{\tilde{B}\bar{B}}\right) + 0\right] . \end{aligned} \tag{4.176} $$

Finally,

$$ {}^{B}\vec{v}^{P} = \frac{{}^{B}d}{dt}\left(\vec{p}^{\tilde{B}\bar{B}}\right) , \tag{4.177} $$

yields

$$ {}^{N}\vec{a}^{P} = {}^{N}\vec{a}^{\bar{B}} + \left[{}^{B}\vec{a}^{P} + {}^{N}\vec{\omega}^{B} \times {}^{B}\vec{v}^{P}\right] + {}^{N}\vec{\omega}^{B} \times \left[{}^{B}\vec{v}^{P}\right] . \tag{4.178} $$

4.3.4.1 EXAMPLE – COMPUTING VELOCITIES WHEN THERE IS A ROLLING POINT OF CONTACT UNDER THE FOOT

When the kinematics of a linkage model are computed, oftentimes a point of ground contact is defined as the starting point of the linkage because it can be assigned a zero velocity. However, if the model includes a rolling point of contact underneath the foot of the stance leg, Equations 4.167 and 4.168

should be used to compute the velocity and acceleration of the point where the ground reaction force is applied. A more precise definition of rolling follows in Chapter 5, but for now it is sufficient to state that the parts *of the foot* in contact with the ground do not slip, and therefore have zero velocity relative to the ground.

Letting $\bar{B}$ be a point on the foot having zero velocity, the velocity and acceleration ${}^{N}\vec{v}^{\bar{B}}$ and ${}^{N}\vec{a}^{\bar{B}}$ are both zero, but the velocity and acceleration of the (moving) contact point P are nonzero,

$$ {}^{N}\vec{v}^{P} = {}^{B}\vec{v}^{P} \quad (4.179) $$

$$ {}^{N}\vec{a}^{P} = {}^{B}\vec{a}^{P} + 2\,{}^{N}\vec{\omega}^{B} \times {}^{B}\vec{v}^{P}\,. \quad (4.180) $$

Although the movement of the ground reaction force is usually neglected in biomechanical modeling studies, the fact that the righthand sides of the above equations are nonzero means that there will be nonzero terms in the dynamic equations of motion arising from this movement (explained in Chapter 6). Therefore, the underfoot movements of the ground reaction force vector should not be neglected.

4.4 EXAMPLE –KINEMATICS OF A 7-DOF ARM MODEL

The derivation of the kinematic equations for an upper extremity model featuring realistic pronation and supination is outlined. The model is shown in Figure 4.13. Because the proximal radius slides over the distal humerus as the elbow is flexed, the velocities and accelerations of points distal to the radiohumeral joint make use of the formulas for the velocity and acceleration of one point moving on a rigid body. The reader should specifically look for this, as he or she might otherwise miss this important point amid the other mathematical equations presented in this comprehensive example. Almost every vector computation described in this chapter is utilized. Where convenient, the mechanical operations of performing vector additions, vector cross products, vector conversions from one reference frame to another, *etc.*, are neglected so that the methodology remains relatively unobscured.

The humerus (rigid body A), ulna (B), radius (C), and hand (D) move with respect to the trunk reference frame N. In anatomical position, the arm is at the side of the trunk with the palm facing anteriorly, the '1' axes of each rigid body reference frame point anteriorly, the '2' axes point laterally to the right, and the '3' axes point distally. Angles q_i, ($i = 1, 2, \ldots, 7$) are all measured positively consistent with the right-hand rule (CCW positive when viewed with the rotation axis pointing out of the paper), and all reference frames are considered to be coincident when the rotation angles are zero.

Rotations q_1 and q_2 are shoulder adductions about the common $\hat{n}_1$, $\hat{a}_1^{'}$ axis, and shoulder flexion about the common $\hat{a}_2^{'}$, $\hat{a}_2^{''}$ axis, where frames $A^{'}$ and $A^{''}$

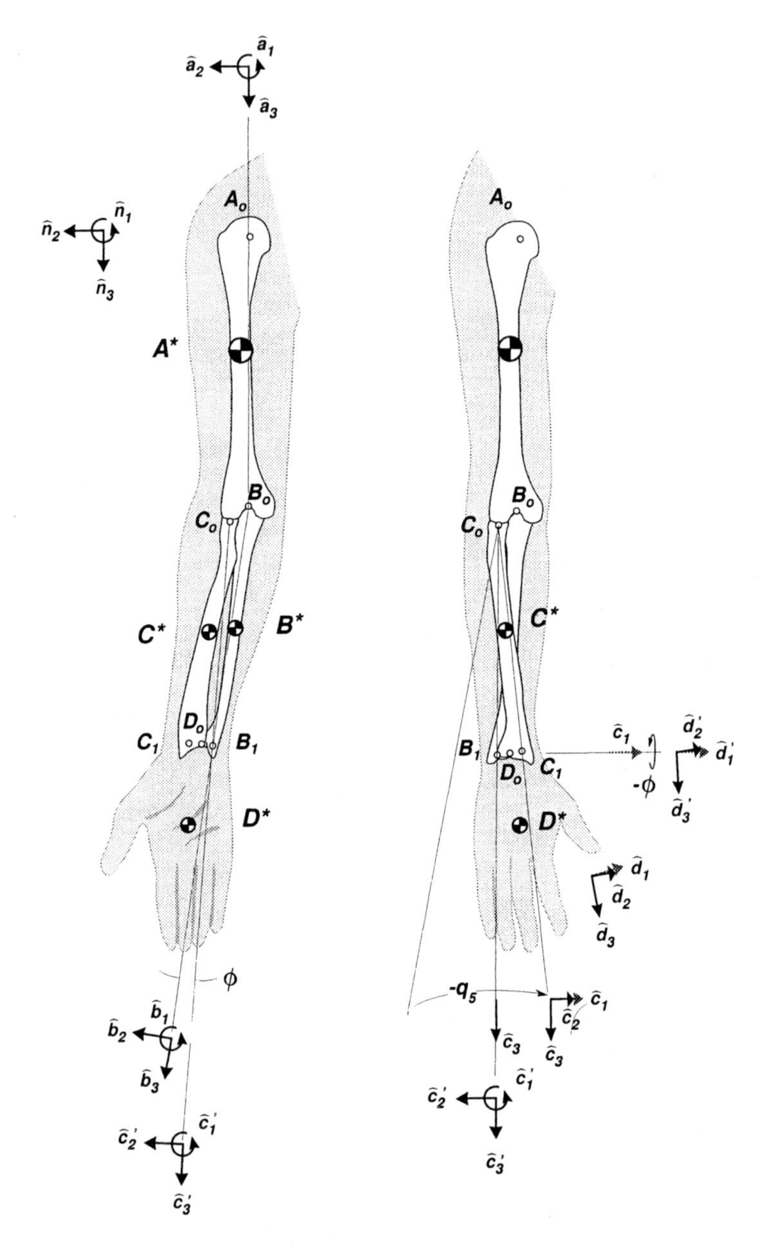

Figure 4.13. A seven degree of freedom model of the upper extremity. This model features the radius and ulna as separate and distinct rigid bodies so that pronation and supination can be realistically modeled. In order to highlight supination angle q_5, the diagrams depict shoulder angles $q_1 = q_2 = q_3 = 0$, elbow angle $q_4 = 0$, and wrist adduction angle $q_6 = 0$.

are two massless "intermediate" reference frames. The third rotation about the shoulder q_3 is external rotation, and occurs about the longitudinal axis of the humerus $\hat{a}_3^{''}$ and $\hat{a}_3$. Direction cosines from N to A are given by

${}^{N}R^{A}$	$\hat{a}_1$	$\hat{a}_2$	$\hat{a}_3$
$\hat{n}_1$	c_2c_3	$-c_2s_3$	s_2
$\hat{n}_2$	$s_1s_2c_3 + c_1s_3$	$-s_1s_2s_3 + c_1c_3$	$-s_1c_2$
$\hat{n}_3$	$-c_1s_2c_3 + s_1s_3$	$c_1s_2s_3 + s_1c_3$	c_1c_2

(4.181)

Elbow flexion is represented as a simple rotation by angle q_4 about the common $\hat{a}_2$, $\hat{b}_2$ axis,

${}^{A}R^{B}$	$\hat{b}_1$	$\hat{b}_2$	$\hat{b}_3$
$\hat{a}_1$	c_4	0	s_4
$\hat{a}_2$	0	1	0
$\hat{a}_3$	$-s_4$	0	c_4

(4.182)

To accurately model the rotation of the radius about the ulna, the *fulcrum of pronation and supination* is determined as the line between the center of the concave proximal end of the radius and the distal end of the ulna (Kapandji, 1982). A positive, *fixed* rotation by angle ϕ about the common $\hat{b}_1$, $\hat{c}_1^{'}$ axis brings unit vector $\hat{c}_3^{'}$ of intermediate reference frame $C^{'}$ parallel to this line.[6] Note that angle ϕ is a property of the geometry of the system, and is not a degree of freedom, as it is a fixed rotation angle. Supination occurs when reference frame C rotates by positive angle q_5 about the common $\hat{c}_3^{'}$, $\hat{c}_3$ axis, and pronation occurs when q_5 is negative. *Sin*(ϕ) and *cos*(ϕ) are represented as s_ϕ and c_ϕ, respectively, in the following tables,

${}^{B}R^{C'}$	$\hat{c}_1^{'}$	$\hat{c}_2^{'}$	$\hat{c}_3^{'}$
$\hat{b}_1$	1	0	0
$\hat{b}_2$	0	c_ϕ	$-s_\phi$
$\hat{b}_3$	0	s_ϕ	c_ϕ

(4.183)

${}^{B}R^{C}$	$\hat{c}_1$	$\hat{c}_2$	$\hat{c}_3$
$\hat{b}_1$	c_5	$-s_5$	0
$\hat{b}_2$	$c_\phi s_5$	$c_\phi c_5$	$-s_\phi$
$\hat{b}_3$	$s_\phi s_5$	$s_\phi c_5$	c_ϕ

(4.184)

In order to realign the wrist flexion axis ($\hat{d}_2$), an inverse rotation by angle $-\phi$ about the common $\hat{c}_1$, $\hat{d}_1^{'}$ axis is required. Because wrist adduction occurs

[6]In practice, and depending on the geometry of the arm segment being modeled, sometimes two fixed rotations are needed to bring $\hat{c}_3^{'}$ parallel, but one rotation is usually sufficiently accurate.

about the same axis, the angles q_6 and $-\phi$ can be combined into one simple rotation. Writing $sin(q_6 - \phi)$ and $cos(q_6 - \phi)$ as $s_{6-\phi}$ and $c_{6-\phi}$, the next direction cosine matrix becomes,

${}^{C}R^{D'}$	$\hat{d}_1'$	$\hat{d}_2'$	$\hat{d}_3'$
$\hat{c}_1$	1	0	0
$\hat{c}_2$	0	$c_{6-\phi}$	$-s_{6-\phi}$
$\hat{c}_3$	0	$s_{6-\phi}$	$c_{6-\phi}$

(4.185)

Finally, wrist flexion angle q_7 is defined as a positive rotation about the common $\hat{d}_2', \hat{d}_2$ axis, yielding the final rotation matrix between rigid bodies C and D,

${}^{C}R^{D}$	$\hat{d}_1$	$\hat{d}_2$	$\hat{d}_3$
$\hat{c}_1$	c_7	0	s_7
$\hat{c}_2$	$s_{6-\phi}s_7$	$c_{6-\phi}$	$-s_{6-\phi}c_7$
$\hat{c}_3$	$-c_{6-\phi}s_7$	$s_{6-\phi}$	$c_{6-\phi}c_7$

(4.186)

Next, the angular velocities are found and expressed in the basis vectors of the moving body. The first two are completely defined.

$$\begin{aligned}
{}^{N}\vec{\omega}^{A'} &= \dot{q}_1\hat{n}_1 && (4.187)\\
&= \dot{q}_1\left(c_2c_3\hat{a}_1 - c_2s_3\hat{a}_2 + s_2\hat{a}_3\right) && (4.188)\\
{}^{A'}\vec{\omega}^{A''} &= \dot{q}_2\hat{a}_2'' && (4.189)\\
&= \dot{q}_2\left(s_3\hat{a}_1 + c_3\hat{a}_2\right) && (4.190)\\
{}^{A''}\vec{\omega}^{A} &= \dot{q}_3\hat{a}_3 && (4.191)\\
{}^{N}\vec{\omega}^{A} &= {}^{N}\vec{\omega}^{A'} + {}^{A'}\vec{\omega}^{A''} + {}^{A''}\vec{\omega}^{A} && (4.192)\\
&= \left(\dot{q}_1c_2c_3 + \dot{q}_2s_3\right)\hat{a}_1 + \left(-\dot{q}_1c_2s_3 + \dot{q}_2c_3\right)\hat{a}_2 \\
&\quad + \left(\dot{q}_1s_2 + \dot{q}_3\right)\hat{a}_3 && (4.193)\\
&= \mathcal{A}_1\hat{a}_1 + \mathcal{A}_2\hat{a}_2 + \mathcal{A}_3\hat{a}_3 && (4.194)
\end{aligned}$$

In the last equation, $\mathcal{A}_i$ $(i = 1,\ 2,\ 3)$ are the $\hat{a}_i$ measure numbers when ${}^{N}\vec{\omega}^{A}$ is expressed in the basis vectors of A.

$$\begin{aligned}
{}^{N}\vec{\omega}^{B} &= {}^{N}\vec{\omega}^{A} + {}^{A}\vec{\omega}^{B} && (4.195)\\
&= \left(\dot{q}_1c_2c_3 + \dot{q}_2s_3\right)\left(c_4\hat{b}_1 + s_4\hat{b}_3\right) \\
&\quad + \left(-\dot{q}_1c_2s_3 + \dot{q}_2c_3\right)\hat{b}_2 \\
&\quad + \left(\dot{q}_1s_2 + \dot{q}_3\right)\left(-s_4\hat{b}_1 + c_4\hat{b}_3\right) + \dot{q}_4\hat{b}_2 && (4.196)
\end{aligned}$$

$$\begin{aligned}
&= [(\dot{q}_1 c_2 c_3 + \dot{q}_2 s_3)\, c_4 - (\dot{q}_1 s_2 + \dot{q}_3)\, s_4]\, \hat{b}_1 \\
&\quad [-\dot{q}_1 c_2 s_3 + \dot{q}_2 c_3 + \dot{q}_4]\, \hat{b}_2 \\
&\quad [(\dot{q}_1 c_2 c_3 + \dot{q}_2 s_3)\, s_4 + (\dot{q}_1 s_2 + \dot{q}_3)\, c_4]\, \hat{b}_3 &(4.197)\\
&= \mathcal{B}_1 \hat{b}_1 + \mathcal{B}_2 \hat{b}_2 + \mathcal{B}_3 \hat{b}_3 &(4.198)
\end{aligned}$$

where $\mathcal{B}_i$ $(i = 1, 2, 3)$ are the $\hat{b}_i$ measure numbers when ${}^N\vec{\omega}^B$ is expressed in the basis vectors of B.

Using the same methodology, the angular velocities of C and D are determined by successively adding simple angular velocities according to the addition theorem for angular velocities.

$$\begin{aligned}
{}^N\vec{\omega}^C &= {}^N\vec{\omega}^B + {}^B\vec{\omega}^C &(4.199)\\
&= \left(\mathcal{B}_1 \hat{b}_1 + \mathcal{B}_2 \hat{b}_2 + \mathcal{B}_3 \hat{b}_3\right) + \dot{q}_5 \hat{c}_3 &(4.200)\\
&= \mathcal{C}_1 \hat{c}_1 + \mathcal{C}_2 \hat{c}_2 + \mathcal{C}_3 \hat{c}_3 &(4.201)
\end{aligned}$$

$$\begin{aligned}
{}^N\vec{\omega}^D &= {}^N\vec{\omega}^C + {}^C\vec{\omega}^D &(4.202)\\
&= (\mathcal{C}_1 \hat{c}_1 + \mathcal{C}_2 \hat{c}_2 + \mathcal{C}_3 \hat{c}_3) + \dot{q}_6 \hat{c}_1 + \dot{q}_7 \hat{d}_2 &(4.203)\\
&= \mathcal{D}_1 \hat{d}_1 + \mathcal{D}_2 \hat{d}_2 + \mathcal{D}_3 \hat{d}_3 &(4.204)
\end{aligned}$$

$\mathcal{C}_i$ $(i = 1, 2, 3)$ are the $\hat{c}_i$ measure numbers when ${}^N\vec{\omega}^C$ is expressed in the basis vectors of C, and $\mathcal{D}_i$ $(i = 1, 2, 3)$ are the $\hat{d}_i$ measure numbers when ${}^N\vec{\omega}^D$ is expressed in the basis vectors of D. The reader should note that ${}^C\vec{\omega}^{D'}$ could have been written as $\dot{q}_6 \hat{c}_1{}'$ instead of $\dot{q}_6 \hat{c}_1$, but it is preferable to use basis vectors of the actual rigid body segments whenever possible.

Velocities of Points. Next, the position vectors must be defined to each important point of the kinematic chain. Important points include points at which a velocity is computed, or points that are helpful in defining the velocities of other points. These are identified beginning at the most proximal point. A_o is the center of shoulder rotation, and hence has either a measured, or zero velocity in N. The derivations will proceed using ${}^N\vec{v}^{A_o} = 0$, but any known velocity can be substituted in its place. B_o is the midpoint of the axis of the humeroulnar joint. C_o is the center of the concave, cup shaped, proximal end of the radius, and is in contact with point $\bar{A}$ of the distal, lateral head of the humerus (the *capitulum humeri*). Note that as the elbow flexes with angle q_4, the location of point C_o moves with respect to body A, while a succession of points $\bar{A}$ in contact with C_o are actually points that are fixed in body A. Points B_1 and C_1 are defined as the centers of the distal ends of the ulna and radius, and are considered to be in line with the curved, centroidal axes of these bones. Point D_o is midway between points B_1 and C_1, and is useful to define both the centroid of the wrist, and a point about which to compute muscle moments. Point P is any point on the hand at which the velocity is computed, and is given three arbitrary coordinates with respect to point D_o.

$$\vec{p}^{A_oB_o} = \ell_A\hat{a}_3 \tag{4.205}$$

$$\vec{p}^{B_o\bar{A}} = \delta\hat{a}_2 + r\hat{c}_3 \tag{4.206}$$

$$\vec{p}^{\bar{A}C_o} = 0 \tag{4.207}$$

$$\vec{p}^{C_oB_1} = \ell_{CB}\hat{c}_3 \tag{4.208}$$

$$\vec{p}^{C_oC_1} = \ell_{C_2}\hat{c}_2 + \ell_{C_3}\hat{c}_3 \tag{4.209}$$

$$\vec{p}^{B_1C_1} = \vec{p}^{C_oC_1} - \vec{p}^{C_oB_1} \tag{4.210}$$

$$\vec{p}^{C_oD_o} = \vec{p}^{C_oB_1} + \frac{1}{2}\vec{p}^{B_1C_1} \tag{4.211}$$

$$\vec{p}^{D_oP} = \ell_{D_1}\hat{d}_1 + \ell_{D_2}\hat{d}_2 + \ell_{D_3}\hat{d}_3 \tag{4.212}$$

The formula relating the velocities of two points fixed in a rigid body are used for the majority of the velocity calculations. The velocity of point C_o in N, however, uses the formula for the velocity of one point moving on a rigid body, since the proximal end of the radius glides upon the *capitulum* with speed $r\dot{q}_4$ and direction $\hat{c}_1^{'}$ $(\hat{b}_1)$. Relative to rigid body A, then,

$$\begin{aligned} {}^A\vec{v}^{C_o} &= r\dot{q}_4\hat{c}_1^{'} && (4.213)\\ &= r\dot{q}_4\hat{b}_1 && (4.214)\\ &= r\dot{q}_4\left(c_4\hat{a}_1 - s_4\hat{a}_3\right) . && (4.215)\end{aligned}$$

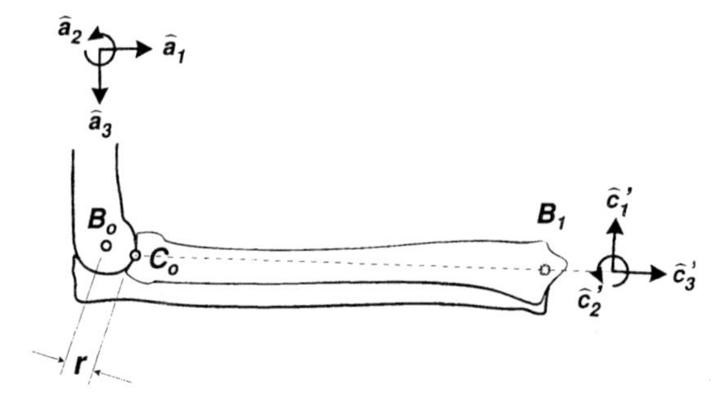

Figure 4.14. The radiohumeral joint and the radius, shown from the lateral side. The proximal end of the radius glides upon the *capitulum* as the elbow is flexed or extended. Thus, point C_o has a nonzero velocity in reference frame A. Its velocity in reference frame N is computed using the formula for one point moving on a rigid body.

Also, the previous equation is differentiated to obtain the acceleration of C_o in A,

$$ {}^{A}\vec{a}^{C_o} = r\ddot{q}_4\left(c_4\hat{a}_1 - s_4\hat{a}_3\right) + r\dot{q}_4\left(-s_4\dot{q}_4\hat{a}_1 - c_4\dot{q}_4\hat{a}_3\right) \quad (4.216) $$

$$ = r\ddot{q}_4\hat{b}_1 - r\dot{q}_4^2\hat{b}_3\,. \quad (4.217) $$

Therefore, the velocities of all points of the system named to this point are,

$$ {}^{N}\vec{v}^{A_o} = 0 \quad (4.218) $$

$$ {}^{N}\vec{v}^{B_o} = {}^{N}\vec{v}^{A_o} + {}^{N}\vec{\omega}^{A} \times \vec{p}^{A_oB_o} \quad (4.219) $$

$$ {}^{N}\vec{v}^{C_o} = {}^{N}\vec{v}^{\bar{A}} + {}^{A}\vec{v}^{C_o} \quad (4.220) $$

$$ = {}^{N}\vec{v}^{B_o} + {}^{N}\vec{\omega}^{A} \times \vec{p}^{B_o\bar{A}} + r\dot{q}_4\hat{c}_1' \quad (4.221) $$

$$ = {}^{N}\vec{v}^{B_o} + {}^{N}\vec{\omega}^{A} \times \left(\delta\hat{a}_2 + r\hat{c}_3\right) + r\dot{q}_4\hat{b}_1 \quad (4.222) $$

$$ {}^{N}\vec{v}^{D_o} = {}^{N}\vec{v}^{C_o} + {}^{N}\vec{\omega}^{C} \times \vec{p}^{C_oD_o} \quad (4.223) $$

$$ {}^{N}\vec{v}^{P} = {}^{N}\vec{v}^{D_o} + {}^{N}\vec{\omega}^{D} \times \vec{p}^{D_oP}\,. \quad (4.224) $$

The accelerations follow the same patterns as the velocities, using the analogous formulas. Typically, however, the accelerations are only desired for the mass centers, as inertial forces are evident in locations with mass. The following are the locations of mass centers A, B, and C, denoted A^*, B^*, and C^*, respectively. In each case, three unspecified coordinates locate the mass centers from the proximal joint of the rigid body.

$$ \vec{p}^{A_oA^*} = \rho_{A_1}\hat{a}_1 + \rho_{A_2}\hat{a}_2 + \rho_{A_3}\hat{a}_3 \quad (4.225) $$

$$ \vec{p}^{B_oB^*} = \rho_{B_1}\hat{b}_1 + \rho_{B_2}\hat{b}_2 + \rho_{B_3}\hat{b}_3 \quad (4.226) $$

$$ \vec{p}^{C_oC^*} = \rho_{C_1}\hat{c}_1 + \rho_{C_2}\hat{c}_2 + \rho_{C_3}\hat{c}_3 \quad (4.227) $$

The angular accelerations may be differentiated in either the N reference frame, or the body frame. Differentiations in the local body reference frames is almost always the most convenient, because the angular accelerations are usually crossed with position vectors defined in the reference frames affixed to the local bodies. Thus, the most convenient expressions for these are,

$$ {}^{N}\vec{\alpha}^{A} = \dot{\mathcal{A}}_1\hat{a}_1 + \dot{\mathcal{A}}_2\hat{a}_2 + \dot{\mathcal{A}}_3\hat{a}_3 \quad (4.228) $$

$$ {}^{N}\vec{\alpha}^{B} = \dot{\mathcal{B}}_1\hat{b}_1 + \dot{\mathcal{B}}_2\hat{b}_2 + \dot{\mathcal{B}}_3\hat{b}_3 \quad (4.229) $$

$$ {}^{N}\vec{\alpha}^{C} = \dot{\mathcal{C}}_1\hat{c}_1 + \dot{\mathcal{C}}_2\hat{c}_2 + \dot{\mathcal{C}}_3\hat{c}_3 \quad (4.230) $$

$$ {}^{N}\vec{\alpha}^{D} = \dot{\mathcal{D}}_1\hat{d}_1 + \dot{\mathcal{D}}_2\hat{d}_2 + \dot{\mathcal{D}}_3\hat{d}_3 \quad (4.231) $$

where the respective measure numbers are differentiated with respect to time using the rules for differentiating products of sines and cosines.

The accelerations are found via a logical progression from point A_o,

$$
\begin{aligned}
{}^N\vec{a}^{A_o} &= 0 && (4.232)\\
{}^N\vec{a}^{B_o} &= {}^N\vec{a}^{A_o} + {}^N\vec{\omega}^A \times \left({}^N\vec{\omega}^A \times \vec{p}^{A_oB_o}\right)\\
&\quad + {}^N\vec{\alpha}^A \times \vec{p}^{A_oB_o} && (4.233)\\
{}^N\vec{a}^{C_o} &= {}^N\vec{a}^{\bar{A}} + {}^A\vec{a}^{C_o} + 2\,{}^N\vec{\omega}^A \times {}^A\vec{v}^{C_o} && (4.234)\\
&= \left({}^N\vec{a}^{B_o} + {}^N\vec{\omega}^A \times \left({}^N\vec{\omega}^A \times \vec{p}^{B_o\bar{A}}\right)\right.\\
&\quad \left. + {}^N\vec{\alpha}^A \times \vec{p}^{B_o\bar{A}}\right)\\
&\quad + {}^A\vec{a}^{C_o} + 2\,{}^N\vec{\omega}^A \times {}^A\vec{v}^{C_o} && (4.235)
\end{aligned}
$$

where the equation for the acceleration of two points fixed on a rigid body was used to expand the acceleration of $\bar{A}$ in reference frame A in terms of known quantities. The remaining accelerations are computed using the accelerations of two points in a rigid body formula,

$$
\begin{aligned}
{}^N\vec{a}^{D_o} &= {}^N\vec{a}^{C_o} + {}^N\vec{\omega}^C \times \left({}^N\vec{\omega}^C \times \vec{p}^{C_oD_o}\right)\\
&\quad + {}^N\vec{\alpha}^C \times \vec{p}^{C_oD_o} && (4.236)
\end{aligned}
$$

$$
\begin{aligned}
{}^N\vec{a}^{P} &= {}^N\vec{a}^{D_o} + {}^N\vec{\omega}^D \times \left({}^N\vec{\omega}^D \times \vec{p}^{D_oP}\right)\\
&\quad + {}^N\vec{\alpha}^D \times \vec{p}^{D_oP}. && (4.237)
\end{aligned}
$$

4.5 EXERCISES

1. Find the potential energy of a four-link planar linkage using the definitions of Figure 4.10. Start by developing direction cosine matrices beginning with the ground (N) reference frame and working upward. Assume that the links have joints separated by lengths ℓ_i and masses m_i ($i = 1, 2, 3, 4$) centered between the joints.

2. A weight W is hung on an inextensible string of length ℓ to form a planar pendulum. If the vertical angle of the string is θ, find the velocity of W by taking the cross product of the pendulum's angular velocity and the appropriate position vector.

3. Prove that the moment of a couple is the same regardless of the point chosen to compute the moments about. It may be helpful to consider the following cases:

   (a) A couple composed of two coplanar forces, with any point P contained in the same plane.

(b) A couple composed of two coplanar forces, with a non-coplanar point P.

(c) A couple composed of any number of non-coplanar forces, with any point P.

4. Estimate the force F_b in the *biceps* muscle required to hold a cup full of liquid in a static posture similar to Figure 4.2. The muscle force should balance the gravitational moment of the forearm and hand about the elbow. Estimate muscle origin and insertion positions and limb dimensions by direct measurement on a volunteer.

5. Find the angular velocity of the pelvis following the three rotations depicted in Figures 3.12, 3.13, and 3.14.

6. This problem refers to Figure 4.7, and uses the $\hat{z}-\hat{x}'-\hat{z}''$ Euler rotations of the pelvis derived in Chapter 3 (Figure 3.15 and Equation 3.48).

   Let point O represent the "origin" of the *gluteus medius* muscle, and point I represent its insertion on the greater trochanter. More precisely, these points are located at the centroids of the attachment areas. Let the origin of pelvic reference frame P be at the centroid of the *acetabulum* (point C), and let the following position vectors be used to locate O and I from point C,

$$\vec{p}^{CO} = o_1\hat{p}_1 + o_2\hat{p}_2 + o_3\hat{p}_3 \tag{4.238}$$
$$\vec{p}^{CI} = i_1\hat{f}_1 + i_2\hat{f}_2 + i_3\hat{f}_3\,. \tag{4.239}$$

   (a) Find the position vector pointing from O to I, or $\vec{p}^{OI}$. Express your answer in either P or F basis vectors.

   (b) Show that the velocity of the muscle, $\vec{v}^{OI}$, can be found by finding the first time derivative of ${}^P\vec{p}^I$.

   (c) Find ${}^P\vec{v}^I$ by direct differentiation of ${}^P\vec{p}^I$ (in the P reference frame).

   (d) Find ${}^P\vec{v}^I$ by relating the time derivatives of ${}^P\vec{p}^I$ in two reference frames,

$$\frac{{}^Pd({}^P\vec{p}^I)}{dt} = \frac{{}^Fd({}^P\vec{p}^I)}{dt} + {}^P\vec{\omega}^F \times {}^P\vec{p}^I\,. \tag{4.240}$$

7. The following are simplified forms of the position vectors locating mass centers A^*, B^*, and C^* of the seven degree of freedom arm model depicted in Section 4.4,

$$\vec{p}^{A_oA^*} = \rho_{A_3}\hat{a}_3 \tag{4.241}$$
$$\vec{p}^{B_oB^*} = \rho_{B_3}\hat{b}_3 \tag{4.242}$$
$$\vec{p}^{C_oC^*} = \rho_{C_3}\hat{c}_3\,. \tag{4.243}$$

(a) Find the velocities of points A^*, B^*, and C^* relative to the N reference frame. Leave your answers in mixed bases for simplicity. What difference is there, if any, between treating point C_o as a point moving on a rigid body versus assigning C_o to be fixed at the centroid of the *capitulum*?

(b) Find the acceleration of A^* in N.

References

Kane, T. R., and Levinson, D. A. (1985) *Dynamics: Theory and Applications.* McGraw-Hill, New York, NY.

Kapandji, I. A. (1982) *The Physiology of the Joints, Volume One – Upper Limb,* Fifth Edition (English). Churchill Livingstone, New York, NY.

Kapandji, I. A. (1985) *The Physiology of the Joints, Volume Two – Lower Limb,* Fifth Edition (English). Churchill Livingstone, New York, NY.

Meriam, J. L., and Kraige, L.G. (1997) *Engineering Mechanics Volume 2 – Dynamics.* John Wiley & Sons, New York, NY.

Strasser, H. (1917) *Lehrbuch der Muskel-und Gelenkmechanik.* Springer, Berlin.

Chapter 5

MODELS OF THE SKELETAL SYSTEM

Objective – To define inertial properties of the body segments and ways in which they are joined together to form kinematic linkage models of the musculoskeletal system.

Linkages are arrangements of materials that move with respect to one another in well defined ways. One can think of the upper extremity, beginning at the shoulder joint, as a linkage because the movements of the humerus, ulna, radius, and hand are constrained to follow particular pathways. Elbow flexion and extension are defined by the humerus and ulna rotating about an axis that is nearly fixed with respect to one another. Pronation and supination occur with the radius (the bone on the thumb side of the forearm) rotating about the ulna (the bone on the side opposite the thumb). The latter motion is not as easy to define as elbow flexion. However, we know from observation and everyday experiences that the distal ends of the radius and ulna do not separate and travel about in random fashion. They follow well defined motion pathways because of their shapes, the architecture of their articulating surfaces, and ligaments which tie the bones together. The ligaments are arranged to maintain precise relationships between the positions, angulations, and motions of the bones comprising the system.

This chapter will explore ways to develop models of the segments, and then ways to join them together in ways that emulate the skeletal system.

5.1 EQUIVALENCE AND REPLACEMENT

In Section 3.3, the motions of a rigid body were presented as being kinematically equivalent to the movements of affixed reference frames. By systematically tracking the component angular rotations of the reference frames, a simple method was developed whereby very complex motions of rigid bodies and points upon or within them could be quantified. Likewise, models of limb segments can be greatly simplified if the following assumptions are made:

1. The collection of molecules, cells, tissues, *etc.*, making up a particular body segment is treated as a rigid body;

2. The inertia of each small volume of tissue making up a particular body segment is included by incorporating its mass and location within a macroscopic description of the inertial properties of the whole rigid body segment;

3. The collection of forces, moments of forces, and torques acting upon a particular body segment is replaced by a single force and a single torque.

Each of these assumptions are made with regard to one another. If the rigid body assumption is not made, then rotational and linear accelerations could distort the body and lead to changes in the distribution of mass, and thus change the inertial properties of the body segment. Furthermore, forces acting upon a particular portion of the body would cause deflections of the tissues, and thus cause additional changes in mass distribution. In the future, these intrasegmental changes will likely be accounted for. Models of the body segments will become more representative of each segment by including the properties of hard *and* soft tissues making up each segment.

For the present, making these assumptions greatly simplifies the task of modeling the motions of the limb segments. It allows us to replace the collections of tissues making up each body segment with dynamically equivalent rigid bodies of equivalent mass and well defined, unchanging inertial properties. The third assumption allows us to group the aggregate sum of forces acting upon a body segment – forces that include the tensions of the ligaments, joint capsules, skin, and every structure we choose not to explicity define in our model – as a single force and a single torque. This simplifies the computational task by orders of magnitude, and allows us to focus upon the actions of the more significant structures (such as muscles) without burdening us with the fine details better left for future analyses.

5.2 SEGMENTAL MASS, PRODUCT OF INERTIA, AND MOMENT OF INERTIA

Each small volume of tissue having density ρ_i and volume dV is summed over the entire volume V to obtain the total mass m of the segment,

$$m = \int_V \rho_i dV \,. \tag{5.1}$$

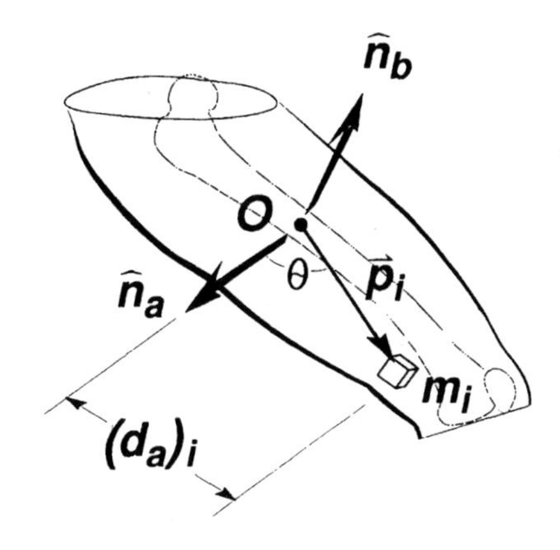

Figure 5.1. Figure for describing products of inertia and moments of inertia. An element of mass m_i is located from the origin O by position vector $\vec{p}_i$. Then, given a unit direction vector $\hat{n}_a$, the cross product $\vec{p}_i \times \hat{n}_a$ creates a vector perpendicular to $\vec{p}_i$ and $\hat{n}_a$ with scalar magnitude $(d_a)_i = |\vec{p}_i| \sin\theta$. The element of mass contributes to the moment of inertia as the product of m_i and ${(d_a)_i}^2$. For a product of inertia, a second unit direction vector $\hat{n}_b$ (usually perpendicular to $\hat{n}_a$) must be defined. $(d_b)_i$ is computed in the same way as $(d_a)_i$, except that $\hat{n}_b$ is used instead of $\hat{n}_a$.

If the position vector $\vec{p}_i$ from any point O to the particular element of tissue dV is known, then the location of the center of mass is defined from O as

$$\vec{p}_{cm} = \frac{\int_V \vec{p}_i \rho_i dV}{\int_V \rho_i dV}. \tag{5.2}$$

In the case of a finite number of N point masses $m_i = \rho_i V_i$ $(i = 1, 2, \ldots, N)$,

$$\vec{p}_{cm} = \frac{\sum_{i=1}^{N} m_i \vec{p}_i}{\sum_{i=1}^{N} m_i}. \tag{5.3}$$

For this equation, the vector $\vec{p}_i$ defines the position of mass element m_i relative to point O. Appendix A contains positions of the centers of mass for body segments in a lower extremity model.

Products of inertia I_{ab} describe another way in which mass is distributed. But instead of being defined only about a point O, I_{ab} is also defined in relation to two directional axes. In Figure 5.1, unit vectors $\hat{n}_a$ and $\hat{n}_b$ are shown passing though the point about which I_{ab} is to be defined (point O). The two vector directions are usually defined perpendicular to each other. The product of inertia I_{ab} about O for axes $\hat{n}_a$ and $\hat{n}_b$ is defined as

$$I_{ab} = \sum_{i=1}^{N} m_i (\vec{p}_i \times \hat{n}_a) \cdot (\vec{p}_i \times \hat{n}_b) \tag{5.4}$$

$$= \sum_{i=1}^{N} m_i (d_a)_i (d_b)_i \tag{5.5}$$

where $(d_a)_i$ and $(d_b)_i$ are the perpendicular distances between the position of mass m_i and the two axes.

The *moment of inertia* I_{aa}, is simply abbreviated as I_a,

$$I_a \equiv I_{aa} \tag{5.6}$$

$$= \sum_{i=1}^{N} m_i \left(\vec{p}_i \times \hat{n}_a\right) \cdot \left(\vec{p}_i \times \hat{n}_a\right) \tag{5.7}$$

$$= \sum_{i=1}^{N} m_i \left(d_a\right)_i^2 \, . \tag{5.8}$$

The moment of inertia is a special type of product of inertia, where only one axial direction is considered. With respect to the rotation axis, I_a quantifies how resistant a system of particles or a rigid body is to changing its angular velocity about that axis. For instance, a torque $\vec{\tau}$ applied to a rigid body will change its angular acceleration $\vec{\alpha}$ in proportion to the moment of inertia,

$$\vec{\tau} = I_a \vec{\alpha} \, , \tag{5.9}$$

just as $\vec{F} = m\vec{a}$ in the linear realm.

When point O is picked to coincide with the center of mass of a particular rigid body, then the moments of inertia are known as *central moments of inertia*. Also, the moment of inertia I_O about axis $\hat{n}_O$ passing through point O may be used to compute the moment of inertia about a parallel axis $\hat{n}_P$ passing through a different point P. If d is the perpendicular distance between $\hat{n}_O$ and $\hat{n}_P$, and M is the total mass of the rigid body, then the moment of inertia about point P is,

$$I_P = I_O + Md^2 \, . \tag{5.10}$$

Central principal moments of inertia for the body segments of a male subject are contained in Appendix A.

5.2.1 INERTIA TENSOR

Defining three mutually perpendicular unit vectors $\hat{b}_1$, $\hat{b}_2$, and $\hat{b}_3$ passing through O, allows three moments of inertia (I_{ii}) and six products of inertia (I_{ij}) to be computed ($i = 1, 2, 3$; $j = 1, 2, 3$; $i \neq j$). Writing these nine values in a matrix format creates the *inertia tensor* for rigid body B about point O. The moments of inertia form the main diagonal, while the products of inertia are written in the off-diagonal positions of the 3×3 matrix,

$$I = \begin{bmatrix} I_{11} & I_{12} & I_{13} \\ I_{21} & I_{22} & I_{23} \\ I_{31} & I_{32} & I_{33} \end{bmatrix} \tag{5.11}$$

If a different set of basis vectors were defined, $\hat{b}_1'$, $\hat{b}_2'$, and $\hat{b}_3'$, the inertia tensor I' will be different from I. Thus, the inertia tensor defined for a rigid body about point O is not unique to that point. I is dependent upon both the position of O and the directions of the basis vectors with which I is defined. Although there are nine elements in the inertia tensor, only six numbers need be computed to completely specify it, because

$$I_{ij} = I_{ji}\,, \tag{5.12}$$

for $i = 1, 2, 3$ and $j = 1, 2, 3$.

When point O is chosen to be at the center of mass of the body, there is at least one set of mutually perpendicular unit vector directions $\hat{b}_1$, $\hat{b}_2$, $\hat{b}_3$, (and combinations of $\hat{b}_1$, $\hat{b}_2$, $\hat{b}_3$, $-\hat{b}_1$, $-\hat{b}_2$ and $-\hat{b}_3$) that yields an inertia tensor with nonzero elements only on the main diagonal,

$$I^* = \begin{bmatrix} I_1^* & 0 & 0 \\ 0 & I_2^* & 0 \\ 0 & 0 & I_3^* \end{bmatrix} \tag{5.13}$$

In these cases, I_i^* is known as a *central principal moment of inertia* about axis $\hat{b}_i$.

The principal moments of inertia contain the largest and smallest values that the moments of inertia can attain about the center of mass. Knowing all three principal moments of inertia is equivalent to knowing the moments and products of inertia about any three mutually perpendicular axes passing through the same point. Conversely, knowing the inertia tensor about any set of perpendicular axes completely describes the rotational inertial characteristics of the rigid body, because inertia tensors can be transformed to other bases. Biomechanical limb segments, however, are almost exclusively described in terms of their principal moments of inertia (see Appendix A), and the subject of tensor transformation will not be discussed.

5.3 LINKED SEGMENT MODELS

5.3.1 "SOFT", VERSUS "HARD" CONSTRAINTS

The configurational and motion constraints discussed in this section are informally called "hard" constraints because they rigidly define relationships between the generalized coordinates and generalized speeds of the system being described. That is, for every constraint equation, a degree of freedom is lost.

For illustration, consider the planar three-link robotic manipulator mounted to a rigid table as depicted in Figure 5.2. Typical robotic manipulators are composed of rigid members joined by rigid revolute joints. When the end-effector is brought into contact with the table at point P, one degree of freedom will be lost (the end-effector is free to slide along the table, changing the

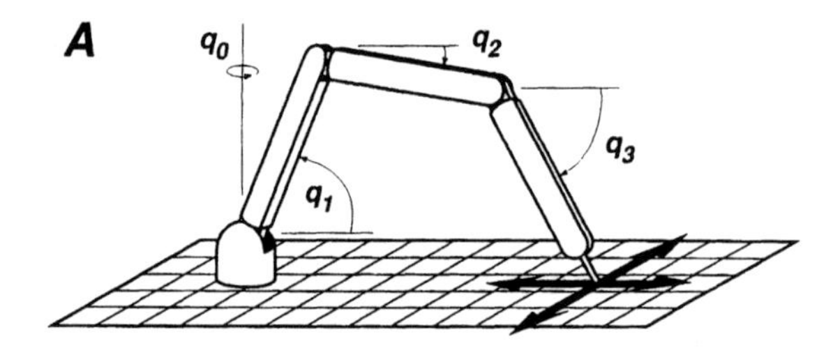

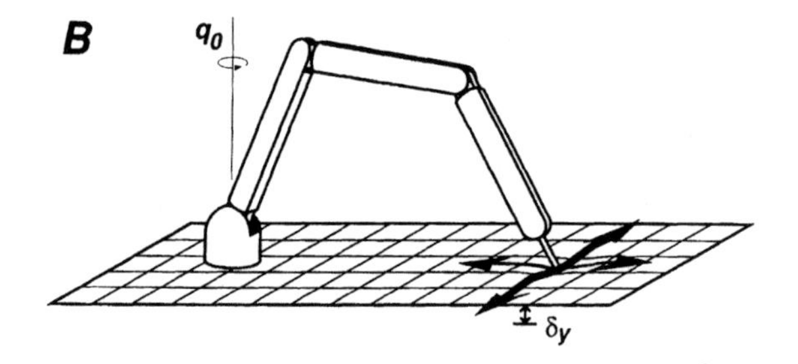

Figure 5.2. A three segment kinematic linkage with hard and soft endpoint constraints. A. When hard constraints are imposed, and the end of the linkage is in contact with the surface, a degree of freedom is lost because the endpoint cannot deform the surface. B. When soft constraints are in effect, the surface is allowed to deform a vertical distance δ_y, which can be made proportional to the vertical force and velocity of deformation.

coordinate of P in the X direction, but the end-effector cannot pass through the table). A loop equation can be written to express the fact that the end-effector is in contact with the table but cannot pass through it,

$$\ell_1 \sin(q_1) + \ell_2 \sin(q_2) + \ell_3 \sin(q_3) = 0 . \tag{5.14}$$

The effect of a soft constraint can be included by replacing the rigid table by a surface with some compliance (Figure 5.2B). This allows the end-effector to deform the surface slightly. Equation 5.14 for a slightly compliant surface would become,

$$\ell_1 \sin(q_1) + \ell_2 \sin(q_2) + \ell_3 \sin(q_3) = \delta_y \tag{5.15}$$

where δ_y is a number representing a small vertical displacement, and can be computed based upon the vertical force (F_y) applied, the stiffness of the surface (k), and any damping (b) provided at the interface,

$$\delta_y = \frac{F_y + b\dot{y}}{k} . \tag{5.16}$$

It is equivalent to think of the end effector displacing the compliant surface, which in turn exerts a force proportional to the displacement δ_y and velocity $\dot{y}$,

$$F_y = k\delta_y - b\dot{y} . \tag{5.17}$$

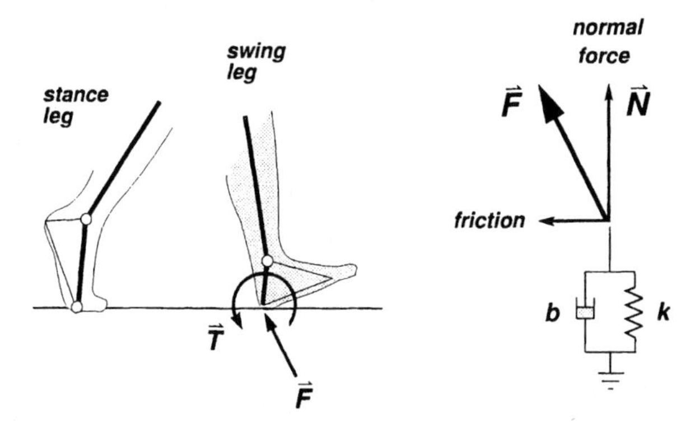

Figure 5.3. The "soft" kinematic constraint utilized in a bipedal human walking model. While the stance-leg foot remains hinged to the ground throughout the simulation of movement, a damped spring exerts an upward force ($\vec{N}$) on the swing-side heel during double-leg support. A frictional force of magnitude $\mu|N|$ prevents the heel from sliding forward excessively. The vector sum of $\vec{N}$ and the frictional force is indicated by $\vec{F}$. An applied torque ($\vec{T}$) keeps the forefoot from falling through the ground should contact be made.

It is likely that biomechanical systems possess a high degree of "softness." By their very nature and construction, the joints have translational and rotational "slop," and thus lend themselves to being modeled well by "soft" constraints. As opposed to "hard" constraints which rigidly constrain the system configuration, soft constraints allow a small amount of motion to occur at the interface between rigid bodies.

Soft constraints can be exploited in simulations when a transition from an open kinematic chain to a closed kinematic arrangement occurs. For example, such a transition occurs during every step of human walking. Simulating the biomechanical transition from single-leg support to double-leg support is simplified through the use of soft constraints (Figure 5.3). Using "hard" constraints at the joints in the model would require dynamic equations to be derived for two separate systems – one set for the single-leg support model and one set for the double-leg support model. This is necessary because the two models have differing numbers of degrees of freedom. On the other hand, if a "soft" constraint were used at one or more of the joints, then the system in double-leg support would retain the same number of degrees of freedom as the unconstrained system. Thus, only one set of dynamic equations needs to be computed.

5.4 SINGLE AND MULTIPLE DEGREE OF FREEDOM JOINTS

Strictly speaking, almost all of the joints in the body have multiple degrees of freedom. However, there are a large number of joints that are well approximated by a joint model with one degree of freedom. To name a few, the *humeroulnar* joint is very well defined by a single rotation axis. The healthy knee, also, has been popularly approximated as a *single-degree-of-freedom*, or *SDOF* joint even though it has been demonstrated by many that it exhibits other motions besides flexion and extension. The key is to realize the degree to which a musculoskeletal model's accuracy will be compromised when certain motions are not represented by the degrees of freedom. Simplifications have always been required in past modeling studies, either because of a lack of computing power or because data analysis has been too tedious. In fact, the whole point of creating a *model* is to *clarify* the behavior of a much more complicated system. Thus, there probably is a limit to the complexity of a *useful* model. As long as one realizes the limitations of the joint models, and the ramifications of eliminating degrees of freedom in a dynamic skeletal model, simplifications should be made. Whenever possible, it is much more convenient computationally to utilize SDOF joints rather than including up to six degrees of freedom for each joint.

Much of the discussion in this section concerns some of the unique features of SDOF joints. However, the reader should be aware that a multiple DOF joint is simply a collection of SDOF joints grouped together and systematically arranged. Therefore, almost all of what is discussed applies to *both* SDOF and multiple DOF joints.

5.4.1 INSTANTANEOUS CENTER OF ROTATION

The term SDOF joint implies that the relationship between the angular and relative positions of two rigid bodies is completely described by a single variable. A simple example of this is a pin joint, which represents a point of interconnection rigidly fixed in each body and about which both bodies can rotate. A more complex example is given in the next section, where a two-joint statically determinate model of the knee is presented.

A common trait among many SDOF joint models is that they are *planar* joints, meaning that all motions (even of three dimensional objects rotating about oblique, tilted axes) are contained in a single reference plane. Equally, in planar rigid body motions having non-zero relative angular velocities, there exists, at every instant in time, a point in the plane of motion that has zero velocity in both rigid body reference frames. This point, called the *instantaneous center of rotation*, or *ICR*, can be thought of as the point of rigid body A about which rigid body B rotates, and simultaneously as the point of body B about which body A rotates. When the positions of the ICR are plotted in time relative to either rigid body reference frame, a pathway of ICRs results.[1] A

[1]Two different pathways, called *poloids*, are obtained depending on whether A is considered as the fixed frame, or whether B is considered as the fixed frame. The two poloidal curves roll on each other, with their

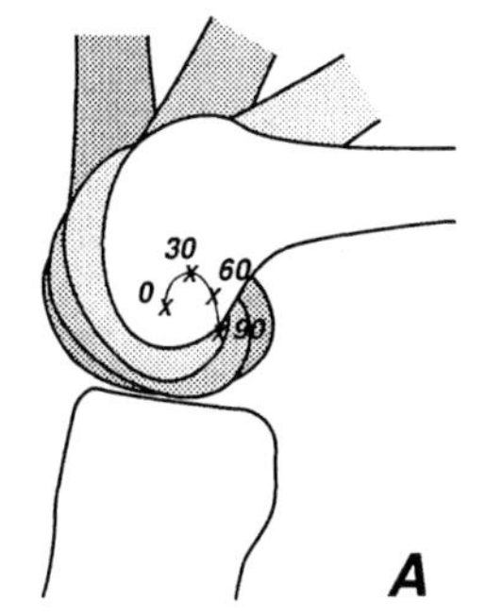

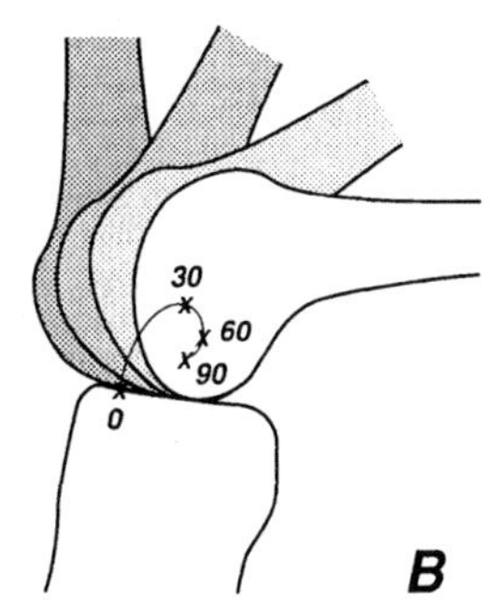

Figure 5.4. An illustration of the difficulties in measuring the ICR via the Method of Reuleaux (see Figure 5.5A). A. When an ICR pathway measured via the Reuleaux Method and reported in the literature was used to rotate a planar representation of the joint, the femur separated from the tibia. B. A prescribed pathway allows the joint surfaces to remain in contact. The ICR was placed at the contact point when the knee was in full extension, to create a relative rolling motion during initial joint flexion.

good working definition for the ICR between reference frames A and B is *the point instantaneously having zero velocity in both reference frames.* Figure 5.4 shows, for example, a pathway of the ICR for the femur moving into flexion, with respect to a fixed tibia.

The reader is cautioned that the ICR is an approximation of relative motion that substitutes a hypothetical (two-dimensional) rotation in place of a (two or three-dimensional) transformation having both rotation *and* translation. In other words, it first presumes that the relative motion is completely planar and revolute. This is okay during a smooth, planar joint rotation, but it is objectionable when translations accompany the joint rotation because the ICR location is often translated outside the physiological joint. In the case where one bone quickly and suddenly slides upon another during the rotational motion, the ICR is translated infinitely far away.

A second objection many biomechanists have with the ICR is that it is a very difficult concept to apply. For example, as the the time sampling interval decreases, it becomes increasingly difficult to draw accurate perpendicular bisectors. Also, no measurement is without some inaccuracy, and the effects of inherent point localization errors magnifies the difficulty of determining the intersection point of the two (probably inaccurate) perpendicular bisectors. Soudan *et al.* (1979) described the difficulty in locating the ICR using the standard *Method of Reuleaux*, and provided a more accurate "tangent method" (see Figure 5.5). However, even the tangent method requires accurate knowledge of

intersections defining the location of the ICR. Knowledge of the poloids is considered to be kinematically equivalent to knowing the relative motions between the two rigid bodies.

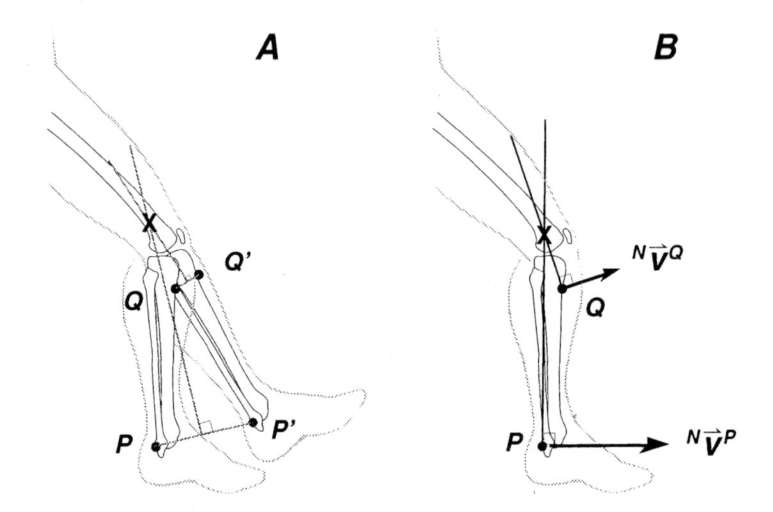

Figure 5.5. Two methods for finding the instantaneous center of rotation (ICR) for a single degree of freedom joint. A. The Method of Reuleaux uses the positions (points P and Q) of two markers or landmarks at time t. At time $t + \Delta t$, P and Q are found at positions P' and Q'. Straight connecting lines are drawn between P and P', and between Q and Q'. Then, perpendicular bisectors are drawn from the connecting lines. The location where the perpendicular bisectors intersect is the approximate, time-averaged location of the ICR ($\times$) for the time interval Δt. B. The Tangent Method uses the instantaneous velocities of points P and Q. Lines are drawn through P and Q in directions perpendicular to the velocity vectors ${}^{N}\vec{v}^{P}$ and ${}^{N}\vec{v}^{Q}$. Where the perpendicular lines cross is the instantaneous location of the ICR ($\times$).

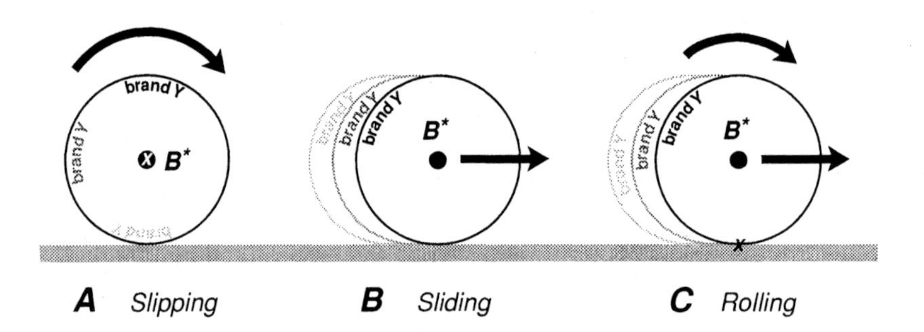

Figure 5.6. ICR locations during slipping (I), sliding (II), and rolling (III), for a wheel moving relative to the ground. A. Slipping is defined when the wheel rotates but its axle B^* remains stationary. The ICR location(s) can be considered to be any point along the wheel axis, here shown at the hub of the wheel ($\times$). B. In sliding, the wheel translates without rotation. In this case, the location of the ICR is undefined, as it is at $\pm\infty$. C. When a wheel rolls upon the ground, the points of the wheel and the ground that are in contact have a common velocity of zero. In this case, the ICR is located at the contact point ($\times$).

the instantaneous velocities of two points. No currently available measurement device makes a direct measurement of velocity, and thus one typically must

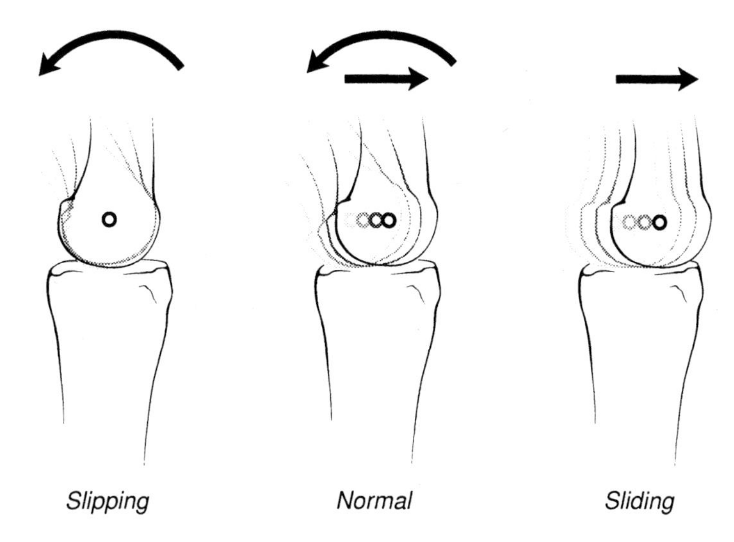

Figure 5.7. ICR locations in the tibiofemoral joint during slipping, a normal mixture of rolling and slipping, and sliding.

settle for a smooth curve drawn between successive marker positions sampled at discrete intervals of time.

Though the ICR is rather limited as a descriptor of joint motion, the ICR has been used to describe gross orthopaedic problems in joints. For instance, Nordin and Frankel (1989) describe how distraction and compression of the knee can be diagnosed based upon the relative positions of the ICR and the articulating joint surface. To illustrate how the ICR defines the kinematic relationships between two contacting bodies, consider two rigid bodies A and B moving relative to each other. For clarity, we will first consider one body B to be a wheel, and the other body A to be a horizontal surface (Figure 5.7A). For the purposes of discussion, B and A contact each other at points $\bar{B}$ of rigid body B and $\bar{A}$ of rigid body A. The center of the wheel B is point B^*. Figure 5.7 provides a physiological analogy to these cases.

- *Case I.* If a wheel was revolving (${}^A\vec{\omega}^B \neq 0$) but its axle remained fixed relative to the ground (${}^A\vec{v}^{B^*} = 0$), then the ICR of the wheel relative to the ground is at the hub of the wheel. This case may be described as *slipping*, and is characterized by *rotation without translation.*

- *Case II.* If a wheel purely translates relative to the ground (${}^A\vec{v}^{B^*} \neq 0$ and ${}^A\vec{\omega}^B = 0$), then the location of the ICR is undefined, as it is located vertically upward or downward at a distance of infinity from the wheel-ground interface. This case is called *sliding*, and is characterized by *translation without rotation.*

- *Case III.* When the velocities of the contacting points $\bar{B}$ and $\bar{A}$ are equal, ${}^{N}\vec{v}^{\bar{B}} = {}^{N}\vec{v}^{\bar{A}}$ where N is a third reference frame, this defines the condition for *rolling without slip* at the interface. When a wheel rolls on the ground without slipping, the ICR is at the point of ground-wheel contact, because the part of the wheel touching the ground must have the same velocity as the part of the ground touching the wheel.

Usually, in biomechanical joints, the ICR will be located away from the area of contact between two bones, but not infinitely far away as in Case II. This means that the bone is not purely rolling, nor purely translating. In smooth joint motions such an ICR pathway is likely to describe a mixture of rolling, slipping and sliding. The farther away the point of contact and the ICR are, the more sliding, slipping, or translating will tend to be evident. In "jerky" joint motions such as a smooth rotation interrupted by an abrupt translation, the ICR will abruptly translate away from the smooth ICR pathway. Thus the ICR is potentially a very sensitive indicator of joint derangement.

It is possible this sensitivity can be exploited for clinical uses provided a long list of details are adequately addressed. For instance, the motions of the bones themselves (not the positions of markers mounted to the skin) should be used to determine the ICR pathway. If the motion is planar, the images must be viewed exactly perpendicular to the plane containing the motion. In the past, X-rays were sometimes used for this and it was difficult to keep the axis of joint rotation parallel to the path of the X-ray beam. Finally, if the motion is not completely planar, the ICR should actually be replaced by an *instantaneous axis of rotation*, or *IAR*. The IAR replaces a 3-D transformation by an equivalent rotation in the same way that the ICR does this for 2-D motions. If a 2-D approximation to 3-D motion is desired, one must first define the plane containing the primary components of the motion. The ICR is the intersection point of the IAR with this plane. Because the IAR is unlikely to remain perpendicular to this plane throughout the motion, the plane's location is usually fixed along the axes of the long bones and along the midline of the joint.

5.4.2 THE MOMENT PRODUCED BY A SPANNING FORCE

When two rigid bodies joined by a SDOF joint are spanned by a tension producing element (muscle-tendon actuator or ligament), the forces applied to each rigid body exert equal and opposite moments on the respective bodies. Consider the case of two bodies A and B in contact at point P (Figure 5.8). If a tension-producing element spans the joint and attaches to point O of body A and I of body B, the moment applied by the tensile force $\vec{T}$ on A about the point P will be equal and opposite to the moment applied by the opposite tensile force $-\vec{T}$ on B about P,

$$\vec{M}_A = \vec{p}^{PO} \times \vec{T} \tag{5.18}$$

$$\begin{aligned} \vec{M}_B &= \vec{p}^{PI} \times \left(-\vec{T}\right) && (5.19)\\ &= \vec{p}^{PO} \times \left(-\vec{T}\right) && (5.20)\\ &= -\vec{M}_A\,. && (5.21) \end{aligned}$$

These moments, $\vec{M}_A$ and $\vec{M}_B$, are not torques because their magnitudes and directions are dependent on the locations of P, O, and I.[2]

Because a tensile spanning element draws the bones together, the muscle tension contributes indirectly to the contact force at the joint. The contact force, the direct muscle tension force, and body forces (gravitational and inertial) together create the torque which causes rotation. Unless the contact force's line of action passes through the ICR, the ICR is a poor choice about which to sum moments when summing up the moments of these forces. This is because the ICR is a point in space that generally has no forces acting through it. If the contact point P is used as the reference point instead of the ICR, then the computation of the joint reaction force and its moment about the ICR can be avoided.

This is one major difference between mechanical versus biomechanical linkages. Mechanical linkages are typically joined together with pins, which serve as fixed axis hinges. *The hinge locations act as ICRs and also are the locations*

[2]It is important to remind the reader that the *torque* on body A would have to include the moments of *every* force acting on body A. This is because torque is the moment of a couple, and a couple is realized when the inertial body force is included along with the sum of *all* other distance, body, and external forces acting on the body (see Section 4.1.4.1).

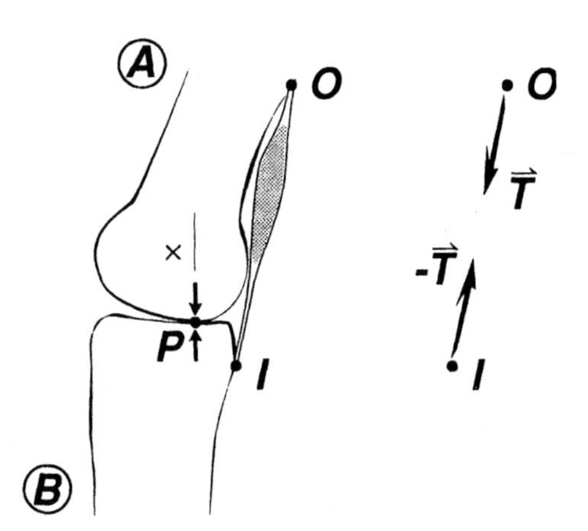

Figure 5.8. An abutting joint with a single degree of freedom. Bone segments A and B are in contact at a single point P though which the joint reaction forces act (arrows). A hypothetical location of the ICR is also shown (×) which does not lie along the line of action of the joint reaction forces. A single muscle creates tension and spans the joint. Force $\vec{T}$ is exerted on body A at the muscle origin (point O). An equal and opposite force $-\vec{T}$ is exerted on body B at the muscle insertion (point I).

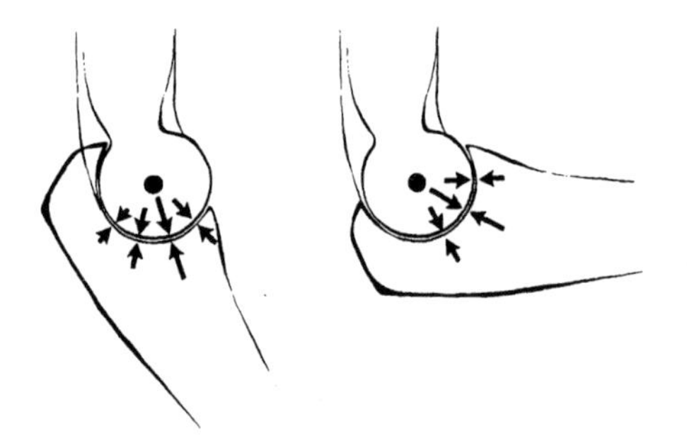

Figure 5.9. Focal point in cylindrical or spherical joints. If friction is minimal, this point serves as the most convenient point to sum joint moments about. The contact forces act locally in a direction perpendicular to the articulating surfaces. If the lines of action of the local contact forces come together in a focal point, the moments of these forces will all be zero about that point.

where the contacting forces between the adjacent bodies are applied. Because the joint reaction forces are applied through the ICR, the moments of the contact forces are zero. In biomechanical joints, the contact forces are applied near the ICR, but their lines of action might not pass *through* the ICR. If the moment is computed about any point other than points along the line of action of the contact force, then the moment of the contact force must be included in the description of the moment created by a spanning element.

If the joint surfaces are cylindrical or spherical in geometry and are essentially frictionless, then the centroid of the joint would serve as the most convenient point to sum moments about. This is because the contact forces are perpendicular to the joint surface. Even when the contact force is distributed across the surface, the lines of action of the local contact forces converge in a *focal point* in symmetrical joints. Computing moments about the focal point eliminates the need to compute the contact forces and the moments. The humeroulnar joint and the hip joint are good examples of cylindrical and spherical geometries, respectively. Further explanation and an illustration of this is provided in Figure 5.9.

5.4.3 EFFECTIVE ORIGIN AND INSERTION POINTS

Consider two rigid segments linked by an n-DOF revolute joint with contact point C and a musculotendon actuator not passing through C (Figure 5.10). If tension is present, the musculotendon actuator will exert a moment about the joint spanned. The moment $\vec{M}$ acting on the proximal segment of the joint C is the vector cross product of two vectors, the position vector $\vec{p}$ from the joint to a point on the line of action of the musculotendon, and the tensile force $\vec{F}$ acting on the segment

$$\vec{M} = \vec{p} \times \vec{F} . \tag{5.22}$$

The line of action of $\vec{F}$ passes through both the "effective origin" EO and the "effective insertion" EI. The effective origins and insertions are defined to be the first points of contact that the spanning tension element encounters on the adjacent rigid bodies. Here, $\vec{F}$ is considered to point away from EO and toward EI.

Sometimes, additional points along the musculotendon pathway are defined, called "via points." Via points are defined to route the pathway along the outside of bony surfaces, particularly for joint extensors. While via points are not required to determine the muscle moment applied to the segments adjacent to the joint, the via points do help to define the length of the musculotendon actuator. The muscle-tendon path is usually approximated by straight lines between the via points, so that the resulting pathway is piecewise linear. Pandy (1999) provides additional details regarding via points and the general computation of joint moments. Good examples of musculotendon paths requiring via point definitions are found in Appendix B.

If multiple muscle activation levels are used to calculate $\vec{M}$ for a single joint configuration (as they are here), it is more convenient to define the "unit moment" $\vec{m}_u$ (the moment due to a unit force), so computations of the cross product above can be minimized. Let $\vec{p}^{EO\,EI}$ refer to the vector from the effective origin to the effective insertion. Then, the unit moment is,

$$\vec{m}_u = \vec{p} \times \frac{\vec{p}^{EO\,EI}}{|\vec{p}^{EO\,EI}|} \,. \tag{5.23}$$

Once $\vec{m}_u$ is known, the joint moment due to the musculotendon force is easily calculated given any magnitude of $\vec{F}$,

$$\vec{M} = \pm|\vec{F}|\vec{m}_u \,. \tag{5.24}$$

The sign of $\vec{M}$ is defined to be consistent with the angular definitions, with a positive moment causing a positive segmental or joint acceleration. $\vec{m}_u$ can also be used to define the (scalar) *moment arm* ρ of $\vec{F}$ about C, which is a commonly used measure of the effectiveness of $\vec{F}$ in producing moments of

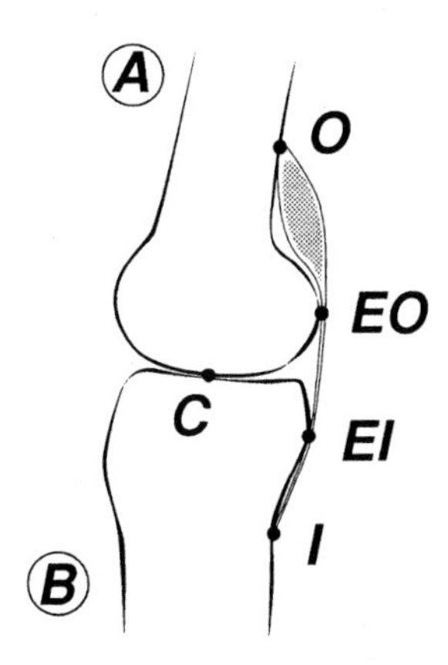

Figure 5.10. A muscle pathway spanning two rigid bodies A and B is represented by 3-segment line-of-action. The bodies are assumed to be in contact at a single point C. The origin is labeled point O, and the insertion, point I. Because the musculotendon pathway traverses around the bony prominences of the joint, an effective origin EO and an effective insertion point EI is defined for the purpose of computing muscle moments.

force. In three dimensions,

$$\rho = |\vec{m}_u|, \tag{5.25}$$

which can be thought of as the perpendicular distance from the point to the line of action defined by $\vec{F}$. ρ may be calculated in any plane of motion by finding the magnitude of the projection of $\vec{m}_u$ in the desired plane.

The magnitude of $\vec{F}$ is the tendon force F^T, which must be computed individually for each musculotendon actuator. The method outlined in Section 2.4 should be used to solve for F^T, which requires that the instantaneous length of each muscle-tendon compartment (ℓ^{MT}) be supplied at the outset. To find ℓ^{MT}, the coordinates of the origin are first transformed to the reference frame containing the insertion point. Then, musculotendon length is easily computed as the sum of three vector magnitudes,

$$\ell^{MT} = |\vec{p}^{O\,EO}| + |\vec{p}^{EO\,EI}| + |\vec{p}^{EI\,I}|, \tag{5.26}$$

where O and I refer to muscle origin and insertion, and EO and EI are the coordinates of the effective origin and effective insertion.

In general, the moments produced by each muscle crossing a joint are calculated in three dimensions at desired instants in time and summed to obtain the total joint moment. If the joint has multiple degrees of freedom, the components of the total moment will cause motions to occur properly without any additional effort. A component of the total joint moment can be used in place of the joint torque in dynamic simulations of movement if planar, SDOF joints are employed.

5.4.4 RELATIONSHIP OF MOMENT ARM AND MUSCLE EXCURSION

In a single degree of freedom planar joint, it is often useful to estimate the change in length of a muscle by *integrating the moment arm.* To derive this useful approximation, we start by assuming that our SDOF planar joint adjoining rigid bodies A and B has no energy dissipation or storage elements, *i.e.*, there is no friction, damping, or elastic storage of energy. (In fact, there is some damping, and some elastic energy storage in the stretch of tendons and ligaments and the compression of cartilage. Also, there is a small amount of friction between the articulating surfaces. Most of these effects are small in normal physiological movements.) If energy is conserved, and not stored anywhere in the system, then the shortening of the spanning muscle by a change in length $\Delta\ell$ produces work on the system, since

$$W_{shortening} = \int_{\ell}^{\ell+\Delta\ell} \vec{F} \cdot d\vec{\ell}. \tag{5.27}$$

Work, of course, is a scalar quantity because the dot product produces a scalar. The change in length of the muscle produces a relative change in the angular position of body B relative to body A. This can be thought of as the rotational analog of work, much like the work done in twisting a torsional spring,

$$W_{rotating} = \int_{\theta}^{\theta+\Delta\theta} |\vec{\tau}| \, d\theta \, . \tag{5.28}$$

Since energy is not stored or dissipated in the system, all of the work done in shortening the muscle will go into rotation, and these two work quantities will be equal to each other. For small changes in length $\Delta\ell$ and small changes in angle $\Delta\theta$, we can simplify these expressions to

$$\left|\vec{F}\right| \Delta\ell = |\vec{\tau}| \, \Delta\theta \, , \tag{5.29}$$

or,

$$\frac{\Delta\ell}{\Delta\theta} = \frac{|\vec{\tau}|}{\left|\vec{F}\right|} \, . \tag{5.30}$$

In the limit, as $\Delta\ell$ and $\Delta\theta$ approach infinitesimally small values, we can express Equation 5.30 using differentials,

$$\frac{d\ell}{d\theta} = \frac{|\vec{\tau}|}{\left|\vec{F}\right|} \, . \tag{5.31}$$

At this point, another set of simplifying assumptions must be made. First, it is acknowledged that the effects of gravity are significant. Yet, gravitational contributions to the torque are ignored because only the muscle's contributions to the torque are desired. Second, the muscle contraction is assumed to be performed under *quasistatic* conditions, so that any inertial forces due to accelerations of the limbs A and B are negligibly small. Making these assumptions simplifies the computation of the torque, which is computed by finding the moment of the muscle force and the joint reaction force. Third, it is assumed that the moments of the contact forces are zero about some point (either a focal point or a single contact point). By computing all moments about that point, the moment due to the joint reaction force is also eliminated. Only the muscle force remains, and the magnitude of the torque $|\vec{\tau}|$ can be expressed as,

$$|\vec{\tau}| = \left|\vec{r} \times \vec{F}\right| \tag{5.32}$$

$$= |\vec{r}| \left|\vec{F}\right| sin\beta \, , \tag{5.33}$$

where $\vec{r}$ is a vector from the contact point to any point along the line of action of the muscle force, and β is the angle between $\vec{r}$ and $\vec{F}$.

Therefore, the equation becomes:

$$\frac{d\ell}{d\theta} = \frac{|\vec{r}| \left|\vec{F}\right| sin\beta}{\left|\vec{F}\right|} \tag{5.34}$$

or,

$$\frac{d\ell}{d\theta} = m\,, \tag{5.35}$$

where m, the *moment arm* has been written in place of the quantity $|\vec{r}|\,sin\beta$. Another way to think of the moment arm in the planar SDOF joint, or any joint for that matter, is that *the moment arm of a muscle spanning a joint is the perpendicular distance from "the joint" to the line of action of the muscle force.*

This brings up the subject of what exactly constitutes "the joint". For the case of a pin joint, the definition of the moment arm is easy to apply. The pin is the place where the two bodies are joined together and apply forces on one another. The pin also serves as the joint rotation axis. In the case of a physiological joint where two bones touch over an area of contact, one might be tempted to substitute the instantaneous center or axis of rotation (ICR or IAR) in place of the fixed pin of the analogous pin-jointed system. While this idea is intellectually attractive, usage of a moment arm defined from one of these fictitious points would lead one to an erroneous result. This is because the joint torque calculation must then include the moment of the joint reaction force in addition to the moment of the muscle force, according to the discussion in the preceding section. To resolve this issue, *the "joint" is defined as the point through which the net joint contact force is applied.* The joint is not the instantaneous center of joint rotation.

Equation 5.35 shows the moment arm of the joint can be estimated by dividing the muscle excursion (over some small distance $d\ell$) by the resulting change in angle of the joint ($d\theta$). Again, taking approximations allows us to rewrite the excursion of the muscle in terms of an integral quantity:

$$\Delta\ell = \int_{\ell}^{\ell+\Delta\ell} d\ell \tag{5.36}$$

$$= \int_{\theta}^{\theta+\Delta\theta} m\,d\theta\,, \tag{5.37}$$

which relates the approximate muscle excursion length to the integral of the moment arm.

These latter two expressions are extremely valuable for estimating joint moments and muscle excursions.

5.5 PASSIVE CONSTRAINTS AT THE JOINTS

Moments due to passive elastic structures at the joints (*e.g.*, ligaments, joint capsules, and surrounding tissues) can be modeled as simple nonlinear springs (Audu and Davy, 1985; Davy and Audu, 1987) which provide strong restoring joint moments near the extremes of joint motion. Damping provides a measure

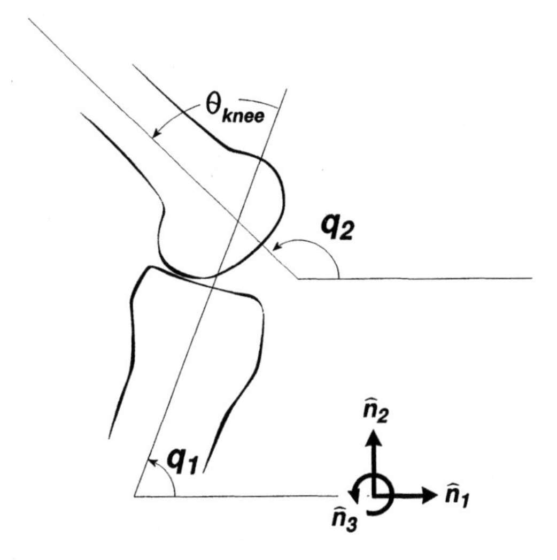

Figure 5.11. Segmental angles (q_1, q_2) must be converted into joint angles for passive joint moment calculations. In this example, the knee angle $\theta_{knee} = q_2 - q_1$.

of resistance to the joint velocities, and makes intuitive sense as the joint capsules are infused with synovial fluids. These should be added any time a particular joint approaches the limits of its range of motion.[3] If the n degrees of freedom are defined in terms of segmental angles q_i, the segmental angles must be converted into joint angles (θ) and joint angular velocities ($\dot{\theta}$) by performing the proper calculations, $\theta = q_i - q_j$ and angular velocities $\dot{\theta} = \dot{q}_i - \dot{q}_j$ (see Figure 5.11). The user should also take care to include the contributions of *joint moments* properly within the set of dynamic equations expressed in the form of *segmental torques*.[4]

[3] For inverse dynamic analyses, the joint ranges of motion are known *a-priori*, and thus it is easy to determine whether the passive moments should be added to a joint model. For dynamic simulations, they should always be added!

[4] See Section 6.6.6 in the next chapter for more on this subject.

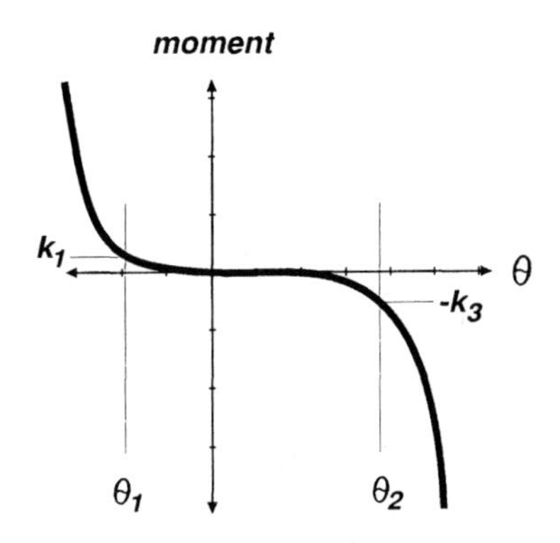

Figure 5.12. Double exponential function used to generate the passive joint moment. "Breakpoints" θ_1 and θ_2 are prescribed well inside the joint's range of motion. Torque magnitudes increase rapidly when θ lies outside of the range $\theta_1 < \theta < \theta_2$. The odd symmetry of these curves serves to apply moments which restore the joint angle to its normal range of motion.

Using joint angles in radians and joint angular velocities in radians per second, the passive joint moments take the form,

$$M_{i,pass}(\theta,\,\dot{\theta}) = k_1 e^{-k_2(\theta-\theta_1)} \;-\; k_3 e^{-k_4(\theta_2-\theta)} \;-\; c_1\dot{\theta}\,, \tag{5.38}$$

where $M_{i,pass}(\theta)$ is the measure number of the joint flexion or extension moment expressed in $N-m$. For joint angle θ_i and joint axis $\hat{k}_i$, the scalar contributions are defined as,

$$M_{i,pass}(\theta,\,\dot{\theta}) = \vec{M}_{i,pass}(\theta)\cdot\hat{k}_i\,. \tag{5.39}$$

With one equation, this function approximates the passive moment contributions provided by the joint structures at both extremes of joint motion. It provides for large restoring moments (*i.e.*, the joint is stiff) when θ falls outside the range set by the breakpoints θ_1 and θ_2 (when $\theta < \theta_1$ or $\theta > \theta_2$). The exponents in the double exponential must be negative between the breakpoints, when $\theta_1 < \theta < \theta_2$. Otherwise, the passive moment will attain unrealistically large values within the joint range of motion. Note that k_2 and k_4 govern the sharpness of the break, and that θ_1 and θ_2 set the location of the breakpoints in θ but θ_1 and θ_2 are well *inside* of the limits of the joint range of motion. Generally speaking, at $\theta = \theta_1$, the moment is approximately equal to k_1, and at $\theta = \theta_2$, the moment has an approximate magnitude of $-k_3$.

Figure 5.12 shows a typical curve of passive moment and the constants used to obtain it. When damping coefficients are not known, typical models have used a value of $c_1 = 0.1$, which provides 1 unit of torque for every 10 radians per second.

Example: Passive moment curves for flexion/extension in the lower extremity. Figure 5.13 shows the resultant curves of passive torque developed under quasistatic conditions ($\dot{\theta} \approx 0$) from Equation 5.38 for the ankle, knee, and hip joints.

5.5.1 KNEE JOINT EXTENSION

The structure and function of the knee is a complex topic, because there are *three* bones which contribute to the mechanics of knee extension. Many excellent studies have been reported in the literature, and each has contributed to our understanding (see the Reference list for a selection). However, the mechanical interactions of the patella, tibia, and femur can be condensed into a single degree of freedom model with a few simplifications (Figure 5.14, Yamaguchi and Zajac, 1989). Because the quasistatic knee extension model serves as a good example of the preceding material, it is included briefly here. For more details, the reader is referred to the original paper.

Modification of the procedure for computing joint moments is required at the knee joint, because of the interactions of the patellofemoral and tibiofemoral

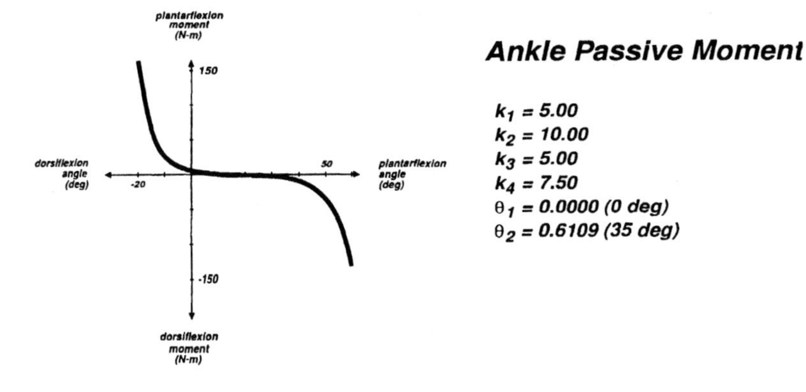

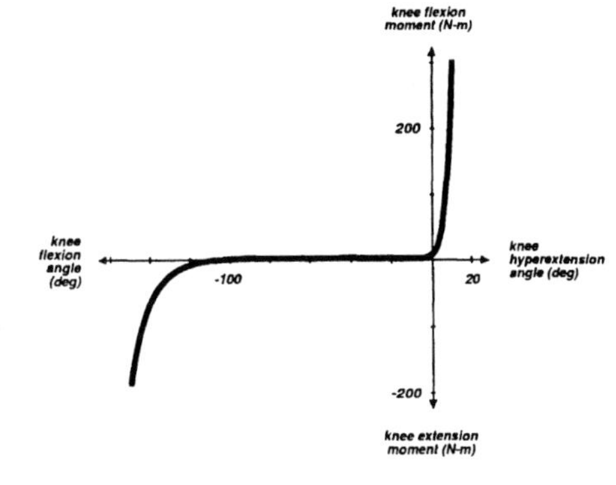

Knee Passive Moment

$k_1 = -3.10$
$k_2 = 5.90$
$k_3 = -14.00$
$k_4 = 22.00$
$\theta_1 = -1.9199$ (-110 deg)
$\theta_2 = 0.0349$ (+2 deg)

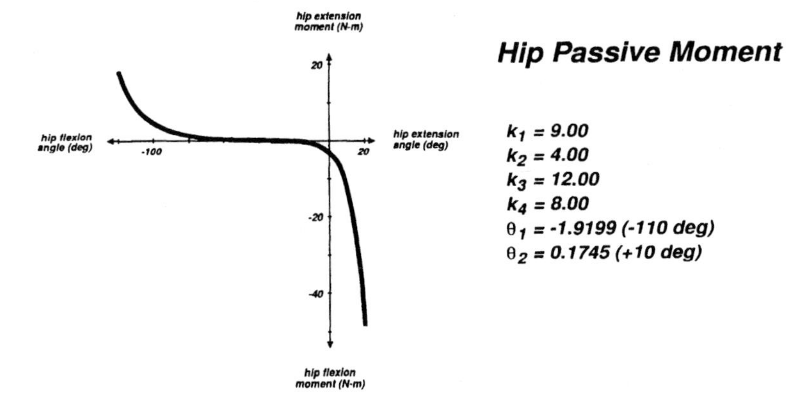

Figure 5.13. Passive moment curves for flexion and extension of the ankle, knee, and hip joints.

joints. Accuracy of the computed extensor moment arm is considered to be critical at the knee because large muscular forces are exerted across small extensor moment arms. Small absolute errors in moment arm thus yield large errors in extensor moment of force. In developing the SDOF knee model, what is desired is a model that incorporates the essential mechanics of knee extension without being too complex to use in whole-body simulations of movement.

Despite its simplicity, the model is able to reproduce the internal configuration of the joint (angles α, β, and the positions of the tibiofemoral and patellofemoral contact points) at various flexion angles (θ). Also, the model uses a static force balance to compute the reduction in the force transmitted from the quadriceps tendon to the patellar ligament. This allows the model to compensate for the differences between the "actual moment arm" (m_{act}) and the "effective extensor moment arm" (m_{eff}). As shown in Figure 5.14, m_{act} is defined as the perpendicular distance from the centroid of the patellar ligament force to the tibiofemoral contact point. m_{eff} is the apparent moment arm of the quadriceps muscle when tension is produced. m_{eff} is the moment arm that is desired during movement simulation studies, because knowledge of the effective moment arm enables the extension moment to be computed more easily from the quadriceps muscle tension $F_q = |\vec{F}_q|$.

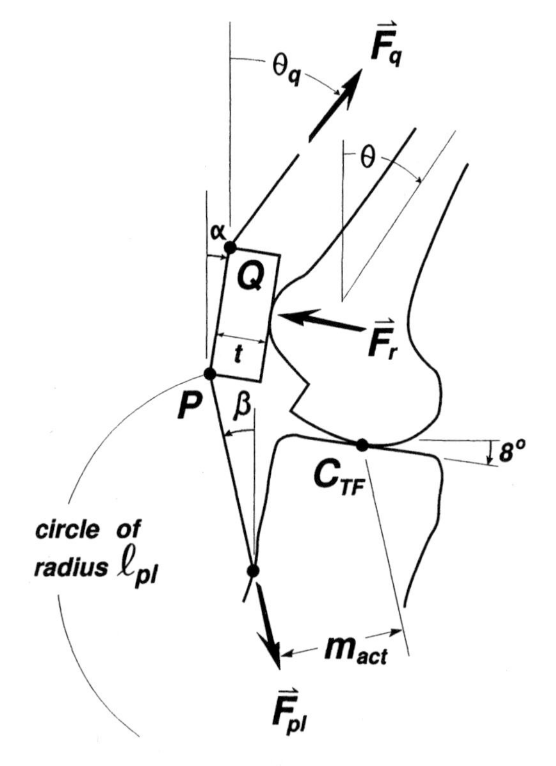

Figure 5.14. The planar, static model of the knee used to compute the knee effective extensor moment arm. Orientations of the patellar and patellar ligament axes (α, β) are determined in order to balance the forces $\vec{F}_r$, $\vec{F}_{pl}$) and moments governing patellar equilibria given an applied quadriceps force and direction ($\vec{F}_q$ and θ_q). Tibiofemoral joint motions are defined by a prescribed pathway of instantaneous centers of joint rotation to yield proper locations of tibiofemoral contact with joint flexion. Once the actual moment arm (m_{act}) is determined, the ratio of force transmitted through the patella ($|\vec{F}_{pl}|/|\vec{F}_q|$) is used to express the effective extensor moment arm in terms of the applied quadriceps force, $m_{eff} = m_{act}|\vec{F}_{pl}|/|\vec{F}_q|$.

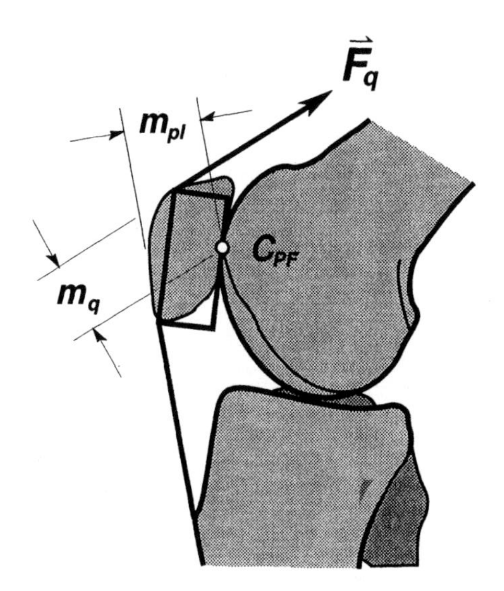

Figure 5.15. Xerogram tracing of the knee, showing how the dimensions of the rectangular patella model were defined. m_{pl} and m_q define the moment arms of the patellar ligament and quadriceps tendon, respectively, about the patellofemoral contact point (C_{PF}).

Defining precisely how the knee extends is complicated by an axis of tibiofemoral joint rotation that moves and changes orientation with flexion and extension. It is generally agreed in the literature that the femur rolls backward on the tibial plateau during the initiation of flexion from an extended position. Because the length of the tibial plateau is only about half that of the corresponding articulating surface on the femoral condyles, the femur begins to slip increasingly until its backward motion is halted by the anterior and posterior cruciate ligaments. Because of this, the ICR should be located at the tibiofemoral contact at full extension, and should rise off of the surface with increasing knee flexion. Experimental data measuring the ICR location is not useful due to inaccuracies. An example of this was illustrated previously in Figure 5.4. The solution *prescribes* the tibiofemoral motion by creating a pathway of the ICR that delivers proper contact locations and interface conditions.

The key to understanding the action of the patellofemoral joint is realizing that the patella functions both as a spacer and a lever. The spacing function increases the displacement of the patellar ligament away from the tibiofemoral contact point C_{TF} and increases m_{act}. However, because the patella rocks about its contact with the femur, it also functions as a force reducing lever to alter the force transmitted from the quadriceps tendon to the patellar ligament (Grood *et al.*, 1984).

In the knee model, the patellar leveraging effect was computed by numerically solving the equations governing static equilibria of the patellar mechanism. The force balance equation includes the unknown patellofemoral joint reaction ($F_r = |\vec{F}_r|$), Figure 5.14) and patellar ligament ($F_{pl} = |\vec{F}_{pl}|$) tensions as functions of the unknown patellar and patellar ligament orientations (α and

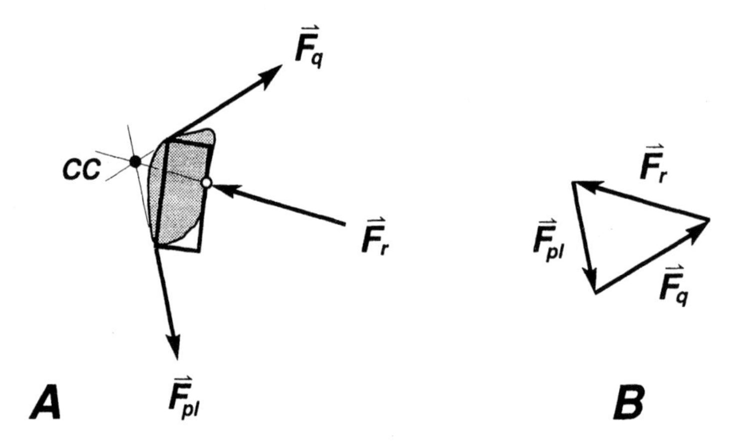

Figure 5.16. Graphical solution of the patellar moment and force balance equations when forces $\vec{F}_r$, $\vec{F}_{pl}$ and $\vec{F}_q$ are considered to be coplanar. A. A three-force coplanar system must have a point of concurrence (CC) to be in static moment equilibrium. If the lines of action of two forces ($\vec{F}_{pl}$ and $\vec{F}_q$) are known, point CC can be found. Then force $\vec{F}_r$ is drawn along the line through the patellofemoral contact point and point CC. B. The three force vectors added together tip to tail must sum to zero to satisfy force equilibrium. A perfect "vector triangle" is formed using the three force vectors.

β, respectively), given a known quadriceps force and direction (θ_q),

$$\begin{bmatrix} cos\alpha & sin\beta \\ -sin\alpha & cos\beta \end{bmatrix} \begin{bmatrix} F_r \\ F_{pl} \end{bmatrix} = \begin{bmatrix} sin\theta_q \\ cos\theta_q \end{bmatrix} F_q \,. \tag{5.40}$$

In addition to satisfying the force equilibrium equation above, F_{pl} also has to satisfy moment equilibrium,

$$F_{pl} m_{pl} = F_q m_q \,, \tag{5.41}$$

where m_{pl} and m_q are the moment arms of the patellar ligament and quadriceps tendon forces about the patellofemoral contact point (Figure 5.15).

At this point, the model still is not determinate given a value of flexion angle θ (the degree of freedom). There are four unknowns α, β, F_r, and F_{pl} and only three equations. Making the further assumption that the patellar ligament is inextensible allows the patellar ligament orientation (β) to be determined once a value for α is specified. This is because the lower anterior edge of the patella (point P) is constrained by the inextensible ligament to follow a circular arc. If α was known, the equation of a line adjoining points P and Q of Figure 5.14 could be developed, and the coordinates of point P found by solving for the upper intersection of the line and the circle. Since m_{pl} and m_q are also known functions of α, the system of three equations could be solved for the (three) unknowns α, F_r, and F_{pl}. An approach using Newton-Raphson iteration is suggested to iteratively solve Equations 5.40 and 5.41 based on an initial guess for α.

A graphical method can also be performed. Because the patella is acted upon by three relatively coplanar forces, the three forces must pass through a common *point of concurrence* for moment equilibrium to be satisfied (Figure 5.16). Therefore, given an initial guess for patellar angle α, one could find the locations of points P and Q and the lines of action of $\vec{F}_q$ and $\vec{F}_{pl}$. Where these intersect is the point of concurrence, CC. Patellofemoral joint reaction force $\vec{F}_r$ must then lie along the line containing CC and the point of patellofemoral contact, C_{PF}. Next, a vector diagram summing forces $\vec{F}_q$, $\vec{F}_{pl}$, and $\vec{F}_r$ is made. If the three forces do not have the right lengths and directions to form a closed triangle when added tip to tail, then force equilibrium is not satisfied, and a new value for α must be tried.

5.6 EXERCISES

1. Estimate the spring constant k for a model which incorporates a soft constraint underneath the heel of a runner. The soft constraint takes the form depicted in Figure 5.3.

2. Using a motion tracking system, a videotape, or a series of digital photographs, compute the position of the knee ICR by direct measurement. Place two markers on each body segment of a volunteer over areas that have a minimum of overlying soft tissue. A two dimensional measurement is sufficient, provided the plane containing the motions of the thigh and shank is perpendicular to the field of view. The subject should flex and extend the knee as smoothly as possible without changing the apparent axis of rotation. Try both methods described in Figure 5.5.

3. Using muscle fiber lengths, and origin and insertion data from the literature or from the Appendix, estimate the moment arm of a muscle averaged over the full range of flexion and extension. Use what you know about the tension-length properties of muscle, and assume that each muscle operates over the full range of flexion and extension. Compare your estimates with measurements on your own body, on a skeleton, or published in the literature.

4. Placing one's own limbs in different positions changes the gravitational moment exerted about the proximal joint. An estimate of the passive joint moment can be statically obtained in this way.

   (a) Select a particular joint, and estimate the passive joint moment in two positions near a maximally flexed position, and two positions near a maximally extended position.

(b) Using a commercially available spreadsheet program, generate a double exponential curve using Equation 5.38.

(c) Adjust the k values so that the double exponential curve passes through all four data points.

5. Using the profiles of the tibia, femur, and patella from Section 5.5.1, select a knee flexion angle θ. Replace the anterior articulating surface of the femur with a circular curve of appropriate radius in the sagittal plane (this can be done for a small range of knee flexion angles). Starting from an initial guess for patella angle α, perform the calculations described in Section 5.5.1. If the equations for balancing the forces and moments on the patella are not satisfied, try another guess for α. Two or three iterations are sufficient.

   What are your best answers for angles α, β, and the force ratios F_r/F_q, and F_{pl}/F_q?

6. A rigid body is maintained in static equilibrium by the action of three nonparallel forces. Prove that any such three force system must be coplanar, and must exhibit a point of concurrence.

References

Audu, M. L., and Davy, D. T. (1985) "The influence of muscle model complexity in musculoskeletal motion modeling." *Journal of Biomechanical Engineering*, V. 107, pp. 147-157.

Davy, D. T., and Audu, M. L. (1987) "A dynamic optimization technique for predicting muscle forces in the swing phase of gait." *Journal of Biomechanics*, V. 20, n. 2, pp. 187-201.

Denham, R. A., and Bishop, R. E. D. (1978) "Mechanics of the knee and problems in reconstructive surgery." *Journal of Bone and Joint Surgery*, V. 60-B, n. 3, pp. 345-351.

Goodfellow, J., Hungerford, D. S., and Zindel, M. (1976) "Patello-femoral joint mechanics and pathology." *Journal of Bone and Joint Surgery* V. 58-B, n. 3, pp. 287-290.

Grood, E. S., Suntay, W. J., Noyes, F. R., and Butler, D. L. (1984) "Biomechanics of the knee-extension exercise." *Journal of Bone and Joint Surgery* V. 66-A, n. 5, pp. 725-734.

Hungerford, D. S., and Barry, M. (1979) "Biomechanics of the patello-femoral joint." *Clinical Orthopaedics and Related Research*, n. 144, pp. 9-15.

Kaufer, H. (1971) "Mechanical function of the patella." *Journal of Bone and Joint Surgery* V. 53-A, n. 8, pp. 1551-1560.

Lindahl, O., and Movin, A. (1967) "The mechanics of extension of the knee-joint." *Acta Orthopaedica Scandinavica*, V. 38, pp. 226-234.

Nisell, R., Nemeth, G., and Ohlsen, H. (1986) "Joint forces in extension of the knee." *Acta Orthopaedica Scandinavica*, V. 57, pp. 46-41.

Pandy, Marcus G. (1999) "Moment arm of a muscle force." In *Exercise and Sport Science Reviews*, J. O. Holloszy (ed.), V. 27, pp. 79-118. joint paper

Soudan, K., Van Audekercke, R., and Martens, M. (1979) "Methods, difficulties and inaccuracies in the study of human joint kinematics and pathokinematics by the instant axis concept. Example: The knee joint." *Journal of Biomechanics*, V. 12, pp. 27-33.

Van Eijden, t. M. G. J., De Boer, W., and Weijs, W. A. (1985) "The orientation of the distal part of the quadriceps femoris muscle as a function of the knee flexion-extension angle." *Journal of Biomechanics*, V. 18, n. 10, pp. 803-809.

Van Eijden, t. M. G. J., De Boer, W., and Weijs, W. A. (1986) "A mathematical model of the patellofemoral joint." *Journal of Biomechanics*, V. 18, n. 10, pp. 803-809.

Wismans, J., Veldpaus, F., Janssen, J., Huson, A., and Struben, P. (1980) "A three-dimensional mathematical model of the knee joint." *Journal of Biomechanics*, V. 13, pp. 677-685.

Yamaguchi, G.T. and Zajac, F.E. (1989) "A planar model of the knee joint to characterize the knee extensor mechanism." *Journal of Biomechanics*, V. 22, n. 1, pp. 1-10.

III

DYNAMIC EQUATIONS OF MOTION

Chapter 6

DYNAMIC EQUATIONS OF MOTION

Objective – To introduce the reader to Kane's Method as a means of deriving the dynamic equations of motion for a musculoskeletal system.

6.1 GENERALIZED SPEEDS AND MOTION CONSTRAINTS

In this book, the generalized speed u_i is defined as the first time derivative of the ith generalized coordinate q_i, or,

$$u_i \equiv \dot{q}_i \,. \tag{6.1}$$

q_i can be either a linear coordinate, such as x, y, or z, or an angular measure such as θ, α, or β. There are occasions when dynamicists do not follow this convention, and define, say $u_1 \equiv \dot{q}_3$. A practical example will be described later in Section 6.9. However, these occasions are rare and are done for specific purposes that are outside of the scope of most biomechanics problems.[1]

If an equation relating the generalized speeds can be written,

$$u_j = \mathcal{F}\left(u_1,\, u_2,\, u_3,\, \ldots,\, u_n\right)\,, \tag{6.2}$$

where f is some function of the generalized speeds, or, more generally, for two functions $\mathcal{F}$ and $\mathcal{G}$ of the generalized speeds,

$$\mathcal{F}\left(u_1,\, u_2,\, u_3,\, \ldots,\, u_n\right) = \mathcal{G}\left(u_1,\, u_2,\, u_3,\, \ldots,\, u_n\right)\,, \tag{6.3}$$

[1]Another example is given here. Normally, n generalized coordinates create n generalized speeds and n dynamic equations. The jth dynamic equation is usually developed from q_j (and u_j) and hence is strongly associated with this degree of freedom ($j = 1, \ldots, n$). However, it is possible that a specific order of the dynamic equations is desired that differs from the order of the generalized coordinates. The order is actually dictated by the order of the generalized speeds u_j, and thus generalized speeds and coordinates can be swapped to change the ordering of the equations.

a *motion constraint* is said to be in effect.

If a particular system is subject to a configurational constraint, where one or more constraint equations can be written that relate the generalized coordinates, one could argue that the configurational constraint equations can be differentiated with respect to time to obtain a motion constraint. While this is certainly true, it would imply that every differentiable configurational constraint equation could be described either as a configurational constraint or as a "motion constraint." To avoid any confusion in this area, the term motion constraint will not be used if the constraint equations are obtainable from configurational constraints.

6.2 PARTIAL VELOCITY VECTORS

To obtain dynamic equations via Kane's Method, *partial velocity* and *partial angular velocity* vectors must be obtained. As their names might imply, these quantities are select portions of the actual velocity and angular velocity vectors. Generally, the partial velocities must be obtained at points where forces act, and partial angular velocities must be obtained for rigid bodies on which torques or moments act. To obtain the partial velocities, one needs only to inspect the velocity expression of the point in question. It is a lot like picking cherries - once a person is shown the cherry tree, all they need to learn is to differentiate the cherries from the leaves and branches. Picking out the partial velocities from a velocity vector is almost as easy! Likewise, the partial angular velocities are obtained via a simple inspection of the angular velocity vectors defining the rotational motions of each rigid body of the system.

Definition: A system S has n degrees of freedom relative to an inertial reference frame N. Given the velocity of any point P in N as a function of the generalized coordinates q_i, $(i = 1, 2, 3, \ldots, n)$, the generalized speeds $u_1, u_2, u_3, \ldots, u_n$, and some scalar function of time $f(t)$, the velocity of P in N can be expressed in terms of n quantities ${}^N\vec{v}_i^P$, $(i = 1, 2, 3, \ldots, n)$, called the partial velocities of P in N,

$$ {}^N\vec{v}^P = {}^N\vec{v}_1^P u_1 + {}^N\vec{v}_2^P u_2 + {}^N\vec{v}_3^P u_3 + \cdots + {}^N\vec{v}_n^P u_n + f(t) . \qquad (6.4) $$

The quantity ${}^N\vec{v}_i^P$ is referred to as the "i^{th} partial velocity of P in N." Similarly, partial angular velocities ${}^N\vec{\omega}_i^B$, $(i = 1, 2, 3, \ldots, n)$ may be defined for the rotation of a rigid body B in an inertial reference frame N,

$$ {}^N\vec{\omega}^B = {}^N\vec{\omega}_1^B u_1 + {}^N\vec{\omega}_2^B u_2 + {}^N\vec{\omega}_3^B u_3 + \cdots + {}^N\vec{\omega}_n^B u_n + g(t) . \qquad (6.5) $$

where $g(t)$ is a scalar function of time. The quantity ${}^N\vec{\omega}_i^B$ is referred to as the "i^{th} partial angular velocity of B in N."

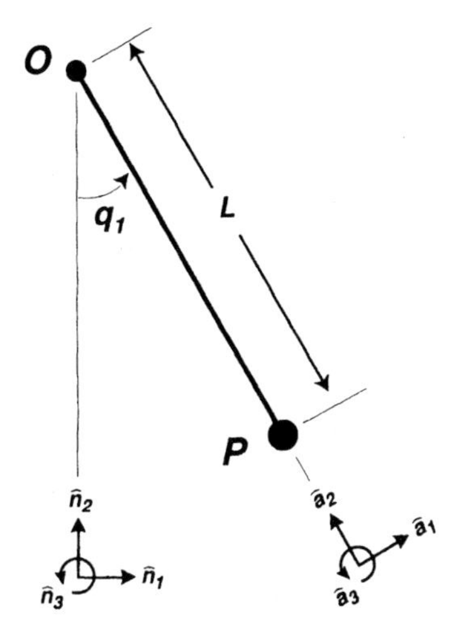

Figure 6.1. An inextensible pendulum of length L.

Exactly what is a partial velocity, or a partial angular velocity? In the inextensible pendulum model shown in Figure 6.1, the center of mass P moves with velocity

$$ {}^{N}\vec{v}^{P} = L\dot{q}_1\hat{a}_1 \, . \tag{6.6} $$

In terms of generalized speeds, this velocity can be rearranged in the form of Equation 6.4,

$$ {}^{N}\vec{v}^{P} = (L\hat{a}_1)\,u_1 \tag{6.7} $$

$$ = {}^{N}\vec{v}_1^{P}\,u_1 \tag{6.8} $$

where the first partial velocity of P in N is simply the multiplier of the first generalized speed u_1,

$$ {}^{N}\vec{v}_1^{P} = L\hat{a}_1 \, . \tag{6.9} $$

In this simple case, the reader should note that the first partial velocity ${}^{N}\vec{v}_1^{P}$ is a *vector quantity*, not a scalar quantity, which is parallel to the actual velocity vector ${}^{N}\vec{v}^{P}$ but has a different magnitude. In a more complex example having n degrees of freedom as in Equation 6.4, one can extend this analogy by thinking of ${}^{N}\vec{v}^{P}$ as the vector sum of n component velocities ${}^{N}\vec{v}_i^{P}\,u_i$, with every component velocity having a partial velocity vector ${}^{N}\vec{v}_i^{P}$ parallel to it, but having different magnitude.

6.2.1 EXAMPLE – PARTIAL VELOCITIES FOR AN ELASTIC PENDULUM

Elasticity will now be added to the string of the pendulum model (Figure 6.2). Let the length L be the sum of the relaxed length L_o and the generalized coordinate q_2,

$$L = L_o + q_2 \,. \tag{6.10}$$

In this case, the velocity of P in N can be found either by inspection, or by differentiating the position vector from O to P,

$$\vec{p}^{OP} = -L\hat{a}_2 \tag{6.11}$$

$$\begin{aligned}
{}^{N}\vec{v}^{P} &= \frac{{}^{N}d}{dt}\left(\vec{p}^{OP}\right) && (6.12)\\
&= \frac{{}^{A}d}{dt}\left(\vec{p}^{OP}\right) + {}^{N}\vec{\omega}^{A} \times \left(\vec{p}^{OP}\right) && (6.13)\\
&= -\dot{L}\hat{a}_2 + \dot{q}_1\hat{a}_3 \times (-L\hat{a}_2) && (6.14)\\
&= -\dot{L}\hat{a}_2 + L\dot{q}_1\hat{a}_1 && (6.15)\\
&= -\dot{q}_2\hat{a}_2 + L\dot{q}_1\hat{a}_1 && (6.16)\\
&= (-\hat{a}_2)\,u_2 + (L\hat{a}_1)\,u_1 \,. && (6.17)
\end{aligned}$$

Therefore, the first and second partial velocities of P in N are

$${}^{N}\vec{v}_1^{P} = L\hat{a}_1 \tag{6.18}$$

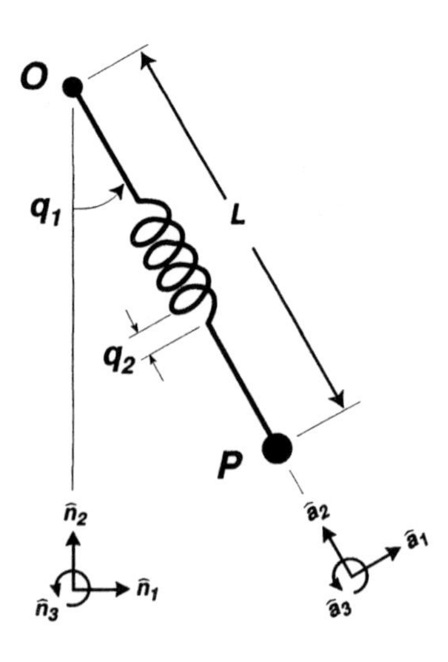

Figure 6.2. An extensible pendulum.

$$^{N}\vec{v}_2^P = -\hat{a}_2 \,. \tag{6.19}$$

A physical interpretation of partial velocities and partial angular velocities is unimportant compared to developing an understanding of their utility. This is best presented by comparing partial velocities to the perturbations used in the classical methods of years past.

The classical methods of deriving dynamic equations of motion often required the definition of a small, imaginary perturbation in the position of a mass. Because the perturbation changed the energy of the system, energy functions incorporating the perturbation were often formed. Then, a series of mathematical operations was performed. Small perturbations multiplied together yielded smaller terms, which could then be ignored in subsequent calculations if they were negligibly small compared to other terms in the equations. When a term could be ignored, and when it could not, was a matter of judgment which could only come about through many years of practice under the mentorship of a master dynamicist. To make matters even more difficult, depending upon the type of system one was analyzing, there were a variety of methodologies one could choose among that would deliver a closed form solution to the dynamic equations of motion. Some methods were easier to implement than others, and most were arduous and impractical when the number of coupled degrees of freedom exceeded three.

The invention of partial velocities and partial angular velocities in Kane's Method eliminated the need to define and use small perturbations. Furthermore, the partial quantities allowed the dynamic equations for simple and complex systems alike to be derived in exactly the same way, every time. So they are the essential keys to Kane's Method!

Example. Let's say that a force $\vec{F}$ acts on a rigid body B at point P, which necessitates the calculation of the velocity $^{N}\vec{v}^P$, for instance,

$$^{N}\vec{v}^P = (u_1 + u_2)\,\hat{n}_1 + u_2\hat{n}_2 \,. \tag{6.20}$$

$^{N}\vec{v}^P$ can be refactored in terms of the generalized speeds into

$$^{N}\vec{v}^P = \hat{n}_1 u_1 + (\hat{n}_1 + \hat{n}_2)\,u_2 \,. \tag{6.21}$$

Then, the first partial velocity of P in N is found from inspection as,

$$^{N}\vec{v}_1^P = \hat{n}_1 \,, \tag{6.22}$$

and the second partial velocity of P in N is:

$$^{N}\vec{v}_2^P = \hat{n}_1 + \hat{n}_2 \,. \tag{6.23}$$

It is easy to see that the ith partial velocity ($i = 1, 2$ in this example) is simply the multiplier of the ith generalized speed u_i.

6.3 GENERALIZED ACTIVE FORCES

Generalized active forces are scalar quantities that embody the contributions of active forces (all forces except inertial forces) to the dynamic equations of motion.

Definition: The n generalized active forces F_r for a system S in reference frame N having n degrees of freedom, ν points at which forces act, and μ rigid bodies upon which torques act are defined as

$$F_r \equiv \sum_{i=1}^{\nu} {}^N\vec{v}_r^{P_i} \cdot \vec{R}_i + \sum_{j=1}^{\mu} {}^N\vec{\omega}_r^{B_j} \cdot \vec{\tau}_j \tag{6.24}$$

for $r = 1, 2, \ldots, n$. $\vec{R}_i$ is the sum of all distance and body forces (except for inertial forces) acting at each point P_i, ($i = 1, 2, \ldots, \nu$) and $\vec{\tau}_j$ is the torque acting on each body B_j ($j = 1, 2, \ldots, \mu$) of the system S. The inertial forces at centers of mass and the inertial torques for rigid bodies are accounted for separately as explained in the following section. Restated, the first summation refers to *all points at which forces act* in the system S, whereas the second summation refers to *all rigid bodies upon which torques act.* Note also that the entire system S is considered. There is no need to subdivide S into a series of isolated component subsystems or a set of free body diagrams.

6.4 GENERALIZED INERTIA FORCES

Generalized inertia forces are scalar quantities that embody the contributions of inertial quantities (inertial forces and torques) to the dynamic equations of motion.

Definition: The n generalized inertia forces F_r for a system S having n degrees of freedom in reference frame N are defined as

$$F_r^* \equiv \sum_{i=1}^{\nu} {}^N\vec{v}_r^{P_i^*} \cdot \vec{R}_i^* + \sum_{j=1}^{\mu} {}^N\vec{\omega}_r^{B_j} \cdot \vec{\tau}_j^* \tag{6.25}$$

for $r = 1, 2, \ldots, n$. $\vec{R}_i^*$ is defined to be the *inertia force* acting at each mass center P_i^* ($i = 1, 2, \ldots, \nu$),

$$\vec{R}_i^* \equiv -m_{P_i^*} \, {}^N\vec{a}^{P_i^*} , \tag{6.26}$$

where $m_{P_i^*}$ is the mass concentrated at point P_i^*, and ${}^N\vec{a}^{P_i^*}$ is the acceleration of P_i^* in reference frame N. $\vec{\tau}_j^*$ is defined to be the *inertia torque* for body B_j in N ($j = 1, 2, \ldots, \mu$),

$$\vec{\tau}_j^* \equiv -I_B \, {}^N\vec{\alpha}^{B_j} . \tag{6.27}$$

I_{B_j} is the moment of inertia of B_j about an axis parallel to its angular acceleration ${}^N\vec{\alpha}^{B_j}$ and passing through the center of mass of body B_j.

In Equation 6.25, the first summation refers to each of the ν *points* at which an inertial force acts in the system S, whereas the second summation refers to each of the μ *rigid bodies* contained in the system S.

In biomechanical modeling, separate "rigid" bodies usually represent each body segment. The inertial characteristics of each rigid body segment are further approximated by a single mass center located at the body's centroid and its moment of inertia. It is assumed that the vector sum of all inertial forces acting on each particle of mass contained within each body segment can be grouped and applied at the center of mass of the body. Therefore, in biomechanical situations meeting this criteria, the number of points having mass equals the number of rigid bodies,

$$\mu = \nu\,, \tag{6.28}$$

so that the indices i and j in Equation 6.25 can be combined. In this case, the F_r^* equation simplifies to

$$F_r^* \equiv \sum_{j=1}^{\nu} \left({}^N\vec{v}_r^{B_j^*} \cdot \vec{R}_j^* + {}^N\vec{\omega}_r^{B_j} \cdot \vec{T}_j^* \right), \tag{6.29}$$

where the mass centers B_j^* for rigid bodies B_j ($j = 1, 2, \ldots, \mu$), has been used to replace the notation for an arbitrary point P_j^* possessing mass.

6.5 DYNAMIC EQUATIONS OF MOTION

For a system S possessing n degrees of freedom, there exist n scalar dynamic equations of motion which describe the accelerations of the system as a function of the generalized coordinates and the generalized speeds. Unless there are configurational or motion constraints, there are n equations because there are n generalized coordinates. They are defined as

$$F_r + F_r^* = 0\,, \tag{6.30}$$

for $r = 1, 2, \ldots, n$. When the equations are rearranged and stacked row upon row, they become,

$$F_1^* = -F_1 \tag{6.31}$$

$$F_2^* = -F_2 \tag{6.32}$$

$$\vdots = \vdots \tag{6.33}$$

$$F_n^* = -F_n\,. \tag{6.34}$$

The left hand sides contain the second derivatives of the generalized coordinates, $\ddot{q}_k$, $k = 1, 2, \ldots, n$. Rewritten with the $\ddot{q}_k$ terms placed into vector $\ddot{\vec{Q}}$, the n dynamic equations of motion can be written in matrix form,

$$M\ddot{\vec{Q}} = \vec{T} + \vec{V} + \vec{G} + \vec{E}\,. \tag{6.35}$$

M is a square ($n \times n$) matrix called the mass matrix, and $\ddot{\vec{Q}}$, $\vec{T}$, $\vec{V}$, $\vec{G}$, and $\vec{E}$ are all "tall" (dimension $n \times 1$) matrices.[2] $\ddot{\vec{Q}}$ contains the accelerations of the n generalized coordinates defining the configuration of the system. $\vec{T}$, $\vec{V}$, $\vec{G}$, and $\vec{E}$ contain torques, moments, and force terms which arise from a number of different forces. If the degrees of freedom are all revolute, then the $\vec{T}$, $\vec{V}$, $\vec{G}$, and $\vec{E}$ terms are the torques and moments arising from muscular and ligament forces acting across the joints, centrifugal forces, gravitational forces, and external forces acting on the system, respectively. If not all of the degrees of freedom are revolute, then the $\vec{T}$, $\vec{V}$, $\vec{G}$, and $\vec{E}$ terms will also contain forces relating to the linear accelerations of the masses which are tracked by the non-revolute degrees of freedom.

Although the mass matrix M and the vectors $\vec{T}$, $\vec{V}$, $\vec{G}$, and $\vec{E}$ are all functions of the current time t (and may also be functions of the generalized coordinates and/or generalized speeds), the notation "(t)" is usually dispensed with in the interest of simplicity.

6.5.1 EXAMPLE – EQUATIONS OF MOTION FOR A BALL MOVING DOWN AN INCLINED PLANE

Although this example is mechanically oriented, it illustrates the process of obtaining dynamic equations and shows how different contact conditions are reflected in the final result. A ball B is moving down an inclined plane along the line of steepest descent (Figure 6.3A). An enlarged view of points $\bar{A}$ and $\bar{B}$ in contact at time t is shown in Figure 6.3B. Point $\bar{B}$ is fixed in body B, and point $\bar{A}$ is fixed in body A.

If B is *not* rolling, then ${}^{N}\vec{v}^{\bar{B}} \neq {}^{N}\vec{v}^{\bar{A}}$, q_1 and q_2 are unrelated, and a frictional force $\vec{f}$ tangent to the surface at $\bar{B}$ acts on the ball at $\bar{B}$. If $\vec{\mathcal{N}} = \mathcal{N}\hat{a}_2$ is the normal force acting at $\bar{B}$, then,

$$\vec{f} = \mu\mathcal{N}\hat{a}_1\,, \tag{6.36}$$

where μ is the coefficient of kinetic friction between A and B.

In total, there are normal and frictional force components acting at $\bar{B}$, a gravitational force $m_B\vec{g} = -m_B g\hat{n}_2$, and an inertial force $-m_B\,{}^{N}\vec{a}^{B^*}$ acting

[2] In the context of talking about matrices, the tall $n \times 1$ matrices (or their transposes) are usually referred to as "vectors", because they are convenient forms for numerically expressing the measure numbers of a vector quantity. However, they are actually composed entirely of scalar values and must be associated with directional components to become actual vectors.

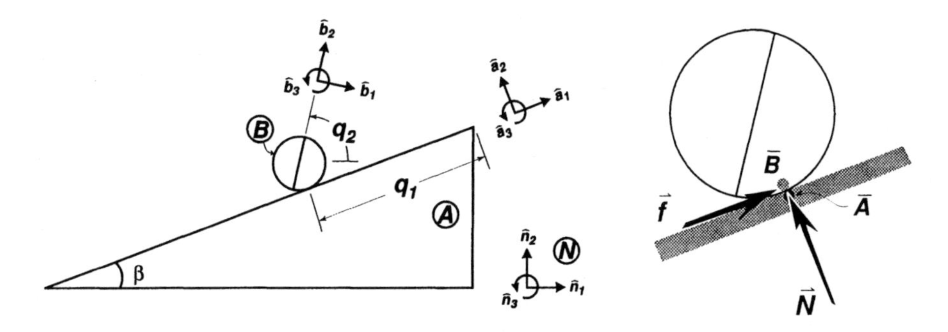

Figure 6.3. A ball (B) moves down an inclined plane (A), as shown in a cross-sectional view (left figure). Reference frame A is fixed at an incline (angle β) with respect to an inertial reference frame N. The ball is considered to be rigid and spherical, and has radius R, mass m_B, and central moment of inertia I_B^*. When released at rest from an arbitrary starting point, gravitational force $-m_B g\hat{n}_2$ causes the ball to move down the plane with some slip at the interface. Generalized coordinate q_1 is the distance between the starting and current contact points, and q_2 is the angular deviation. On the right, an enlarged view of the points of the ball $\bar{B}$ and plane $\bar{A}$ at the instant under consideration. $\bar{B}$ and $\bar{A}$ are points fixed in bodies B and A, respectively, that happen to be in instantaneous contact. A force with normal ($\vec{N}$) and frictional ($\vec{f}$) components acts on body B.

at $\vec{B}^*$. There are no applied torques, but there is an inertial torque $-I_B^* \, {}^N\vec{\alpha}^B$ which accounts for the body's rotational inertia.[3]

There are two degrees of freedom and therefore there will be two dynamic equations. The derivation proceeds below, as briefly as possible.

Angular velocities and accelerations:

$$
\begin{aligned}
{}^N\vec{\omega}^B &= \dot{q}_2\hat{a}_3 && (6.37)\\
&= (0)u_1 + (\hat{a}_3)u_2\,, && (6.38)
\end{aligned}
$$

where the generalized speeds are defined as $u_1 \equiv \dot{q}_1$ and $u_2 \equiv \dot{q}_2$.

$$
\begin{aligned}
{}^N\vec{\alpha}^B &= \frac{{}^N d}{dt}\left({}^N\vec{\omega}^B\right) && (6.39)\\
&= \ddot{q}_2\hat{n}_3 && (6.40)\\
&= \ddot{q}_2\hat{a}_3\,. && (6.41)
\end{aligned}
$$

Partial Angular velocities:

$$
{}^N\vec{\omega}_1^B = 0 \qquad (6.42)
$$

[3]The inertial torque of form $-I\vec{\alpha}$ is analogous to an inertial force of form $-m\vec{a}$ in the linear realm. It can be defined by rearranging the dynamic equation $\sum\vec{\tau} = I\vec{\alpha}$ as $\sum\vec{\tau} - I\vec{\alpha} = 0$, and defining $\vec{\tau}^* = -I\vec{\alpha}$.

$$ {}^N\vec{\omega}_2^B = \hat{a}_3 \tag{6.43} $$

Velocities and accelerations:

$$ \begin{aligned} {}^N\vec{v}^{B^*} &= -\dot{q}_1\hat{a}_1 & (6.44)\\ &= (-\hat{a}_1)u_1 + (0)u_2\,. & (6.45) \end{aligned} $$

$$ \begin{aligned} {}^N\vec{v}^{\bar{B}} &= {}^N\vec{v}^{B^*} + {}^N\vec{\omega}^B \times \vec{p}^{B^*\bar{B}} & (6.46)\\ &= -\dot{q}_1\hat{a}_1 + \dot{q}_2\hat{a}_3 \times R(-\hat{a}_2) & (6.47)\\ &= -\dot{q}_1\hat{a}_1 + R\dot{q}_2\hat{a}_1 & (6.48)\\ &= (-\hat{a}_1)u_1 + (R\hat{a}_1)u_2\,. & (6.49) \end{aligned} $$

Partial velocities:

$$ \begin{aligned} {}^N\vec{v}_1^{B^*} &= -\hat{a}_1 & (6.50)\\ {}^N\vec{v}_2^{B^*} &= 0 & (6.51)\\ {}^N\vec{v}_1^{\bar{B}} &= -\hat{a}_1 & (6.52)\\ {}^N\vec{v}_2^{\bar{B}} &= R\hat{a}_1 & (6.53) \end{aligned} $$

Generalized active forces:

Equation 6.24 is used to formulate these. First determine the number of points through which contributing forces act ($\nu = 2$) and next the number of bodies on which contributing torques are in evidence ($\mu = 1$). Then, for $r = 1$,

$$ \begin{aligned} F_1 &= {}^N\vec{v}_1^{B^*} \cdot m_B\vec{g} + {}^N\vec{v}_1^{\bar{B}} \cdot \left(\vec{\mathcal{N}} + \vec{f}\right) + {}^N\vec{\omega}_1^B \cdot (0) & (6.54)\\ &= -\hat{a}_1 \cdot m_B g(-\hat{n}_2) \;-\hat{a}_1 \cdot (\mathcal{N}\hat{a}_2 + \mu\mathcal{N}\hat{a}_1) + 0 & (6.55)\\ &= m_B g \sin\beta - \mu\mathcal{N} & (6.56)\\ &= m_B g \sin\beta - \mu\,(m_B g \cos\beta) & (6.57)\\ &= m_B g\,(\sin\beta - \mu\cos\beta)\;. & (6.58) \end{aligned} $$

Note that the normal force component $\mathcal{N}\hat{a}_2$ is eliminated because it is perpendicular to ${}^N\vec{v}_1^{\bar{B}} = -\hat{a}_1$. This means that the normal force will not contribute to the first dynamic equation of motion. For $r = 2$, we get the contributions of active forces and torques to the second dynamic equation of motion,

$$ \begin{aligned} F_2 &= {}^N\vec{v}_2^{B^*} \cdot m_B\vec{g} + {}^N\vec{v}_2^{\bar{B}} \cdot \left(\vec{\mathcal{N}} + \vec{f}\right) + {}^N\vec{\omega}_2^B \cdot (0) & (6.59)\\ &= 0 + R\hat{a}_1 \cdot (\mathcal{N}\hat{a}_2 + \mu\mathcal{N}\hat{a}_1) + 0 & (6.60)\\ &= R\mu\mathcal{N} & (6.61)\\ &= R\mu m_B g \cos\beta\,. & (6.62) \end{aligned} $$

Generalized inertia forces:

Equation 6.25 is used to formulate these. In this case, ν is the number of points at which inertial forces act ($\nu = 1$) and μ is the number of bodies having rotational inertias ($\mu = 1$). Then, for $r = 1$,

$$\begin{aligned} F_1^* &= {}^N\vec{v}_1^{B^*} \cdot \left(-m_B\, {}^N\vec{a}^{B^*}\right) + {}^N\vec{\omega}_1^B \cdot \left(-I_B\, {}^N\vec{\alpha}^B\right) && (6.63) \\ &= -\hat{a}_1 \cdot -m_B\left(-\ddot{q}_1\hat{a}_1\right) + 0 \cdot \left(-I_B\, {}^N\vec{\alpha}^B\right) && (6.64) \\ &= -m_B\ddot{q}_1 \,. && (6.65) \end{aligned}$$

For $r = 2$, we get the contributions of inertial forces and torques to the second dynamic equation of motion,

$$\begin{aligned} F_2^* &= {}^N\vec{v}_2^{B^*} \cdot \left(-m_B\, {}^N\vec{a}^{B^*}\right) + {}^N\vec{\omega}_2^B \cdot \left(-I_B\, {}^N\vec{\alpha}^B\right) && (6.66) \\ &= 0 + \hat{a}_3 \cdot \left(-I_B\ddot{q}_2\hat{a}_3\right) && (6.67) \\ &= -I_B\ddot{q}_2 \,. && (6.68) \end{aligned}$$

Dynamic equations of motion:

The first equation, for $r = 1$ is,

$$\begin{aligned} F_1 + F_1^* &= 0 && (6.69) \\ m_B g\left(\sin\beta - \mu\cos\beta\right) - m_B\ddot{q}_1 &= 0 && (6.70) \\ m_B\ddot{q}_1 &= m_B g\left(\sin\beta - \mu\cos\beta\right) \,. && (6.71) \end{aligned}$$

The acceleration terms are written on the left hand side of the equation to facilitate placing the dynamic equations into matrix form. Note that the term m_B is *not* factored out of the equation or the dynamic equation will look peculiar. The second equation is, setting $r = 2$,

$$\begin{aligned} F_2 + F_2^* &= 0 && (6.72) \\ R\mu m_B g\cos\beta + \left(-I_B\ddot{q}_2\right) &= 0 && (6.73) \\ I_B\ddot{q}_2 &= m_B g\left(R\mu\cos\beta\right) \,. && (6.74) \end{aligned}$$

Dynamic equations in matrix form:

$$\begin{bmatrix} m_B & 0 \\ 0 & I_B \end{bmatrix} \begin{bmatrix} \ddot{q}_1 \\ \ddot{q}_2 \end{bmatrix} = m_B g \begin{bmatrix} \sin\beta - \mu\cos\beta \\ R\mu cos\beta \end{bmatrix} \qquad (6.75)$$

Before we leave this example, it is instructive to consider the differences if ball B was instead considered to be rolling down the inclined plane. Following the definition of rolling from Section 5.4.1, then ${}^N\vec{v}^{\bar{B}} = {}^N\vec{v}^{\bar{A}}$. Because the surface of the plane is fixed in reference frame N, and point $\bar{A}$ is a point fixed in A, ${}^N\vec{v}^{\bar{A}} = 0$. Therefore, ${}^N\vec{v}^{\bar{B}} = 0$ and *the normal and frictional forces acting* *on $\bar{B}$ will have no contribution to the dynamic equation of motion.* Secondly, q_1 and q_2 would be related by a configurational constraint equation,

$$q_1 = Rq_2 \,. \tag{6.76}$$

Hence, there would be only one dynamic equation of motion.

6.6 FURTHER DESCRIPTION OF THE MATRICES WHICH FORM THE DYNAMIC EQUATIONS OF MOTION

6.6.1 MASS MATRIX: M

This matrix of dimension $(n \times n)$ is called the *mass matrix* or *inertia matrix* because it contains the mass moments of inertia (I_k), products of masses and distances squared $(m_k r_k^2)$, and instantaneous values of the direction cosines (products of $sin(q_i)$ and $cos(q_i)$). The mass matrix contains information regarding the instantaneous mass distribution of the system. In its most basic form, M is symmetric and positive semi-definite so M^{-1} always exists. The rows of M may appear as linear combinations of the rows from the symmetric form, and thus M might not appear to be symmetric. The individual terms of this matrix are functions of the generalized coordinates, not their first or second time derivatives.

6.6.2 ACCELERATION VECTOR: $\vec{\ddot{Q}}$

The $\vec{\ddot{Q}}$ vector of dimension $(n \times 1)$ contains the second time derivatives of all n generalized coordinates of the system. In a forward dynamic simulation, these are unknown values that must be obtained, usually by numerical solution.

6.6.3 VECTOR OF MOMENTS FROM CENTRIFUGAL FORCES: $\vec{V}$

Vector $\vec{V}$ has dimension $(n \times 1)$. The terms of the vector contain products of angular speeds ($\dot{q}_i^2$ and $\dot{q}_i \dot{q}_j$; $i = 1, 2, \ldots, n$ and $j = 1, 2, \ldots, n$). When the degrees of freedom are rotational, the terms of this vector represent the moments of centrifugal forces.

6.6.4 VECTOR OF MOMENTS FROM GRAVITATIONAL FORCES: $\vec{G}$

This vector has dimension $(n \times 1)$. The terms of this matrix contains the moments of gravitational forces exerted on each body. Because the gravitational moments change as the configuration of the system changes, the terms of this matrix are functions of the generalized coordinates (q_i terms; $i = 1, 2, \ldots, n$) and the gravitational acceleration constant g.

6.6.5 VECTOR OF MOMENTS FROM EXTERNAL FORCES AND TORQUES: $\vec{E}$

Vector $\vec{E}$ has dimension $(n \times 1)$. These contain the effects of external forces on the system. For instance, if forces are applied at the endpoint of a linkage, the forces propagate from segment to segment, and create moments on each link. This usually means a single endpoint force will accelerate every body segment. However, if the dynamic equations of motion have been created by Kane's Method, only the forces applied at points with nonzero velocity and the torques applied to bodies having non-zero angular velocity will appear.[4]

6.6.6 VECTOR OF APPLIED TORQUES: $\vec{T}$

Dimension $(n \times 1)$. This vector contains the torques and moments arising from muscular activations and passive elastic structures surrounding the joints,

$$\vec{T} = [T_1 \;\; T_2 \;\; T_3 \;\; \cdots \;\; T_n]^T \,, \tag{6.77}$$

where T_i simply stands for the "i^{th}" scalar element contained within the vector $\vec{T}$. If the n degrees of freedom q_i $(i = 1, 2, 3, \ldots, n)$ are defined as segmental angles (with respect to a fixed inertial reference such as a horizontal plane), then the T_i terms will be segmental torques. If, however, the degrees of freedom are defined to be joint angles, then the T_i terms will be in the form of joint torques. It is important to make this distinction because the moments applied by muscular activations and passive elastic structures are usually supplied in the form of joint torques. *If segmental angles and segmental torques are used instead of joint torques, then the joint moments provided by the muscles and passive elastic structures must be converted into segmental torques before their contributions can be used in the dynamic equations of motion.* This is illustrated in Section 6.7.

If the dynamic equations are written in the form of joint or segmental torques, the contributions of the active and passive structures to each element of the torque vector may be simply added together to obtain the total torque. For degree of freedom i,

$$T_i = T_{i,passive} + T_{i,active} \,. \tag{6.78}$$

More specifically, if the dynamic equations are written in the form of *joint torques,* this equation takes the form,

$$T_i = M_{i,pass}(\theta_i, \dot{\theta}_i) + \sum_{j=1}^{m} M_{j,act}(\theta_i, \dot{\theta}_i) \,. \tag{6.79}$$

[4]Some forces do not appear in the dynamic equations because they do not create accelerations. See Section 6.8 for a more specific discussion of this topic.

$M_{i,pass}(\theta_i, \dot{\theta}_i)$ is the contribution to the joint torque T_i produced by the passive joint moments according to Equation 5.38, and $M_{j,act}(\theta_i, \dot{\theta}_i)$ is the muscle moment produced by muscle j which contributes to T_i. In this case, m is the number of muscles spanning joint angle θ_i. The passive elastic contribution and the muscle moments are functions of both the joint angle θ_i and joint angular velocity $\dot{\theta}_i$ because of damping in both the joint and the muscular structures.

If, however, the dynamic equations are written in the form of *segment torques*, this equation takes the form,

$$T_i = \sum_{i=1}^{n} \left\{ M_{i,pass}(\theta_i, \dot{\theta}_i) + \sum_{j=1}^{m} M_{j,act}(\theta_i, \dot{\theta}_i) \right\} . \qquad (6.80)$$

In the segmental case, the joint angle θ_i and its time derivative will be obtained from the segmental angles. For instance, a common formulation is $\theta_i = q_i - q_{i-1}$ and $\dot{\theta}_i = \dot{q}_i - \dot{q}_{i-1}$. The equation is the same as Equation 6.79 except that the contributions of all n joints bordering segment i sum together to obtain the total segmental torque T_i.

6.7 EXAMPLE – SEGMENTAL TORQUE CONTRIBUTIONS FOR UNIARTICULAR, BIARTICULAR, AND MULTIARTICULAR MUSCLES

Let body segments A, B, C, and D form a kinematic chain with joints at points B_o, C_o, and D_o as shown in Figure 6.4. Segmental angles q_1 to q_4 define angles of these segment with respect to a horizontal line defined in the N reference frame. For simplicity, only these four bodies will be shown. However, the discussion is applicable to any four consecutive segments of a larger kinematic chain. Segmental torques will be denoted with single subscripts, *e.g.*, $\vec{\tau}_B$ for the torque on segment B, while joint torques will be denoted with dual subscripts separated by a slash, *e.g.*, $\vec{\tau}_{A/B}$ for the torque exerted *by* body A *on* body B.

6.7.1 CASE I – UNIARTICULAR MUSCLE

A uniarticular muscle spans a single joint, as shown in Figure 6.4A. Whether the muscle follows a straight path from its origin O to its insertion I as shown, or goes through a series of via points, the moment exerted by a uniarticular muscle on segments A and B about joint B_o will always be equal in magnitude and opposite in direction.[5] This is because a straight path can always be defined between the first contact points encountered on bodies A and B by the spanning

[5]See Section 5.4.2 for a discussion on the moment produced by a spanning force.

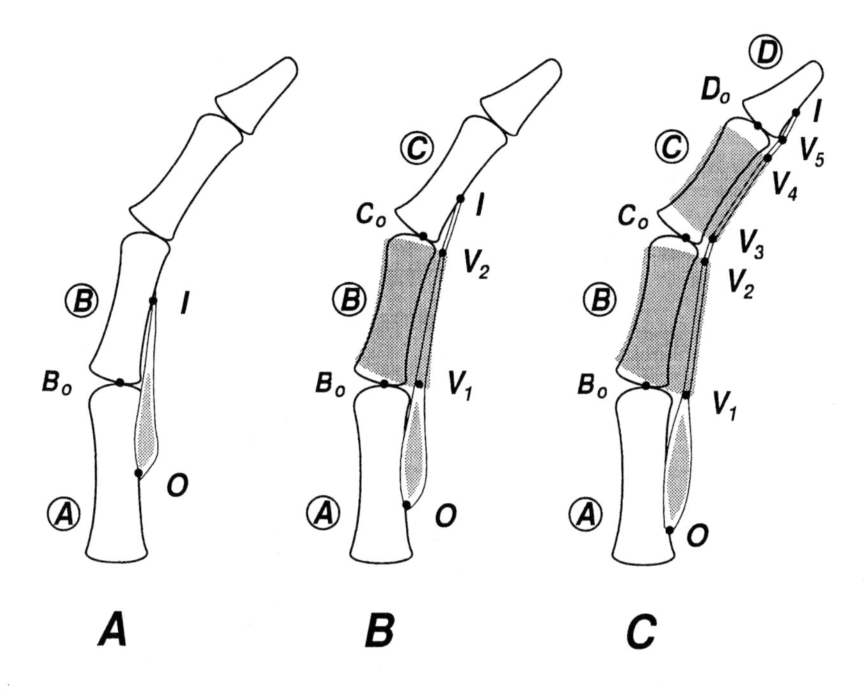

Figure 6.4. A. A uniarticular muscle spans a single joint, and exerts equal and opposite moments on two adjacent segments. B. A biarticular muscle spans two joints. The moments exerted across each joint are equal and opposite, so the muscle moment created on the spanned segment is equal and opposite to the sum of the moments exerted on the segments to which the muscle attaches. C. While less common, a triarticular muscle spans three joints. Again, the moments exerted across the outer joints are equal and opposite to the moments exerted by the muscles on the segments to which they attach. Via points V_i are shown, as muscles spanning multiple joints tend to have sheaths (shaded areas) which redirect the tendon pathways and prevent "bowstringing." The closest two via points to each joint contact location are used to compute the local joint moment.

element.[6] In this case, the joint moment and the segmental moments are equal in magnitude,

$$\vec{\tau}_{A/B} = \vec{\tau}_B = \vec{p}^{B_oI} \times F\frac{\vec{p}^{IO}}{|\vec{p}^{IO}|} \tag{6.81}$$

$$\vec{\tau}_{B/A} = \vec{\tau}_A = \vec{p}^{B_oO} \times F\frac{\vec{p}^{OI}}{|\vec{p}^{OI}|} \tag{6.82}$$

where F is the magnitude of the muscle force. Since $\vec{p}^{IO} = -\vec{p}^{OI}$,

$$\vec{\tau}_{B/A} = -\vec{\tau}_{A/B} \tag{6.83}$$

[6]See the discussion on effective origins and insertions in Section 5.4.3.

and the contributions of the joint torques to the segmental torque vector are

$$\vec{T}' = \begin{bmatrix} -\vec{\tau}_{A/B} & \vec{\tau}_{A/B} & 0 & 0 \end{bmatrix}^T . \tag{6.84}$$

6.7.2 CASE II – BIARTICULAR MUSCLE

A biarticular muscle spans two adjacent joints, as shown in Figure 6.4B. In this case, the moments applied on segments A and C are computed in a fashion similar to that described above. One end of a biarticular muscle usually exhibits some wrapping around the convex side of a joint, or on the concave side a tendon sheath often changes the tendon path to prevent "bowstringing," as shown in the figure. The calculation is modified to use "effective" origin and insertion points EO and EI instead of the actual origin and insertion.

$$\vec{\tau}_A = \vec{p}^{\,B_oO} \times F \frac{\vec{p}^{\,O\,V_1}}{|\vec{p}^{\,O\,V_1}|} \tag{6.85}$$

$$\vec{\tau}_C = \vec{p}^{\,C_oI} \times F \frac{\vec{p}^{\,I\,V_2}}{|\vec{p}^{\,I\,V_2}|} \tag{6.86}$$

Two uniarticular muscles spanning the adjacent joints can be made functionally equivalent to a biarticular muscle. Therefore, the joint moments exerted about joints B_o and C_o will equal the segmental moments on segments A and C, respectively, *as if the joint moments were exerted by uniarticular muscles emanating from segment* B,

$$\vec{\tau}_{B/A} = \vec{\tau}_A \tag{6.87}$$

$$\vec{\tau}_{B/C} = \vec{\tau}_C . \tag{6.88}$$

The contributions to the segmental torque on body B include the joint moments exerted by bodies A and C on body B,

$$\vec{\tau}_B = \vec{\tau}_{A/B} + \vec{\tau}_{C/B} \tag{6.89}$$

$$= -\vec{\tau}_{B/A} - \vec{\tau}_{B/C} \tag{6.90}$$

$$= -\left(\vec{\tau}_A + \vec{\tau}_C\right) . \tag{6.91}$$

In this case, the contributions of the joint torques to the segmental torque vector are

$$\vec{T}' = \begin{bmatrix} -\vec{\tau}_{A/B} & \left(\vec{\tau}_{A/B} - \vec{\tau}_{B/C}\right) & \vec{\tau}_{B/C} & 0 \end{bmatrix}^T \tag{6.92}$$

$$= \begin{bmatrix} \vec{\tau}_A & -\left(\vec{\tau}_A + \vec{\tau}_C\right) & \vec{\tau}_C & 0 \end{bmatrix}^T . \tag{6.93}$$

6.7.3 CASE III – MULTIARTICULAR MUSCLE

A multiarticular muscle spans more than two adjacent joints, as shown in Figure 6.4C. For instance, muscles such as the *biceps* can be considered to be *tri*articular if the humeroulnar joint and the radioulnar joint of the elbow are considered separately. The moments applied on segments A and D are computed in a fashion similar to that described above in the biarticular case, except that separate sets of effective origins and insertions are needed across joint B_o and C_o. Rather than adding subscripts to the EO and EI designations, it is simpler to indicate junctions between straight line segments of the tendon path by V_1, V_2, *etc.*, where the V_i's stand for "via points."

The closest two via points and/or the muscle origin and insertion are used to compute the local joint moments. For instance,

$$\vec{\tau}_A = \vec{p}^{B_oO} \times F\frac{\vec{p}^{OV_1}}{|\vec{p}^{OV_1}|} \tag{6.94}$$

$$\vec{\tau}_D = \vec{p}^{D_oV_5} \times F\frac{\vec{p}^{V_5V_4}}{|\vec{p}^{V_5V_4}|}. \tag{6.95}$$

Another difference between the biarticular case and the triarticular case is that the equal and opposite moment contributions exerted across joints B_o and D_o are applied to segments B and C instead of a single segment.

In this case, the contributions of the joint torques to the segmental torque vector are

$$\vec{T}' = \begin{bmatrix} -\vec{\tau}_{A/B} & \vec{\tau}_{A/B} & -\vec{\tau}_{C/D} & \vec{\tau}_{C/D} \end{bmatrix}^T. \tag{6.96}$$

6.8 CONTRIBUTING AND NONCONTRIBUTING FORCES AND TORQUES

By now, the reader should realize that Kane's Method offers a distinct advantage over other classical dynamical methods. In the formation of the generalized active forces, as described in Section 6.3, the resultant (vector) force at each point is dotted with a partial velocity vector to form a scalar quantity. Also, torques acting on rigid bodies are dotted with partial angular velocities to form additional quantities. Any such quantities that are nonzero contribute directly to the dynamic equations of motion via Equation 6.24. By the same token, some dot products of forces and partial velocity vectors yield a null result, and consequently do not appear in the dynamic equations of motion. Thus, the process of forming the generalized active forces simplifies the dynamic equations before the components making up each equation are formulated! We can take advantage of this knowledge to reduce the amount of work necessary in formulating the dynamic equations, because forces and torques that we anticipate as noncontributory can be neglected.

In general, a force $\vec{F}$ will contribute to the dynamic equations of motion if there are velocity components at the point of force application parallel to $\vec{F}$.

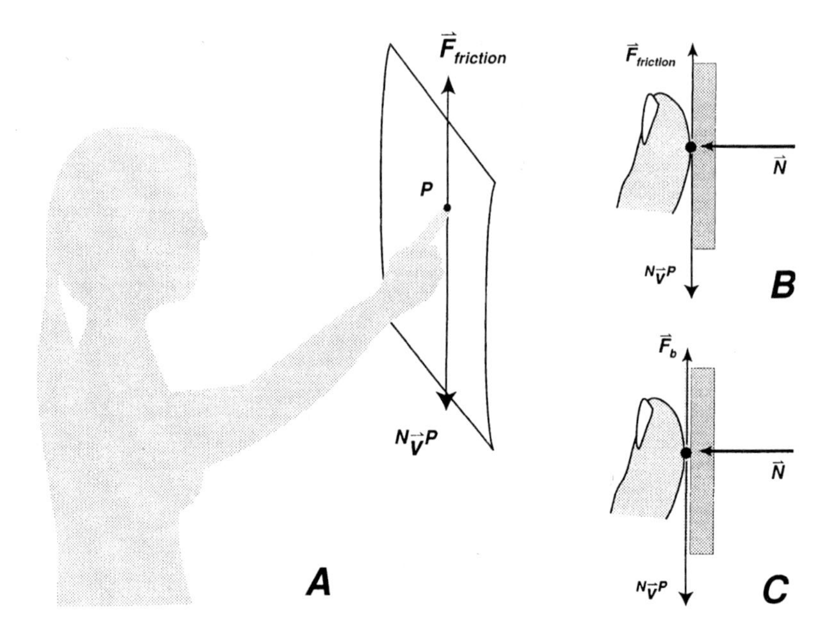

Figure 6.5. A. The frictional force caused by dragging a finger across a computer screen will appear in the dynamic equations of motion, because the velocity of the fingertip at the point of contact P is parallel to the direction of the shear force. B. Other than creating the constraint surface, the normal force component $\vec{N}$ does not contribute, as it is perpendicular to the velocity of the fingertip. C. A fluid lubricant inserted between the fingertip and the computer screen creates a velocity dependent force $\vec{F}_b$ which is contributory.

That is, if it is possible for movements or deflections to occur as a result of that force, then that force will have influence on the positions, velocities, and accelerations of the system and is considered to be contributory. In an angular sense, the same holds true when a rigid body has angular velocity components that are parallel to the resultant torque. When in doubt as to whether a force or torque will contribute to the dynamical equations or not, it should be treated as if it does contribute, and one should include it in the formulation of the generalized active forces. Proper application of Kane's Methods will automatically add or negate any contributions.

For a dynamical system S that is well defined by a system boundary, some examples of forces that contribute, and that do not contribute to the dynamic equations of motion follow.

6.8.1 CONTRIBUTING FORCES

Frictional forces from external sources. In general, anytime there is energy dissipation or energy storage, a contribution to the dynamical equations will result. This is because when friction is present, a force component in the

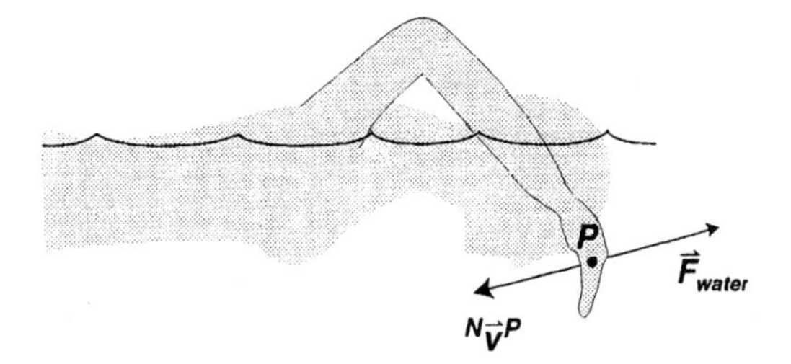

Figure 6.6. A viscous fluid creates a resistive force in a direction opposite to the velocity of the surface or object within the fluid. There is always a contribution to the dynamic equations because the force and the velocity are parallel.

direction of relative motion between the contacting bodies is present, called the shear force. Because the shear force and the velocity of the body creating it are parallel, the dot product of the shear force with the partial velocity will yield a contribution (Figure 6.5). In the case of a viscous fluid interface between adjacent bodies, or a damping element, a velocity dependent force is present that acts parallel to the relative motion, and hence a contribution to the dynamic equations will result (Figure 6.6).

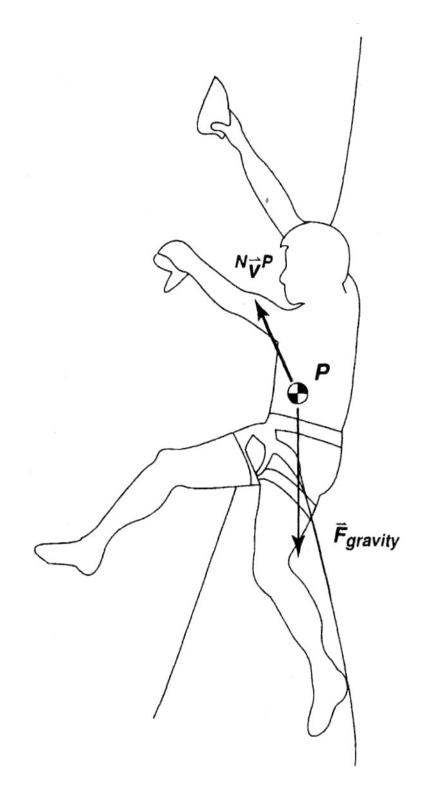

Figure 6.7. A rock climber moving his center of mass upward against the downward pull of gravity. Because the velocity of the center of mass has a component in the vertical direction, the dot product of the gravitational force and the partial velocity vector will yield a contribution.

Body forces (e.g. gravitation). Unless the system S is entirely constrained to horizontal motions, gravitational forces dotted with partial velocity vectors will yield nonzero quantities. Hence, contributions to the dynamical equations will result (Figure 6.7).

External forces and torques. Forces and torques applied across the system boundary on specific points of rigid bodies in S generally will yield contributions, because externally applied forces generally cause motions, or some motion component, to occur in the same direction as the force or torque.

6.8.2 NONCONTRIBUTING FORCES

Internal forces within rigid bodies. The total contribution of all contact and distance forces exerted by all particles of rigid body B of system S on each other is zero. The proof is adapted from Kane and Levinson (1985).

Proof. Body B of system S has two points P and Q representing particles of body B that exert distance and/or contact forces on each other. Point Q is located from point P by position vector $\vec{p}^{PQ}$ as shown in Figure 6.8. The force exerted by point P on point Q is $\vec{F}_{P/Q}$ and the force exerted by point Q on P is $\vec{F}_{Q/P}$. Newton's third law in the "strong" form states that the forces $\vec{F}_{P/Q}$ and $\vec{F}_{Q/P}$ are not only equal in magnitude and opposite in direction, but are also directed along the line joining the two points P and Q. Therefore, vectors $\vec{F}_{P/Q}$, $\vec{F}_{Q/P}$, and $\vec{p}^{PQ}$ are parallel to each other. The contribution of $\vec{F}_{P/Q}$ and $\vec{F}_{Q/P}$ to the generalized active forces is

$$F_r^{'} = \left({}^{N}\vec{v}_r^{P} \cdot \vec{F}_{Q/P}\right) + \left({}^{N}\vec{v}_r^{Q} \cdot \vec{F}_{P/Q}\right) \tag{6.97}$$

$$= \left({}^{N}\vec{v}_r^{P} - {}^{N}\vec{v}_r^{Q}\right) \cdot \vec{F}_{Q/P} \,. \tag{6.98}$$

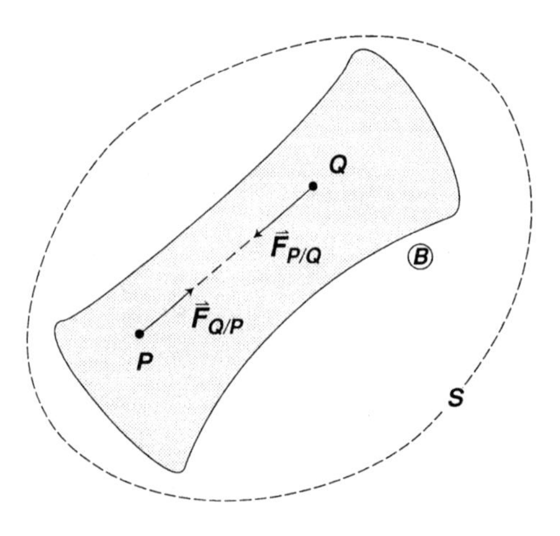

Figure 6.8. Contact and distance forces exerted by particles within rigid bodies of the system S are noncontributing. Here, points P and Q exert a small gravitational attraction on one another.

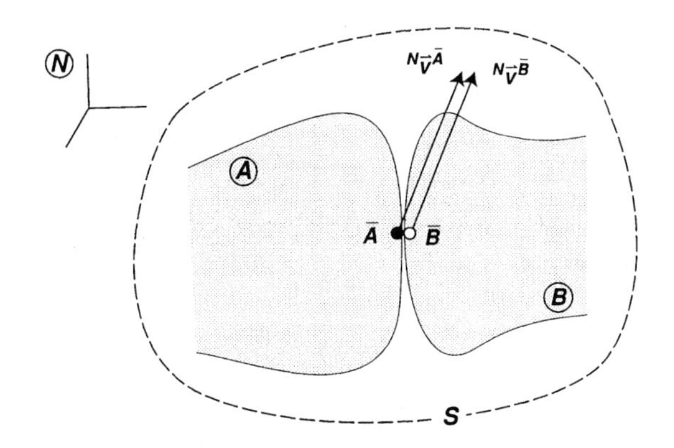

Figure 6.9. Contact forces exerted by rigid bodies of the system S are noncontributing when they are exerted across points of rolling contact. Rolling occurs when the two points of bodies A and B in instantaneous contact have a common velocity, ${}^N\vec{v}^{\bar{A}} = {}^N\vec{v}^{\bar{B}}$.

Next, Equation 4.125 relating the velocities of two points in a rigid body can be extended to partial velocities, because the relationship between the velocities ${}^N\vec{v}^Q$ and ${}^N\vec{v}^P$ can be rewritten in terms of the partial velocities and three scalar functions of time, $f(t)$, $g(t)$, and $h(t)$.

$$ {}^N\vec{v}^Q = {}^N\vec{v}^P + \left({}^N\vec{\omega}^B \times \vec{p}^{PQ}\right) \tag{6.99} $$

$$ \begin{aligned} \sum_{i=1}^{n} {}^N\vec{v}_i^Q u_i + f(t) &= \sum_{i=1}^{n} {}^N\vec{v}_i^P u_i + g(t) \\ &\quad + \left(\left\{\sum_{j=1}^{n} {}^N\vec{\omega}^B u_i + h(t)\right\} \times \vec{p}^{PQ}\right) \end{aligned} \tag{6.100} $$

Therefore, the n equations relating the partial velocities of Q and P can be extracted and rearranged,

$$ {}^N\vec{v}_r^Q = {}^N\vec{v}_r^P + \left({}^N\vec{\omega}_r^B \times \vec{p}^{PQ}\right), \tag{6.101} $$

$$ {}^N\vec{v}_r^P - {}^N\vec{v}_r^Q = -\left({}^N\vec{\omega}_r^B \times \vec{p}^{PQ}\right). \tag{6.102} $$

Equation 6.102 allows Equation 6.98 to be expressed in a different form, and solved,

$$ F_r' = -\left({}^N\vec{\omega}_r^B \times \vec{p}^{PQ}\right) \cdot \vec{F}_{Q/P} = 0, \tag{6.103} $$

because the quantity ${}^N\vec{\omega}_r^B \times \vec{p}^{PQ}$ is perpendicular to $\vec{p}^{PQ}$ and hence is also perpendicular to $\vec{F}_{Q/P}$.

Contact forces exerted within the system S. As a general rule, *all contact forces (including frictional forces)* exerted by a rigid body A on another rigid body B, when A *rolls without slipping* on B and both bodies are elements of system S, are noncontributing.

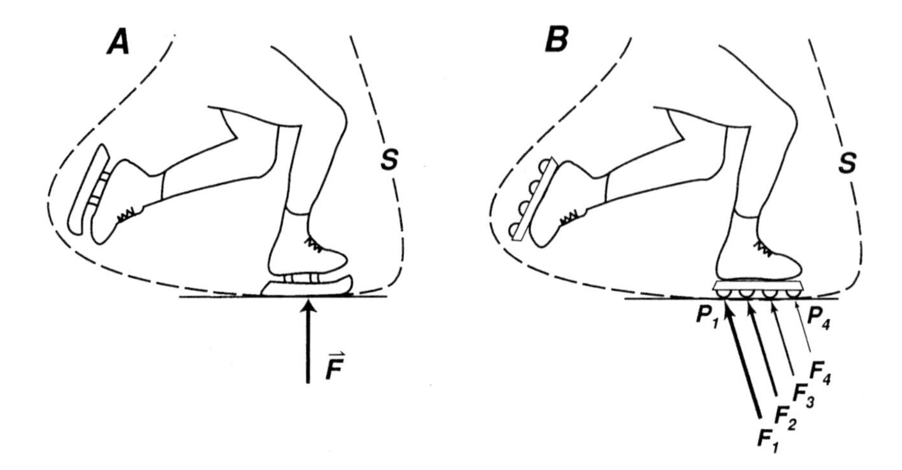

Figure 6.10. Contact forces exerted across the system boundary on rigid bodies of the system S are noncontributing when they are exerted across points having (A) frictionless or (B) rolling contact.

Proof. A definition of *pure rolling* is that there is *no slip.* Kinematically, this means that the points of contact, depicted as point $\bar{A}$ of body A and point $\bar{B}$ of body B in Figure 6.9, have a common velocity. In equation form, this becomes

$$ {}^{N}\vec{v}^{\bar{A}} = {}^{N}\vec{v}^{\bar{B}} \,, \tag{6.104}$$

with similar expressions for the partial velocities ($r = 1,\ 2,\ \ldots$),

$$ {}^{N}\vec{v}_r^{\bar{A}} = {}^{N}\vec{v}_r^{\bar{B}} \,. \tag{6.105}$$

From Newton's third law, there are equal and opposite reaction forces $\vec{R}_{\bar{A}}$ and $\vec{R}_{\bar{B}}$ acting on bodies A and B, respectively, across the contacting surfaces,

$$ \vec{R}_{\bar{A}} = -\vec{R}_{\bar{B}} \,. \tag{6.106}$$

Hence the contribution to the generalized active forces for the case of rolling contacts is

$$ {}^{N}\vec{v}_r^{\bar{A}} \cdot \vec{R}_{\bar{A}} + {}^{N}\vec{v}_r^{\bar{B}} \cdot \vec{R}_{\bar{B}} = {}^{N}\vec{v}_r^{\bar{A}} \cdot \left(\vec{R}_{\bar{A}} + \vec{R}_{\bar{B}}\right) \tag{6.107}$$

$$ = {}^{N}\vec{v}_r^{\bar{A}} \cdot \left(-\vec{R}_{\bar{B}} + \vec{R}_{\bar{B}}\right) \tag{6.108}$$

$$ = 0 \,. \tag{6.109}$$

Contact forces exerted across the system boundary. There are several varieties of noncontributing contact forces exerted across the boundary of system S. *Contact forces exerted across "smooth" (i.e., frictionless) surfaces of rigid*

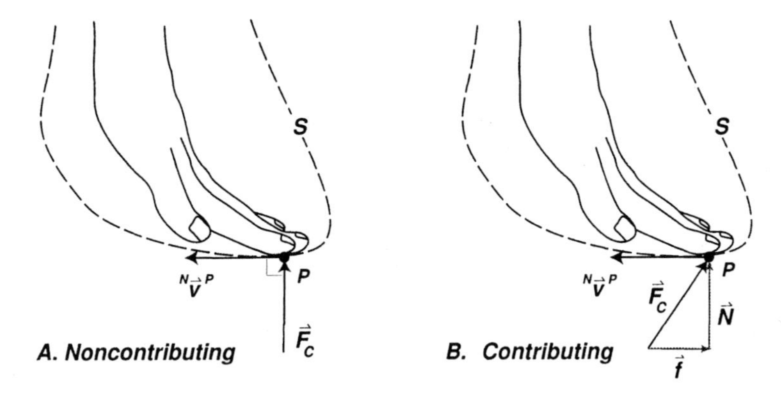

Figure 6.11. Non-contributing, and contributing contact forces $\vec{F}_C$ exerted by an external influence across the system boundary upon a body contained within system S.

bodies on the system S do not contribute because forces exerted across such contacts are necessarily perpendicular to the relative motions between the bodies (Figure 6.10A). Similarly, *contact forces exerted by rolling contacts at the system boundary* do not contribute to the dynamical equations of motion because generalized coordinates can always be chosen in such a way that the velocity of the point of system S in contact at the boundary instantaneously has a zero velocity in the external reference frame (Figure 6.10B).

Proof. Let there be a point P of system S upon which an externally imposed contact force $\vec{F}_C$ acts (Figure 6.11). To be in contact with something outside of S, P necessarily must be on the surface of one of the bodies comprising the system. If friction is present, the frictional force $\vec{f}$ is the component of $\vec{F}_C$ that is directed parallel to the surface tangent at the point of contact and in a direction opposite to the velocity of P in N,

$$\vec{f} = \vec{F}_C \cdot \frac{-{}^N\vec{v}^P}{\left|{}^N\vec{v}^P\right|} \tag{6.110}$$

and has magnitude equal to

$$\left|\vec{f}\right| = \mu \left|\vec{N}\right| . \tag{6.111}$$

$\vec{N}$ is the component of $\vec{F}_C$ perpendicular to the surface. Note that $\vec{f}$ is always parallel to ${}^N\vec{v}^P$ as long as there is no compression of either body nor penetration of one body into another at P. Hence, the dot product of $\vec{f}$ and the partial velocities ${}^N\vec{v}_r^P$ are likely to be nonzero. On the other hand, if the surfaces in contact are considered to be rigid and frictionless, then $\vec{f} = 0$ and $\vec{F}_C = \vec{N}$. In this case, the velocity ${}^N\vec{v}^P$ and the force $\vec{F}_C$ will always be perpendicular and thus noncontributory.

6.9 JOINT REACTION FORCES

Now that the subject of contributing and noncontributing forces has been covered, it is appropriate to discuss ways of solving for the joint reaction forces. Using Kane's Method, these tend to be eliminated early in the process of deriving dynamic equations. As long as there is no dissipation or storage of energy at the contact locations, the joint reaction forces do not appear because equal and opposite forces are applied at points having either common velocities, or velocities perpendicular to the forces in question. Kane's Method has sometimes been criticized by members of the biomechanics community because most joint reaction forces are automatically eliminated. However, it is easy to introduce these joint reaction forces, and to do so at specific joints.

Kane and Levinson (1985) call this "bringing noncontributing forces into evidence." In the beginning of the chapter, Equation 6.1 defined n generalized speeds for n generalized coordinates. The method of bringing a force into evidence simply defines additional generalized speeds *without* defining corresponding generalized coordinates. The additional generalized speeds are introduced at the joints where the otherwise noncontributing forces are applied. Direction vectors are associated with these additional generalized speeds to create nonzero dot products with the noncontributing forces. The definition of each additional generalized speed introduces an additional "dynamic" equation, which allows one noncontributing force component to be computed.

The method may be used to bring any noncontributing force, or any noncontributing torque, into evidence. It is best shown by demonstration in the example following.

One might think that adding a highly constrained generalized coordinate would accomplish the same thing, but this adds an unnecessary degree of freedom and introduces additional complexities. For instance, one could add a small amount of joint compliance via a "soft" constraint, and use a damped spring to limit the allowable displacement at the interface. However, this would introduce oscillations at the natural frequency of the spring which would propagate throughout the system and could cause instabilities.

To solve for more than a few joint reaction forces, it is often easier to revert to a free body diagram *after* the dynamic equations are solved, and the segment accelerations are known. Then, D'Alembert's Principle and Newton's equations can be applied to determine the joint reaction forces.

6.9.1 EXAMPLE – HUMERORADIAL JOINT REACTION FORCE

Suppose we wished to compute the force exerted by the *capitulum* of the distal humerus on the proximal end of the radius. We will use the definitions set forth in the seven degree-of-freedom arm model of Section 4.4. To clarify

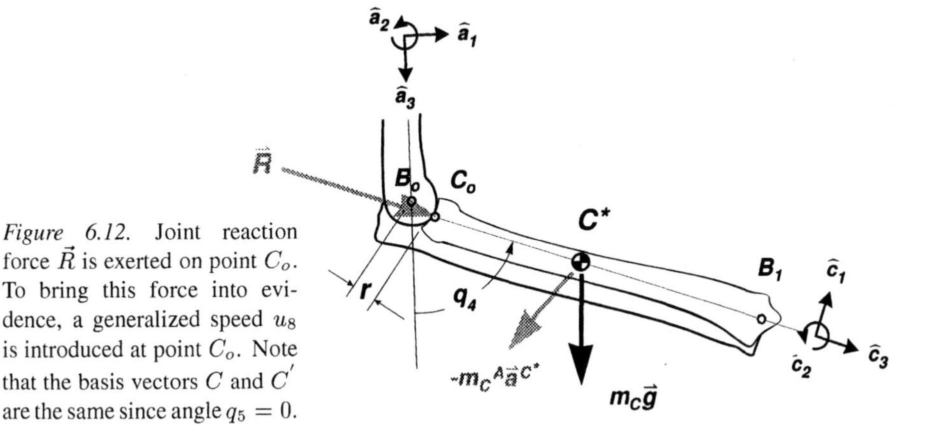

Figure 6.12. Joint reaction force $\vec{R}$ is exerted on point C_o. To bring this force into evidence, a generalized speed u_8 is introduced at point C_o. Note that the basis vectors C and C' are the same since angle $q_5 = 0$.

the illustration, the discussion here will consider the humerus (A) to be fixed, and the radius to be moving in flexion and extension only ($q_5 = 0$; see Figure 6.12). Another simplification locates the center of mass C^* at position $\rho_{C_3}\hat{c}_3$ from point C_o. Furthermore, only the axial component of the joint reaction force $\vec{R} = R\hat{c}_3$ will be found.

To begin, one generalized speed is introduced for every component of $\vec{R}$ desired. Since only the component in the $\hat{c}_3$ direction is to be found, only one additional generalized speed is needed. There are already seven generalized speeds, so the new one is designated as u_8. This generalized speed is introduced at the point of application of $\vec{R}$, at point C_o. One pretends that $\vec{R}$ deforms the proximal radius in the $\hat{c}_3$ direction to bring out the $\hat{c}_3$ component of $\vec{R}$. There is no need to introduce a generalized coordinate q_8 to describe this deformation, as doing so requires an additional dynamic equation to be defined.

The generalized speed u_8 changes the velocities of point C_o and many points distal to it. For instance, the velocity of C_o and the velocity and acceleration of C^* are normally expressed as,[7]

$$
\begin{aligned}
{}^A\vec{v}^{C_o} &= r\dot{q}_4\hat{c}_1' = ru_4\hat{c}_1 && (6.112)\\
{}^A\vec{v}^{C^*} &= \rho_{C_3}\dot{q}_4\hat{c}_1 = \rho_{C_3}u_4\hat{c}_1 && (6.113)\\
{}^A\vec{a}^{C^*} &= \rho_{C_3}\ddot{q}_4\hat{c}_1 - \rho_{C_3}\dot{q}_4^2\hat{c}_3 && (6.114)\\
&= \rho_{C_3}\dot{u}_4\hat{c}_1 - \rho_{C_3}u_4^2\hat{c}_3\,. && (6.115)
\end{aligned}
$$

After introducing u_8, these become,

$$
{}^A\vec{v}^{C_o} = ru_4\hat{c}_1 + u_8\hat{c}_3 \qquad (6.116)
$$

[7]The velocities are expressed here with respect to reference frame A. If the shoulder was allowed to rotate, these velocities should be expressed relative to the N reference frame.

$$ {}^{A}\vec{v}^{C^*} = \rho_{C_3} u_4 \hat{c}_1 + u_8 \hat{c}_3 \quad (6.117) $$
$$ {}^{A}\vec{a}^{C^*} = \rho_{C_3} \dot{u}_4 \hat{c}_1 - \rho_{C_3} u_4^2 \hat{c}_3 + \dot{u}_8 \hat{c}_3 \,. \quad (6.118) $$

Because of the additional generalized speed, additional partial velocities are defined,

$$ {}^{A}\vec{v}_8^{C_o} = \hat{c}_3 \quad (6.119) $$
$$ {}^{A}\vec{v}_8^{C^*} = \hat{c}_3 \quad (6.120) $$

Partial velocities ${}^{A}\vec{v}_8^{P_i}$ must be found for every point P_i ($i = 1, \ldots, \mu$) at which contributing forces $\vec{F}_i$ are applied to the system. Contributions to the generalized active and generalized inertia forces will result if the forces and partial velocities have nonzero dot products. An 8th generalized active force F_8, and an 8th generalized inertia force F_8^* are computed as,

$$ F_8 = \left({}^{A}\vec{v}_8^{C_o} \cdot \vec{R}\right) + \left({}^{A}\vec{v}_8^{C^*} \cdot m_C \vec{g}\right) + \sum_{i=1}^{\mu} {}^{A}\vec{v}_8^{P_i} \cdot \vec{F}_i \quad (6.121) $$

$$ F_8^* = \left({}^{A}\vec{v}_8^{C^*} \cdot (-m_C \, {}^{A}\vec{a}^{C^*})\right) + \sum_{j=1}^{\nu} {}^{A}\vec{v}_8^{B_i^*} \cdot \left(-m_{B_i} \, {}^{A}\vec{a}^{B_i^*}\right) \quad (6.122) $$

In the second equation, B_i^* are the ν points having mass (in addition to C^*) at which inertial forces $-m_{B_i} \, {}^{A}\vec{a}^{B_i^*}$ are developed. Normally, only masses distal to point C_o would have nonzero contributions to the generalized inertia forces.

If we consider only the contributions resulting from the forces at points C_o and C^*, then the additional "8th dynamic equation" is

$$ F_8 + F_8^* = 0 \quad (6.123) $$

$$ \left(\hat{c}_3 \cdot R\hat{c}_3\right) + \left(\hat{c}_3 \cdot m_C \, \vec{g}\right) + \hat{c}_3 \cdot \left(-m_C \left(\rho_{C_3} \dot{u}_4 \hat{c}_1 - \rho_{C_3} u_4^2 \hat{c}_3 + \dot{u}_8 \hat{c}_3\right)\right) = 0 \quad (6.124) $$
$$ R + m_C \, g \, c_4 + \rho_{C_3} m_C u_4^2 - m_C \dot{u}_8 = 0 \,. \quad (6.125) $$

c_4 is the cosine of angle q_4, which is the result of taking the dot product $\hat{c}_3 \cdot \hat{n}_3$ ($\hat{n}_3$ points downward in the direction of the gravitational force $\vec{g} = g\hat{n}_3$). The final step is to solve for R by setting $\dot{u}_8$ to zero, as the system cannot actually move or accelerate with speed u_8,

$$ R = -m_C \, g \, c_4 - \rho_{C_3} m_C u_4^2 \,. \quad (6.126) $$

It may appear that one must go through a lot of extra trouble to bring a joint reaction force, or any noncontributing force, into evidence. If only one force component is required, the extra work is minimal as only one additional generalized speed, one set of partial velocities, and one "dynamic" equation are introduced.

If the external forces and accelerations of the system are known, it is often easier to solve for the joint reactions directly from a free body diagram such as Figure 1.2. The force equation,

$$\sum_k \vec{F}_k = \sum_i m_i \, {}^N\vec{a}^{B_i^*} \,, \tag{6.127}$$

where $\vec{F}_k$ are forces applied across the system boundaries, can be applied to the system as a whole, or to any subset of the system. For instance, the humeroradial force $\vec{R}$ above could have been solved for all three components of $\vec{R}$ by the summation of forces on an isolated radius,

$$\vec{R} + m_C \vec{g} + \sum_k \vec{F}_k = m_C \, {}^A\vec{a}^{C^*} \,, \tag{6.128}$$

where $\vec{F}_k$ are all the forces acting on the radius, including other joint reactions, muscular forces, *etc.* Typically, however, this methodology is performed using the standard Newton-Euler approach, solving for the joint reactions beginning with the outmost link, and working inward. The outermost link is done first, as it has only one set of joint reactions to be solved for (interior links have two sets of joint reactions). This direct method is advantageous because three force components (and moments) acting across the system boundary may be found any place where the boundary is drawn.

For instance, if we hypothetically assumed that the radius is at the very end of the linkage, and that $\sum_k \vec{F}_k = 0$, then we could easily solve for $\vec{R}$,

$$\vec{R} \;=\; m_C \, {}^A\vec{a}^{C^*} - m_C \vec{g} \,. \tag{6.129}$$

If the component of $\vec{R}$ in the $\hat{c}_3$ direction is desired, the situation is analogous to the solution starting from Equation 6.123,

$$\vec{R} \cdot \hat{c}_3 \;=\; \left(m_C \, {}^A\vec{a}^{C^*} - m_C \vec{g} \right) \cdot \hat{c}_3 \tag{6.130}$$

$$\begin{aligned} \left(R\hat{c}_3 \right) \cdot \hat{c}_3 \;&=\; m_C \left(\rho_{C_3} \dot{u}_4 \hat{c}_1 - \rho_{C_3} u_4^2 \hat{c}_3 \right) \cdot \hat{c}_3 \\ &\quad - m_C \left(g \hat{n}_3 \right) \cdot \hat{c}_3 \end{aligned} \tag{6.131}$$

$$R \;=\; -\rho_{C_3} m_C u_4^2 - m_C \, g \, c_4 \,. \tag{6.132}$$

This is the same answer obtained previously in Equation 6.126.

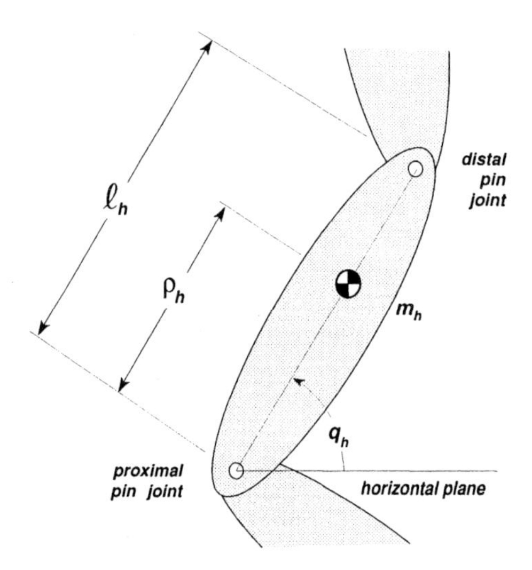

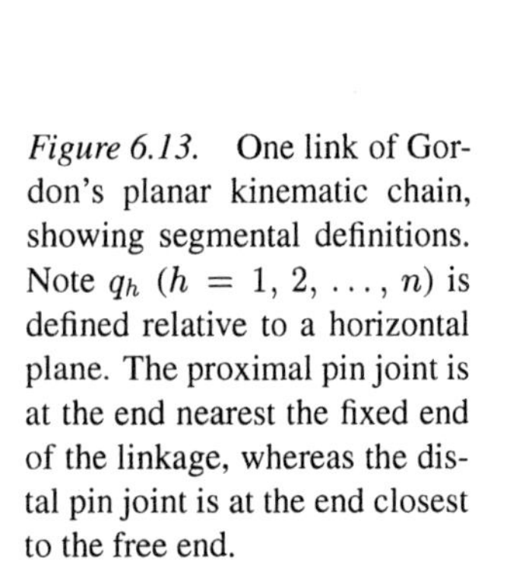

Figure 6.13. One link of Gordon's planar kinematic chain, showing segmental definitions. Note q_h ($h = 1, 2, \ldots, n$) is defined relative to a horizontal plane. The proximal pin joint is at the end nearest the fixed end of the linkage, whereas the distal pin joint is at the end closest to the free end.

6.10 GORDON'S METHOD

By painstaking derivation, Gordon (1990) used Kane's Method to formulate the dynamic equations of motion for a planar linkage having an arbitrary number of links. The result was a "recipe" that allowed the n dynamic equations for an n-link planar linkage to be written down without having to derive them.

What follows are the highlights of his work, and not the actual derivation. His method is *constrained to open, unbranched kinematic chains,* and *requires segmental angles as the generalized coordinates* (Figure 6.13). An added benefit is that the dynamic equations are written in their most compact form, which is where the mass matrix M is symmetric, positive semi-definite (see Homework Problem 3).

Vint and Yamaguchi (1993) added in the contributions of an arbitrary planar endpoint force $\vec{F} = f_1\hat{n}_1 + f_2\hat{n}_2$ to extend the application of Gordon's method to a wide variety of biomechanical tasks. $\vec{F}$ can be specified as any magnitude and direction with the proper choice of parameters f_1 and f_2.

The description begins with the formulation of variables used by subsequent equations. It is suggested that the recipe be followed precisely in the order given. If the equations are to be computerized, this order is recommended as the most efficient way to develop a computer program.

Auxiliary variables:

These are formed using indices $j = 1, 2, \ldots, n-1$, and $h = 1, 2, \ldots, n$,

$$\mu_j = \ell_j \sum_{i=j+1}^{n} m_i \tag{6.133}$$

$$\mu_n = 0 \tag{6.134}$$

$$\delta_h = \mu_h + \frac{I_h + m_h\rho_h^2}{\ell_h} \quad (6.135)$$

$$\gamma_h = \mu_h + m_h\rho_h \quad (6.136)$$

Mass Matrix M:

$$M_{hh} = \ell_h\delta_h c_{hh} \quad (6.137)$$

$$= \ell_h\delta_h \quad (6.138)$$

for the diagonal elements, $h = (1, 2, \ldots, n)$, and

$$M_{hk} = \ell_h\gamma_k c_{hk} \quad (6.139)$$

where $c_{hk} \equiv cos(q_h - q_k)$ and $s_{hk} \equiv sin(q_h - q_k)$ for the elements where $k > h$. Strictly speaking, these are the upper triangular elements. The mass matrix M is also symmetric, so the lower triangular elements are found by reflection,

$$M_{kh} = M_{hk} \quad (6.140)$$

for the off diagonal elements, $h = (1, 2, \ldots, n)$ and $k = (1, 2, \ldots, n)$.

Matrix N:

$$N_{hh} = \ell_h\delta_h s_{hh} \quad (6.141)$$

$$= 0 \quad (6.142)$$

for the diagonal elements, $h = (1, 2, \ldots, n)$, and

$$N_{hk} = \ell_h\gamma_k s_{hk} \quad (6.143)$$

for $k > h$ (upper triangular elements). In contrast to M, the matrix N is antisymmetric, so that

$$N_{kh} = -N_{hk} \quad (6.144)$$

for $h = (1, 2, \ldots, n)$ and $k = (1, 2, \ldots, n)$.

Vectors on the right-hand side of the dynamic equations:

The elements of vectors $\vec{T}$, $\vec{G}$, $\vec{V}$, and $\vec{E}$ can now be written as,

$$T_h = \vec{\tau}_{B_i} \cdot \hat{n}_3 \quad (6.145)$$

$$G_h = -g\gamma_h c_h \quad (6.146)$$

$$V_h = -\sum_{k=1}^{n} N_{hk}\dot{q}_k^2 \quad (6.147)$$

$$E_h = \ell_h\left(f_2 c_h - f_1 s_h\right), \quad (6.148)$$

for $h = (1, 2, \ldots, n)$. The term $\vec{\tau}_{B_i}$ of Equation 6.145 is the torque on body B_i, where $i = (1, 2, \ldots, n)$.

The final step is to write down the dynamic equations of motion. Individually, the equations in scalar form are,

$$\sum_{k=1}^{n} M_{hk}\ddot{q}_k = T_h + G_h + V_h + E_h \tag{6.149}$$

for $h = (1, 2, \ldots, n)$. However, it is most convenient to express the dynamic equations in matrix form. From Equation 6.147, the vector $\vec{V}$ can be obtained by performing the following matrix-vector multiplication,

$$\vec{V} = -N \begin{bmatrix} \dot{q}_1^2 \\ \vdots \\ \dot{q}_n^2 \end{bmatrix} . \tag{6.150}$$

Finally, the equations themselves may be written down,

$$M\ddot{\vec{Q}} = \vec{T} + \vec{G} + \vec{V} + \vec{E} . \tag{6.151}$$

6.10.1 EXAMPLE – DYNAMIC EQUATIONS FOR A FOUR-LINK PLANAR LINKAGE

If the generalized coordinates for a four link planar linkage are defined using segmental angles measured from a horizontal plane, the dynamic equations can simply be written down following the recipe developed above. This is by no means a derivation! With practice using Gordon's method, the variables and equations for n degree of freedom planar systems can be written within minutes.

Auxiliary variables:

$$\mu_1 = \ell_1 (m_2 + m_3 + m_4) \tag{6.152}$$

$$\mu_2 = \ell_2 (m_3 + m_4) \tag{6.153}$$

$$\mu_3 = \ell_3 (m_4) \tag{6.154}$$

$$\mu_n = 0 . \tag{6.155}$$

$$\delta_1 = \mu_1 + \frac{I_1 + m_1\rho_1^2}{\ell_1} \tag{6.156}$$

$$\delta_2 = \mu_2 + \frac{I_2 + m_2\rho_2^2}{\ell_2} \tag{6.157}$$

$$\delta_3 = \mu_3 + \frac{I_3 + m_3\rho_3^2}{\ell_3} \tag{6.158}$$

$$\delta_4 = \mu_4 + \frac{I_4 + m_4\rho_4^2}{\ell_4}. \tag{6.159}$$

$$\gamma_1 = \mu_1 + m_1\rho_1 \tag{6.160}$$

$$\gamma_2 = \mu_2 + m_2\rho_2 \tag{6.161}$$

$$\gamma_3 = \mu_3 + m_3\rho_3 \tag{6.162}$$

$$\gamma_4 = \mu_4 + m_4\rho_4 . \tag{6.163}$$

Mass Matrix M:

$$M_{11} = \ell_1\delta_1 \tag{6.164}$$

$$M_{22} = \ell_2\delta_2 \tag{6.165}$$

$$M_{33} = \ell_3\delta_3 \tag{6.166}$$

$$M_{44} = \ell_4\delta_4 \tag{6.167}$$

$$M_{12} = \ell_1\gamma_2 c_{12} \tag{6.168}$$

$$M_{13} = \ell_1\gamma_3 c_{13} \tag{6.169}$$

$$M_{14} = \ell_1\gamma_4 c_{14} \tag{6.170}$$

$$M_{23} = \ell_2\gamma_3 c_{23} \tag{6.171}$$

$$M_{24} = \ell_2\gamma_4 c_{24} \tag{6.172}$$

$$M_{34} = \ell_3\gamma_4 c_{34} \tag{6.173}$$

$$M_{21} = M_{12} \tag{6.174}$$

$$M_{31} = M_{13} \tag{6.175}$$

$$M_{32} = M_{23} \tag{6.176}$$

$$M_{41} = M_{14} \tag{6.177}$$

$$M_{42} = M_{24} \tag{6.178}$$

$$M_{43} = M_{34} . \tag{6.179}$$

Matrix N:

$$N_{11} = 0 \tag{6.180}$$

$$N_{22} = 0 \tag{6.181}$$

$$N_{33} = 0 \quad (6.182)$$
$$N_{44} = 0 \quad (6.183)$$
$$N_{12} = \ell_1 \gamma_2 s_{12} \quad (6.184)$$
$$N_{13} = \ell_1 \gamma_3 s_{13} \quad (6.185)$$
$$N_{14} = \ell_1 \gamma_4 s_{14} \quad (6.186)$$
$$N_{23} = \ell_2 \gamma_3 s_{23} \quad (6.187)$$
$$N_{24} = \ell_2 \gamma_4 s_{24} \quad (6.188)$$
$$N_{34} = \ell_3 \gamma_4 s_{34} \quad (6.189)$$
$$N_{21} = -N_{12} \quad (6.190)$$
$$N_{31} = -N_{13} \quad (6.191)$$
$$N_{32} = -N_{23} \quad (6.192)$$
$$N_{41} = -N_{14} \quad (6.193)$$
$$N_{42} = -N_{24} \quad (6.194)$$
$$N_{43} = -N_{34}\,. \quad (6.195)$$

Vectors on the right-hand side of the dynamic equations:

The elements of vectors $\vec{T}$ and $\vec{G}$ can now be written as,

$$T_1 = \vec{\tau}_{B_1} \cdot \hat{n}_3 \quad (6.196)$$
$$T_2 = \vec{\tau}_{B_2} \cdot \hat{n}_3 \quad (6.197)$$
$$T_3 = \vec{\tau}_{B_3} \cdot \hat{n}_3 \quad (6.198)$$
$$T_4 = \vec{\tau}_{B_4} \cdot \hat{n}_3 \quad (6.199)$$

$$G_1 = -g\gamma_1 c_1 \quad (6.200)$$
$$G_2 = -g\gamma_2 c_2 \quad (6.201)$$
$$G_3 = -g\gamma_3 c_3 \quad (6.202)$$
$$G_4 = -g\gamma_4 c_4\,. \quad (6.203)$$

The vector $\vec{V}$ is most easily obtained by performing the following matrix-vector multiplication,

$$\begin{bmatrix} V_1 \\ V_2 \\ V_3 \\ V_4 \end{bmatrix} = -N \begin{bmatrix} \dot{q}_1^2 \\ \vdots \\ \dot{q}_n^2 \end{bmatrix}, \quad (6.204)$$

and the external force terms in vector $\vec{E}$ are,

$$E_1 = \ell_1 (f_2 c_1 - f_1 s_1) \quad (6.205)$$

$$E_2 = \ell_2 \left(f_2 c_2 - f_1 s_2 \right) \tag{6.206}$$
$$E_3 = \ell_3 \left(f_2 c_3 - f_1 s_3 \right) \tag{6.207}$$
$$E_4 = \ell_4 \left(f_2 c_4 - f_1 s_4 \right) . \tag{6.208}$$

The final step is to simply write down the dynamic equations of motion. Note that all of the elements of the $\vec{M}$ and $\vec{N}$ matrices are defined, as are all of the elements of $\vec{T}$, $\vec{G}$, $\vec{V}$, and $\vec{E}$.

$$M\ddot{\vec{Q}} = \vec{T} + \vec{G} + \vec{V} + \vec{E} . \tag{6.209}$$

6.11 EXERCISES

1. In Section 1.3.1.1, an outline of Kane's Method for deriving the dynamic equations of a two link planar linkage was given. Perform the full derivation of these equations, using the following *joint* torques,

$$\tau_{N/A} = T_1 \hat{a}_3 \tag{6.210}$$
$$\tau_{A/B} = T_2 \hat{b}_3 . \tag{6.211}$$

   Format the resulting equations in matrix form. Compare the result to those obtained by the Newton-Euler Method (Problem 1, Chapter 1).

2. Problem 1 generates the dynamic equations for angle q_2 defined as a joint angle (the angle between the $\hat{a}_1$ and $\hat{b}_1$ basis vectors). When q_2 is defined as a segmental angle (the angle between the $\hat{n}_1$ and $\hat{b}_1$ basis vectors), the equations will change. Using Kane's Method, rederive the dynamic equations for a two link planar linkage where angles q_1 and q_2 are segmental angles. Use the same torque definitions as in Problem 1. Format the equations in matrix form, and compare the result to the equations given in Problem 2, Chapter 1.

   Compare the terms of the torque vector $\vec{T}$ obtained here to those from Problem 1. Explain any similarities or differences.

3. Problem 2 generates the dynamic equations for angle q_2 defined as a segmental angle. The answer can also be developed using Gordon's Method. Write the equations for the two link planar linkage of Chapter 1 using Gordon's "recipe." Expand the auxiliary variables into basic parameters (ℓ's, ρ's, *etc.*), and compare the result with your answer for Problem 2. Then answer the following questions.

   (a) Assuming that you have derived the equations accurately, the formulations should look quite different. Explain how both sets of equations can be correct.

(b) Which form presents the dynamic equations in their most compact, efficient form?

4. Problem 7 of Chapter 4 asks for the velocities of points A^*, B^*, and C^* of the seven degree-of-freedom arm model, using simplified forms for the position vectors to these mass centers. We will ignore body D and the two wrist angles q_6 and q_7 for simplicity.

   (a) Determine the partial velocities for these three points.

   (b) Determine the partial angular velocities for bodies A, B, and C.

   (c) Find the 1st and the 5th generalized active forces, F_1 and F_5. To simplify the problem, define the torques acting on bodies A, B, and C as zero.

5. The two link planar linkage model of Figure 1.1 can be made three dimensional by adding a degree of freedom about point A_o at the base of the linkage. Rotation angle q_o rotates A about the common $\hat{n}_1$, $\hat{a}_1$ directions, and tilts the plane of the linkage from the vertical plane. The torque applied by body N on A is $\tau_{N/A} = T_o\hat{n}_1 + T_1\hat{a}_3$. Likewise, the torque applied by body A on B is $\tau_{A/B} = T_2\hat{b}_3$.

   Derive the dynamic equations for this three degree of freedom system.

6. Write down the dynamic equations for a five link planar linkage using Gordon's Method. Time yourself, and compare your time with what you would anticipate as the time required to derive the equations using Kane's Method and the Newton-Euler Method. (Note: complexity roughly doubles for each additional degree of freedom.)

7. Derive the dynamic equations for a four degree of freedom, non-planar biomechanical linkage of your choice. Keep the algebra as simple as possible by defining mass centers to lie along lines between adjacent joints.

References

Gordon, Michael E. (1990) "Planar N-Link Open Chain". Appendix 8 of *An Analysis of the Biomechanics and Muscular Synergies of Human Standing*. Ph.D. Dissertation, Department of Mechanical Engineering, Stanford University, Stanford, CA, pp. 223-239.

Kane, T. R., and Levinson, D. A. (1985) *Dynamics: Theory and Applications*. McGraw-Hill, New York, NY.

Meriam, J. L., and L. G. Kraige. (1997) *Engineering Mechanics Volume 2 – Dynamics*. John Wiley & Sons, New York, NY.

Vint, P., and Yamaguchi (1993) Unpublished work performed at Arizona State University.

Chapter 7

CONTROL

Objective – To introduce the reader to inverse and forward dynamic methods of determining the forces and activation levels of the muscles and tendons needed to produce a particular movement.

7.1 INVERSE AND FORWARD DYNAMICS

Dynamic musculoskeletal models and equations are useful only when they can be applied to gain additional knowledge about a living system or to extrapolate behavior beyond what has been observed. To initiate a discussion on this topic, consider the natural "flow" of events by which movements are produced by a living organism. When an animal or human desires to move its limbs, a sequence of motor commands emanates from its central nervous system and excites an appropriate set of musculature into developing contractile force. Since these muscles connect directly to bones via tendons, the tensile forces acting on the bones exert moments about the joints of the skeletal system and give rise to reaction forces within the joints. The summation of actively generated moments, and the kinematic constraints and inertial characteristics of the body, govern how the body segments will move. Thus, the natural flow of events proceeds *outward* from the nervous system to the muscular system, and the muscular system to the skeletal system (Zajac and Gordon, 1989).

Dynamic musculoskeletal models are useful in describing the interrelationship between the applied muscular forces and the resulting motions of the body segments. For instance, the mathematical gait model shown in Appendix A and B could be used as a geometric tool to define the position and velocity dependent transformations between the inputs (here, muscle activations) and the outputs (limb trajectories) of the dynamic equations. These equations, which govern the motion of any body model having n degrees of freedom q_i

($i = 1, 2, \ldots, n$), can be written in matrix-vector form as in Equation 6.35, which is reproduced here for convenience,

$$M\ddot{\vec{Q}} = \vec{T} + \vec{V} + \vec{G} + \vec{E}\,. \tag{7.1}$$

This equation can be restructured to emphasize the dependence of the segmental accelerations $\ddot{\vec{Q}} = [\ddot{q}_1, \ddot{q}_2, \ldots, \ddot{q}_n]^T$ upon the mass characteristics and instantaneous configuration of the body (contained in M), and the segmental torques applied by the muscles and joints ($\vec{T}$), inertia ($\vec{V}$), gravity ($\vec{G}$), and external forces ($\vec{E}$), respectively,

$$\ddot{\vec{Q}} = M^{-1}\left[\vec{T} + \vec{V} + \vec{G} + \vec{E}\right]. \tag{7.2}$$

As already discussed, M (and hence M^{-1}) and $\vec{G}$ are functions of the instantaneous mass distribution as reflected by the values of the degrees of freedom ($\vec{Q} = [q_1, q_2, \ldots, q_n]^T$). $\vec{V}$ is a function of $\vec{Q}$ and the speeds $\dot{\vec{Q}} = [\dot{q}_1, \dot{q}_2, \ldots, \dot{q}_n]^T$. The equations, written in this way, clearly show motion outputs as functions of muscular inputs, and are therefore compatible with the natural flow of *neural-to-muscular-to-skeletal* events. *Dynamic simulations of movement*, or *motion syntheses*, integrate Equation 7.2 forward in time to obtain motion trajectories in response to neuromuscular inputs,

$$\dot{\vec{Q}}(t + \Delta t) = \int_t^{t+\Delta t} \ddot{\vec{Q}} dt \tag{7.3}$$

$$= \int_t^{t+\Delta t} M^{-1}\left[\vec{T} + \vec{V} + \vec{G} + \vec{E}\right] dt\,. \tag{7.4}$$

The term "direct dynamics" will be used to describe this type of analysis.

"Inverse dynamic analyses," as opposed to direct dynamic analyses, proceed in a direction opposite to the natural flow of events. Inverse dynamic methods use observed motions as inputs to the process. Specifically, particular movements are quantified as joint or angular trajectories which vary with time. The trajectories are used together with dynamic musculoskeletal models to determine the joint moments or muscular forces that *must have been evident* to have created the observed motion. To perform an inverse analysis, Equation 7.1 is rearranged as,

$$\vec{T} = M\ddot{\vec{Q}} - \left[\vec{V} + \vec{G} + \vec{E}\right], \tag{7.5}$$

which clearly expresses the output torques as a function of the trajectory dependent terms grouped on the right hand side of the equation.

To gather the data needed to perform inverse dynamic analyses, markers are typically placed upon a living subject in areas that have minimal underlying muscle. Markers are placed either over the approximate joint locations, or alternatively, three-marker "triads"are placed in sets on each rigid body to be tracked. If the markers are all sampled simultaneously, the marker positions at each sampled instant of time are used to compute the values of the generalized coordinates (usually, segmental or joint angles). The generalized coordinates are then filtered or fit with a smoothing function, which facilitates taking derivatives and second derivatives (see the example below). These are substituted into the right hand side of Equation 7.5, the right hand side terms are computed, and used to evaluate the elements of the torque matrix $\vec{T}$.

A second step of the inverse dynamic process is done if the muscular forces are desired. Once $\vec{T}$ is determined, the torques are converted into *joint* torques if they are not already in that form. Then, the muscle forces crossing each joint are assigned so the correct torque is delivered at each joint. The muscles at a particular joint are treated as being "local generators of joint torque," as their actions are considered limited to the joints the muscles actually cross. Typically, some form of static optimization is used to distribute the tensile forces among the redundant muscle set at each instant of time.

Clearly, the inverse dynamic approach applies itself readily to clinical and experimental settings where one works backwards from observed motion trajectories. Since the trajectories of only the body segments of interest are needed, the analyst can use inverse dynamic analyses to focus upon as many segments as necessary within a larger musculoskeletal model.

7.1.1 EXAMPLE – CUBIC SPLINE INTERPOLATION OF GENERALIZED COORDINATES

An easy way to smooth and differentiate data points sampled in time is to fit the data with a polynomial function that has continuous first and second derivatives. The simplest of these is the *cubic spline,*

$$q(t) = at^3 + bt^2 + ct + d\,, \tag{7.6}$$

which is a function that represents a continuously varying joint or segmental angle in time. Instead of having a continuous record of $q(t)$, we typically have n values that have been *sampled* at discrete time instants t_k, $k = 0, 1, \ldots, n$.

The following discussion sketches out some algorithms which are very useful, but are becoming less fashionable with the proliferation of software packages for personal data processing and computing. However, this section is pertinent to solving linear systems in general, and provides the key to achieving stunning reductions in computation time when applying the pseudoinverse optimal control method explained in Section 7.5. Unless the reader is already proficient in numerical analysis, he or she is urged to review this material. The algorithm that is described is extremely simple and satisfactory for basic uses. More sophisticated filtering and smoothing routines are available for biome-

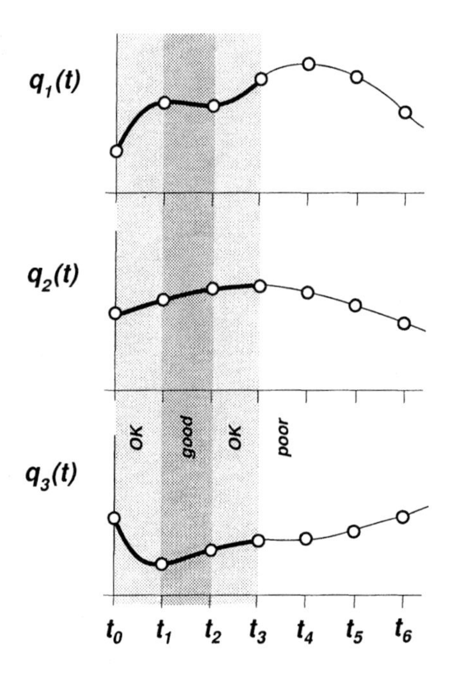

Figure 7.1. A cubic spline fit to the first four points of a longer data series. Each coordinate is parameterized as a function of time, and is interpolated separately. The cubic spline function is the simplest polynomial with continuous first and second derivatives. Except at the beginning and end of the series, the middle of the spline is used for interpolations (dark shaded region). Accuracy decreases for interpolations at either end of the spline (light shaded regions). Once time t proceeds beyond t_{k+2}, the spline is shifted forward in time by one point, k is indexed, and new coefficients are found.

chanical analyses of experimental data, as described by Woltring (1986) and Hatze (1981).

Here's the big idea! There are four unknown coefficients a, b, c, d that can be determined by creating four linearly independent equations. It turns out that these four equations can be formed by assigning time values t_k and functional values $q(t_k)$ for four consecutive data points at a time.[1] Hence, a series of "piecewise" cubic splines is made, and is moved along with time as the interpolation proceeds. Smooth differentiation is accomplished simultaneously because the first and second derivatives can also be formed easily once the coefficients are known,

$$\dot{q}(t) = 3at^2 + 2bt + c \tag{7.7}$$

$$\ddot{q}(t) = 6at + 2b\,. \tag{7.8}$$

Figure 7.1 illustrates the process.

Starting with time $t = t_0 = 0$, the first spline will take points 0, 1, 2, and 3 and can be used to approximate $q(t)$ for times $0 \leq t \leq t_2$. Thus, at the beginning, the first and middle segments of the piecewise cubic spline are used.

[1] Higher ordered spline functions, such as "quintic" splines are also popular, and can be easily made using the same number of points as coefficients to be solved for. The advantage of higher ordered splines is that smoother acceleration functions are derived. Usually, an odd number of points is fit so that the interpolation region is kept in the middle.

As time is incremented, the spline shifts by one sampled point at a time to keep subsequent interpolations in the middle segment of the spline (between the second and third points). Thus, the next piecewise cubic spline is formed using points 1, 2, 3, and 4, and is used to interpolate and smooth the data for times $t_2 \leq t \leq t_3$. The process continues until the last four points of the data series are encountered. Here, the interpolation will use the middle and the last segment of the spline. Extrapolation to extend time values beyond the sampled data (for $t < 0$ or $t > t_n$) is not recommended, as accuracy decreases.

Here's how cubic spline interpolation is actually done: Starting at any time value t_k ($k \leq n-3$), a continuous time spline function $q(t)$, for $t_k < t < t_{k+3}$ can be fit to the four sampled data points $q(t_k)$, $q(t_{k+1})$, $q(t_{k+2})$, and $q(t_{k+3})$ using the following process. The spline function is likely to fit the data best in the middle of the spline ($t_k < t < t_{k+1}$), and less well at the ends of the spline ($t_{k-1} < t < t_k$ and $t_{k+1} < t < t_{k+2}$). If the coefficients can be found, then $q(t)$ will be determined. The coefficients can be solved for using four points, because,

$$
\begin{aligned}
q(t_k) &= at_k^3 + bt_k^2 + ct_k + d && (7.9)\\
q(t_{k+1}) &= at_{k+1}^3 + bt_{k+1}^2 + ct_{k+1} + d && (7.10)\\
q(t_{k+2}) &= at_{k+2}^3 + bt_{k+2}^2 + ct_{k+2} + d && (7.11)\\
q(t_{k+3}) &= at_{k+3}^3 + bt_{k+3}^2 + ct_{k+3} + d\,, && (7.12)
\end{aligned}
$$

or,

$$
\begin{bmatrix} t_k^3 & t_k^2 & t_k & 1 \\ t_{k+1}^3 & t_{k+1}^2 & t_{k+1} & 1 \\ t_{k+2}^3 & t_{k+2}^2 & t_{k+2} & 1 \\ t_{k+3}^3 & t_{k+3}^2 & t_{k+3} & 1 \end{bmatrix} \begin{bmatrix} a \\ b \\ c \\ d \end{bmatrix} = \begin{bmatrix} q(t_k) \\ q(t_{k+1}) \\ q(t_{k+2}) \\ q(t_{k+3}) \end{bmatrix}. \qquad (7.13)
$$

This fits the linear system format,

$$
A\vec{x} = \vec{b} \qquad (7.14)
$$

for which there are many methods of numerically finding the elements a, b, c, d within the unknown vector $\vec{x}$. The most common methodologies are *Gaussian Elimination* and a related process called *LU Decomposition.*

While more complete descriptions of the numerical algorithms are available outside of this book, it is useful to describe the basic schemes behind these methods as they are pertinent to the current discussion and also to the following section describing the pseudoinverse method. Both methods utilize the principle that a system of n linear equations with n unknowns can be solved systematically by transforming the original equations into a different set of equations having an identical solution. The transformation is performed by multiplying the equations, which are contained within the rows of the vector formulation, by scalar coefficients and subtracting the rows from each other. As long as no equation is a multiple of another (this is called "linearly independent"), then the solution to the altered set of equations will be the same as the solution to the original set.

If the right scalar coefficient is chosen, and used to multiply, say the fourth row of Equation 7.13, the resulting row can be subtracted from another row to form a different equation *with one fewer nonzero term* in it. The missing term would be represented in the matrix formulation as a zero. For instance, if every term of the fourth row was multiplied by the scalar value t_{k+2}^3/t_{k+3}^3, and the fourth row was subtracted from the third, then the resulting equation written in the fourth row would have a zero in the leftmost position. By following this process, many more zeros in the A matrix can be obtained. Eventually, the entire square matrix A is formed into a triangular matrix,

$$\begin{bmatrix} a_{11} & a_{12} & a_{13} & a_{14} \\ 0 & a_{22} & a_{23} & a_{24} \\ 0 & 0 & a_{33} & a_{34} \\ 0 & 0 & 0 & a_{44} \end{bmatrix} \begin{bmatrix} a \\ b \\ c \\ d \end{bmatrix} = \begin{bmatrix} b_1 \\ b_2 \\ b_3 \\ b_4 \end{bmatrix}. \tag{7.15}$$

The matrix now has a trivial solution because $d = b_4/a_{44}$, which is then used to find *c*, and so forth. Even though the coefficients of the equations (a_{ij}) have changed through this process, the solution is the same (*a, b, c, d*). We say that matrix A has been transformed into an *upper triangular* form to obtain this solution.

LU Decomposition breaks down the original A matrix into the product of a *lower triangular* matrix L, and an upper triangular matrix U. If $A = LU$, Equation 7.14 can be rewritten as,

$$(LU)\,\vec{x} = \vec{b} \tag{7.16}$$

$$L\,(U\vec{x}) = \vec{b}. \tag{7.17}$$

Now the solution is again trivial but must occur in two steps instead of one,

$$L\vec{y} = \vec{b} \tag{7.18}$$

$$U\vec{x} = \vec{y}. \tag{7.19}$$

It may seem that more work is involved in LU Decomposition than Gaussian Elimination. However, a digital computer needs only a little longer to solve the extra set of equations. LU Decomposition is advantageous because the *same* decomposition (the same L and U matrices) can be used for finding the solutions to *many* different assignments for the right-hand side vector $\vec{b}$. This becomes practical when more than one angle is fit by a spline at one time. Examining Equation 7.13, for instance, for the same time values contained in the A matrix, one might have several different joint angles that need to be interpolated and smoothed.

As a simple example, the $q(t_i)$ values ($i = t_k, \ldots, t_{k+3}$) on the right-hand side might contain four ankle angles in the initial LU Decomposition, and solved for the spline coefficients a through d. Then, reusing the decomposed matrices L and U, four knee angles sampled at the same times t_i could be inserted into the right-hand side and another set of spline coefficients found. Finally, four hip angles can be inserted. When one decomposition is used to solve for many $\vec{b}$ vectors, the computational time savings can be enormous!

7.2 MUSCLE FORCE DISTRIBUTION

One of the classic problems in biomechanics is that of determining the forces or activation levels needed in each of the body's musculotendon actuators to create smooth, coordinated movements (Seireg and Arvikar, 1975; Chao and An, 1978; Crowninshield, 1978; Hardt, 1978; Hatze, 1981; Pedotti *et al.*, 1978; Patriarco *et al.*, 1981). In this chapter, muscle activation levels are considered to be the controls, while the phrase "coordinated movements" refers to *simultaneously* moving the joints through their movement trajectories. Thus, controlling coordinated movements is distinctly different from, and much more difficult than, *sequentially* moving multiple points from their initial to final configurations.[2] Furthermore, if the model is limited to a small number of degrees of freedom, there are many more muscles available than joint movements to produce. Therefore, the muscle set appears to be "redundant." It is fortunate that there are so many muscles, because the muscles are constrained to exert only tensile forces. The force distribution problem becomes one of most efficiently distributing these forces among the muscle set to generate the joint moments required to produce a desired movement.

In the literature, most solutions of the force distribution problem can be grouped into methods that reduce the degree of muscle redundancy in the model, mathematically optimize the solution according to some criterion, or combine the two approaches. In the first method, constraints are introduced until the number of degrees of freedom equals the number of controls, so a determinate problem can be formulated. Since expressing such constraints often cannot be done without adding unphysiological artifacts to the analysis, optimization (the second method) is generally the favored, and most adaptable approach.

Optimization methods search for allowable combinations of the controls to find the "best" one according to some predetermined criterion. The pre-

[2] For instance, sequential movements are commonly used with upper-extremity prostheses due to limitations on the number of available control signals. A sequential movement means that only one joint is moved at a time, and a series of joints are moved one after the other.

sumption is that a mathematical expression called the *cost* or *criterion function* may be formulated to measure how good, say, the resulting movement of a biomechanical model would be in response to a particular control strategy. Typical cost functions penalize deviations from desired movement trajectories and/or excessive use of control effort. The main difficulty with expressing the cost mathematically is formulating physiologically justifiable expressions that reflect a good combination of energy expenditure, metabolic costs, muscle exertion, and fatigue. It is likely that the body naturally "optimizes" at least some aspects of movement production. It remains an attractive idea to emulate the energetics of physiological movement control using mathematical optimization.

Dynamic musculoskeletal models can predict motion patterns. Optimization can be used in these models to determine appropriate muscle sets and their force and activation patterns. The goal is to *predict* the muscle forces that emulate the actual forces *in-vivo*, and by so doing gain some insight into the motor control strategies employed by the central nervous system. Therefore, any optimization algorithms carried over from robotic or industrial uses must deliver physiologically relevant results. Usually, this means that these algorithms require significant modification. For reasons that will be discussed, the *pseudoinverse method* was specifically developed as a fast, and reasonably good way to control the high degree of freedom, multiple muscle models encouraged in this book.

Before proceeding to the pseudoinverse method, other commonly used optimization schemes are described to more fully develop the readers' appreciation of the subject material. By no means have biomechanical control specialists solved the physiological control problem. It is the author's hope that one day the actions of the central nervous system will be reliably predicted. If that day arrives, the control actions, together with realistic musculoskeletal models, will enable neural strategies to be emulated, and musculoskeletal effects to be evaluated. *Neuromusculoskeletal modeling* will become an important tool with which to evaluate and perhaps even solve difficult problems in paralysis and human rehabilitation.

7.3 ALTERNATIVE OPTIMIZATION METHODS

Dynamic processes like walking can be analyzed *quasistatically*, where the events at one instant of time are considered to be unrelated to events at other instants of time. Dynamic processes can also be analyzed *dynamically*, where control decisions are based on how they affect future events and thus are optimized over the entire time interval required to complete the task. Quasistatic analyses have the advantage of being much simpler to implement, but are severely limited in their dynamic modeling capabilities. Dynamic optimal control algorithms based on the calculus of variations are more accurate, but are difficult to use successfully. Unless one has prior knowledge of the form of the optimal control solution, or at least a good initial guess for it, the likelihood of obtaining meaningful results is very small.

A commonly used quasistatic approach is linear programming (Dantzig, 1963). Linear programming is one of the more popular optimization techniques because it requires no prior knowledge of the solution, is easy to understand, and is easily applied to problems in the business world, where costs must be minimized to maximize profits.[3] As long as an optimization problem may be mathematically formulated with linear expressions for the cost and constraints, linear programming can be used. Linear programming has been a popular approach among researchers trying to solve the biomechanical force distribution problem (Seireg and Arvikar, 1975; Chao and An, 1978; Crowninshield, 1978; Hardt, 1978; Hatze, 1981; Pedotti *et al.*, 1978; Patriarco *et al.*, 1981).

Unfortunately, studies suggest that the physiological costs associated with muscular exertions are in fact nonlinear, as suggested by the inverse nonlinear relationship between muscle stress and endurance (Crowninshield and Brand, 1981). Therefore, a linear cost function (as required by linear programming) can only approximate the physiological cost, at best. Also, many of the constraints, especially if they contain geometrical or empirical relationships, may not be readily cast into linear forms. Of course, nonlinear expressions can be linearized about a nominal operating point or trajectory, but such approximations will be valid only if the deviations are small.

Linear programming imposes additional undesirable limitations when used to solve movement and coordination problems. The usual difficulties encountered when applying quasistatic methods to dynamic problems are evident. For instance, the muscle force solutions are optimized at unrelated instants of time, and may not be optimal over the entire movement. Furthermore, the solutions may be unrealistic because no limitations exist to regulate how quickly muscle tension can be developed or dissipated (Hardt, 1978). More seriously, the linear programming method mathematically limits the number of muscles active at any given time to a number equal to the number of constraints. Thus, linear programming would be hard-pressed to predict, for example, muscle synergies requiring the simultaneous control of many coactivated muscles (Herzog and Leonard, 1991; Herzog, 1992).

Furthermore, the way in which muscle actions are interpreted is probably oversimplified in the linear programming approach. Because the focus is on achieving the specific set of torque values delivered by the inverse dynamic calculation, the muscles are treated as local generators of the "torque" (actually, the moments) needed at specific joints. Except for strength and leverage

[3] It was estimated in the late 1980s that one-quarter of the world's computer resources were expended running the Simplex Method of linear programming.

considerations, there is little to distinguish one muscle and its actions from another. The individuality of each muscle's pathway and how it affects the motions of the entire body is ignored in favor of providing the right amount of "joint torque."

This discussion in no way diminishes the many contributions of linear programming to our understanding of musculoskeletal control. Linear programming provided the early basis for formulating the stress based cost function (Hardt, 1978), and illustrated the effects of muscle strength and moment arm. However, because improved methodologies are now available, linear programming is not discussed further.

Another process called "dynamic programming" eliminates many of the limitations imposed by linear programming. Dynamic programming divides the time period of the movement into discrete time stages, and also approximates the continuously varying states (positions and velocities) and controls (muscle forces or activations) as sets having discrete values. For instance, a continuous position $q_i(t)$ which varies between 0 and 5 radians could be represented by 11 discrete levels, $\acute{q}_i(t) \in [0.0, 0.5, 1.0, \ldots, 5.0]$. A continuously varying control, for instance muscle activation $a(t)$, could be varied between 0 and 100% by discrete levels 10% apart, $\acute{a}(t) \in [0.0, 0.1, 0.2, \ldots, 1.0]$. Because the process of discretizing the states and the controls creates a finite number of control combinations to be tried from each of a finite number of states, the process is perfectly suited for digital computation. All of the allowable state and control combinations are evaluated. The process is made efficient by working backwards in time from the desired end state, so that only one set of control histories is saved at each quantized state. The controls that are saved at each quantized state are the controls that will move the system from the current quantized state and time to the desired end state in an optimal way.

Explained in another way, suppose one wished to compile a listing of the best ways to drive to New York City beginning from any roadway position in the continental U.S. The state vector would contain the instantaneous position and velocity of the car. It is likely that the optimal way to get to New York is to remain upon a paved roadway, so this will be a constraint. However, there are still infinitely many states that one could have at any instant of time, because distance measured along a roadway is a continuously variable, analog quantity. To simplify the task, a discrete approximation to the continuously varying states is needed.

One way of simplifying things is to determine the locations of all cities having resident populations exceeding 50,000. These locations would be the quantized states. Because there would be a finite number of these cities, the best way to get from each city to New York could be determined by evaluating all the possible routes. Information about every route is not needed and is discarded. What is stored and published are the driving instructions for the best

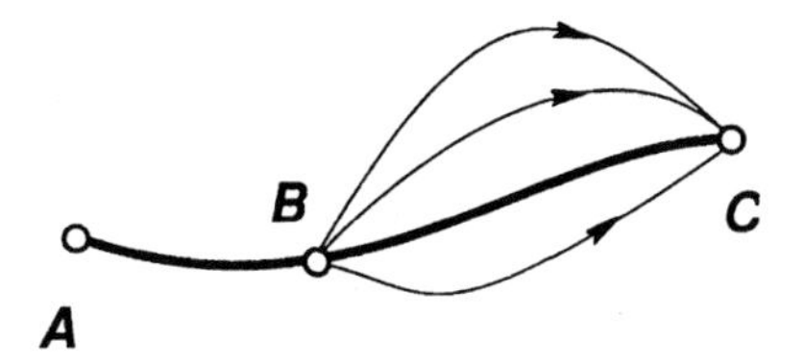

Figure 7.2. The optimal pathway from A to C, passing through B, must include the optimal pathway from B to C. The bold path indicates the optimal pathway between A and C, and B and C.

way to get from each city to the destination. The driving instructions represent the optimal control actions. As long as a driver knew where he or she was, the best set of driving instructions from the nearest city could be looked up.

This process is based upon *Bellman's Principle of Optimality,* which is quoted here without proof (Kirk, 1970),

> *"An optimal policy has the property that whatever the initial state and initial decision are, the remaining decisions must constitute an optimal policy with regard to the state resulting from the first decision."*

To interpret this, suppose we have three states, A, B, and C. By appropriate control actions, the state is moved over time from A to C through state B. Bellman's Principle of Optimality (Bellman, 1957) states that if the optimal state-time pathway from A to C is given as A–B–C, then the optimal path from B to C is the subpath B–C of the optimal path A–B–C (Figure 7.2). In other words, optimal subpaths are contained within optimal paths. Another way to think of this from the biomechanical control point of view is that by optimizing the controls and movements over *intervals* of time, one also optimizes over the total movement, from start to finish. Optimizing over periods of time is called *dynamic optimization* because the states and controls at different times are related. In contrast, linear programming would be considered *static optimization* because events are optimized at instants of time and for which the state and control values at one instant of time are unrelated to events at other instants of time.

Unfortunately, dynamic programming suffers from the "curse of dimensionality," which causes the computation time to increase exponentially with the degree of problem complexity. Letting N be the number of discretized state levels, and M the number of discretized control levels. The total computation time T_{CPU} for a system having n degrees of freedom, m muscles, and traversing K time steps is,

$$T_{CPU} = KN^{2n}M^{m}\Delta t_{CPU}\,. \tag{7.20}$$

In the equation, Δt_{CPU} is the time for one combination of state, control value, and time stage to be evaluated. At current computational speeds, one can easily formulate dynamic programming problems that would take months or years to solve. These limitations were explored by Yamaguchi and Zajac (1990), in controlling a single step of an eight degree-of-freedom skeletal model actuated by 10 muscles. Using laboratory mainframe computers available at that time, each simulation required 14 hours to run, even with an extremely coarse mesh of quantized states and controls. Each successive simulation refined the quantization and advanced the model forward in time slightly. A series of 652 simulations were required to optimize a single step![4]

A type of parameter optimization that improves upon dynamic programming has been useful in biomechanical control problems (Pandy *et al.*, 1992). The control values are initially set at discrete time instants, say at t_k ($k = 0, 1, \ldots, K$) and linearly interpolated at times in between. This provides the advantage of having relatively few control parameters while allowing them to vary continuously. By trying out various combinations of the controls, the computer determines an optimal combination of the controls at the discrete time instants. An optimal solution is determined when the search algorithm "converges" to a solution with minimal cost, which is usually a combination of control effort and a performance measure. Current parameter optimizations require supercomputers, due to the need to search the parameter space with gradient methods.

Another good technique is called the Bremermann Optimizer, which has been applied in biomechanical simulations (Bremermann, 1970). The method apparently performs a semi-random search of control parameter space, checks whether the error and/or cost is increasing or decreasing, and readjusts the search. Gerritsen (1998) has used this method to enable a planar, bipedal human walking model to take 12 steps before accumulating errors caused the model to "fall down." A good feature of the Bremermann Optimizer is that the cost function can be a performance based, or a trajectory based measure. The method is very robust because it does not require the computation of gradients to guide the search, which can be problematic depending upon the stability of the model.

Ultimately, dynamic optimization methods based on the calculus of variations should be used to finely tune the musculotendon forces needed to effect a particular motion. A physiologically based, time integral cost function would have to be formulated. This function could perhaps balance physiological energy expenditure (control efforts) with deviations from desired segment tra-

[4] At the time, it was routine to spend one year deriving, programming, and debugging the dynamic program, and one additional year to optimize the controls. Now each of these tasks can be accomplished in a single day on a personal computer!

jectories (trajectory errors). The dynamic system equations might require linearization about the desired trajectory, which unfortunately would be a difficult and tedious task. Otherwise, a nonlinear optimal control method would have to be employed to predict the changes in cost due to infinitesimal changes in the controls. Gradient, or similar methods of efficiently searching the solution space for optimal solutions would probably be employed to determine the best patterns of control throughout the gait cycle. Certainly, the gains we are experiencing in computer power make such an undertaking increasingly attractive. However, the usage of dynamic, variational optimal control methods is probably premature for a number of reasons.

One requirement difficult to satisfy at this time is the need to know the character of the solution prior to the optimization study. This is because the numerical gradient methods usually employed have difficulty converging to a minimum-cost solution unless they begin searching in the vicinity of the desired solution. If and when a solution is found, one cannot be sure without additional tests that the global, rather than a local, optimum has been reached. Additional difficulties arise because muscles, and particularly paralyzed muscles, are limited in their ability to exert force. Thus, the controls are bounded by nature, which undesirably applies restrictions to variational methods. For these reasons, it is unlikely that a good solution to the redundant problem could be found without significant preliminary study.

The best type of preliminary study to perform would involve direct measurements of musculotendon force within animal or human subjects. Presently, any discussion regarding the best mathematical method to solve the redundant problem in biomechanics would be, at best, open to question. Good physiological measurements of the actual forces within living organisms do not yet exist. Therefore, we cannot evaluate the optimal control solutions with comparable data. The present state of the art compares theoretically predicted muscle activations with patterns of muscle electromyograms. This is the best we can presently do, but it is still unsatisfactory. Before major efforts are undertaken to develop improved optimization techniques, efforts should be made to measure tendon forces *in-vivo*. Unfortunately, this is a difficult experimental undertaking.

Until *in-vivo* muscle force data becomes available, one should consider predictions of internal musculoskeletal forces to be interesting academic exercises. As we wait for this data, undoubtedly other, more powerful dynamic optimization routines will be developed to control multisegmented biomechanical motions. It is hoped that the discussion of the "pseudoinverse method" in Section 7.5 will stimulate efforts to develop muscle force distribution schemes that emphasize efficiency and physiology rather than brute force computing. In any event, the following section will lay the necessary foundation for further, more refined, work.

7.4 FORWARD DYNAMIC SIMULATION

While inverse dynamic methods are more appropriate for analyzing motions well after the movements have been recorded, forward dynamic simulations are useful for predicting what would happen under altered circumstances. *The dynamic equations simply enable accelerations to be computed given velocities and positions, forces and torques.* This is, in fact, what makes all the trouble of deriving dynamic equations worthwhile! Out of the infinite number of possible motions that could occur, the dynamic equations predict the motion that will occur.

To see this, the n dynamic equations for a system having n degrees of freedom are reprinted here,

$$M\ddot{\vec{Q}} = \vec{T} + \vec{G} + \vec{V} + \vec{E}\,. \tag{7.21}$$

All the elements on the right-hand side sum together to form a single $n \times 1$ vector $\vec{b}$. Given the parameters of the system, the current torques and external forces, and the instantaneous values of the generalized coordinates and the generalized speeds, all of the elements of matrices M, $\vec{T}$, $\vec{G}$, $\vec{V}$, and $\vec{E}$ will be fully specified as numerical values. The mass matrix M is positive semidefinite, which means that

$$\vec{x}^T M \vec{x} \geq 0\,, \tag{7.22}$$

for all vectors $\vec{x} \neq \vec{0}$. If the n rows are linearly independent, M is said to have "full rank." One of the properties of a positive semidefinite matrix with full rank is that it is invertible, and therefore, a solution can always be found. Equation 7.21 will assume the linear form exhibited in Equation 7.14, which is easily solved for the unknown accelerations contained in $\ddot{\vec{Q}}$ using algorithms such as the methods introduced in Section 7.1.1.[5]

Given the accelerations, it becomes easy to predict the velocities and positions at a small time Δt in the future. Theoretically, one can do the following,

$$\dot{\vec{Q}}(t+\Delta t) = \int_t^{t+\Delta t} \ddot{\vec{Q}}\, dt \tag{7.23}$$

$$= \int_t^{t+\Delta t} M^{-1}\left[\vec{T} + \vec{G} + \vec{V} + \vec{E}\right] dt \tag{7.24}$$

$$\vec{Q}(t+\Delta t) = \int\!\!\int_t^{t+\Delta t} \dot{\vec{Q}}\, dt \tag{7.25}$$

$$= \int\!\!\int_t^{t+\Delta t} M^{-1}\left[\vec{T} + \vec{G} + \vec{V} + \vec{E}\right] dt\,. \tag{7.26}$$

[5]Gordon (1990) suggests Cholesky Decomposition as the most efficient method of finding the solution $\vec{x}$ for linearly system $A\vec{x} = \vec{b}$, where A is positive semi-definite. However, Gaussian Elimination and LU Decomposition programs are routinely available and tend to be more robust for biodynamic problems.

Of course, one does not need to find M^{-1} to obtain the future positions and velocities because Gaussian Elimination or LU Decomposition is much more efficient than finding the inverse of M and multiplying it with the right-hand side.

The method typically performed is even more efficient. The n second order differential equations are broken down into $2n$ first order equations by way of forming a *state vector* $\vec{U}(t)$, such that,

$$\vec{U}(t) = \begin{bmatrix} \left[\dot{\vec{Q}}(t)\right] \\ \\ \left[\vec{Q}(t)\right] \end{bmatrix} = \begin{bmatrix} \dot{q}_1(t) \\ \vdots \\ \dot{q}_n(t) \\ q_1(t) \\ \vdots \\ q_n(t) \end{bmatrix}, \tag{7.27}$$

containing the known positions and velocities. Its derivative is,

$$\dot{\vec{U}}(t) = \begin{bmatrix} \ddot{q}_1(t) \\ \vdots \\ \ddot{q}_n(t) \\ \dot{q}_1(t) \\ \vdots \\ \dot{q}_n(t) \end{bmatrix}. \tag{7.28}$$

At time $t = 0$, only the first n positions of $\dot{\vec{U}}(t)$ are unknown and need to be solved for using the dynamic equations of motion. If this vector can be integrated forward in time by Δt, the entire state vector is projected forward in time and specifies the future state, $\vec{U}(t + \Delta t)$. Again, at time $t + \Delta t$, all of the elements of $\vec{U}(t + \Delta t)$ and the second n elements of $\dot{\vec{U}}(t + \Delta t)$ are known from the forward time projection. The first n positions of $\dot{\vec{U}}(t + \Delta t)$ are found from the dynamic equations, and the state is again projected forward in time. This process is repeated until the simulation achieves its desired final time.

A common way to project the state vector forward in time is to utilize a *Runge-Kutta integration* computer software program. Many are commercially available and typically project a state vector forward by one time increment on each call to the program. All that needs to be supplied is the current state, the name of the program that computes the derivative of the state, and a specification of the error tolerance as an absolute error, a relative error, or both. Because the accuracy of the forward integration increases as the forward time step decreases, one can achieve forward integration results that approach the accuracy of the computer upon which it is run. The algorithms adjust their own internal time steps to perform the integration and to keep the error within specified limits.

If one incorporates muscle activation dynamics within the musculoskeletal model, it is often convenient to append the time varying activations $a_j(t)$, $(j = 1, \ldots, m)$ for the m muscles into the state vector as well,

$$\vec{U}(t) = \begin{bmatrix} \dot{q}_1(t) \\ \vdots \\ \dot{q}_n(t) \\ q_1(t) \\ \vdots \\ q_n(t) \\ a_1(t) \\ \vdots \\ a_m(t) \end{bmatrix} \quad ; \quad \dot{\vec{U}}(t) = \begin{bmatrix} \ddot{q}_1(t) \\ \vdots \\ \ddot{q}_n(t) \\ \dot{q}_1(t) \\ \vdots \\ \dot{q}_n(t) \\ \dot{a}_1(t) \\ \vdots \\ \dot{a}_m(t) \end{bmatrix}. \tag{7.29}$$

One needs only to supply the initial values $a_j(0)$, a program that computes the derivatives $\dot{a}_j(t)$, adjust the size of the state vector to $2n + m$ throughout the simulation program, and pass it to the *Runge-Kutta* algorithm. The entire state vector, including muscle activation dynamics, will then be projected forward in time. For little extra effort, a forward simulation can be made to effectively match physiological, biodynamic movements.

7.5 PSEUDOINVERSE METHOD

The pseudoinverse method is a computationally efficient way of performing dynamic optimizations of movement without many of the undesirable features of linear programming, dynamic programming, parameter optimization, or gradient searching (Yamaguchi *et al.*, 1995). It can be run on desktop computers, and is especially well suited to problems having very complex forward dynamic musculoskeletal models. The more muscle pathways one includes, the better the solution becomes. However, the pseudoinverse method requires a specific desired motion trajectory to be predefined, and requires a cost function which is expressible in terms of a vector length. Besides these limitations, it neatly circumvents many of the problems associated with other methods of obtaining solutions to the biomechanical force distribution problem. The method also takes advantage of a forward dynamic interpretation of muscle function, which states that each muscle has a unique line of action and therefore generates a unique *global* pattern of segmental angular accelerations.

Three seemingly unrelated factors work together in the pseudoinverse method. These are now discussed in turn.

7.5.1 THE CROWNINSHIELD-BRAND COST FUNCTION

The first is the form of a physiologically related cost function proposed by Crowninshield and Brand (1981),

$$C = \sqrt[p]{\sum_{j=1}^{m} \sigma_j^p}\,. \tag{7.30}$$

This equation creates a scalar value C out of m muscle stresses raised to the pth power. The stress σ_j in muscle j is given by its tensile force f_j divided by its *physiological cross sectional area*, or $PCSA_j$,

$$\sigma_j = \left(\frac{f_j}{PCSA_j}\right)\,. \tag{7.31}$$

The $PCSA$ is the average cross section, and is obtained for fusiform muscles by dividing the muscle volume by its optimal muscle fiber length ℓ_o^M, or via the formula presented by Alexander and Vernon (1975),

$$PCSA_{fusiform} = \frac{m^M}{\rho^M\,\ell_o^M}\,. \tag{7.32}$$

Fusiform muscles are those having fibers that are parallel to the tendons, while pennate muscles have muscle fibers pulling at an angle to the tendons. In this formula, m^M is the muscle mass, and ρ^M is its density (typically $1.05\,g/cm^3$). The superscripts (^{M}) are used to indicate a muscle parameter and are *not* numerical exponents. For pennate muscles the $PCSA$ is corrected to account for pennation angle α,

$$PCSA_{unipennate} = \frac{m^M}{2\rho^M t^M} sin(2\alpha)\,, \tag{7.33}$$

where t^M is the layer thickness of muscle pennation.

In Appendix B, the optimal muscle forces reported for each muscle may be converted to $PCSA$ by dividing F_o^M by the specific muscle strength of 31.39 N/cm^2.

Crowninshield and Brand (1981) determined that the most physiologically relevant muscle force distributions were obtained during maximal endurance tasks when $p = 3$. They also reported that the solutions delivered using $p = 2$ and $p = 4$ were nearly identical to those at $p = 3$. They suggested the even values would enable optimal control methods employing quadratic costs to be used. Though two decades have passed since their original paper was published, this formulation has withstood the test of time. Many researchers still use the cost function in its basic or a slightly modified form.

7.5.2 MUSCLE INDUCED ACCELERATIONS

The phrase "muscle induced accelerations" aptly describes the quantifiable accelerations produced by individual muscle contractions (Zajac and Gordon, 1989). These are developed by first rewriting the dynamic equations of motion (Equation 6.35) in slightly different form,

$$\ddot{\vec{Q}} = M^{-1}\left[\vec{T}+\vec{V}+\vec{G}+\vec{E}\right] \tag{7.34}$$

$$= M^{-1}\left[\vec{T}\right]+M^{-1}\left[\vec{V}+\vec{G}+\vec{E}\right]. \tag{7.35}$$

The premultiplication by M^{-1}, the inverse mass matrix, *transforms the moments on the right-hand side of the dynamic equations into accelerations.* The individual terms of the right-hand side may be transformed separately, but it is usually most convenient to group the velocity, gravitational, and external force terms together. The accelerations produced by this grouping are the accelerations from muscle and joint actions, and the accelerations produced by inertial, gravitational, and external forces,

$$\ddot{\vec{Q}} = \ddot{\vec{Q}}_t + \ddot{\vec{Q}}_{vge}. \tag{7.36}$$

The second term $\ddot{\vec{Q}}_{vge}$ expresses accelerations that are not easily changed, as they depend upon the positions and velocities of the system. In other words, the second term is *state dependent* whereas the first term is *control dependent.* $\ddot{\vec{Q}}_t$ may be decomposed further,

$$\ddot{\vec{Q}}_t = M^{-1}\left[\vec{T}_{act}+\vec{T}_{pass}\right], \tag{7.37}$$

where $\vec{T}_{act}$ contains the active contributions of muscle moments, and $\vec{T}_{pass}$ includes the passive contributions of the joint moments created by ligaments and soft tissues (*e.g.*, Equation 5.38). Continuing, the total active torque $\vec{T}_{act}$ can be decomposed into the vector sum of m muscles' individual moment contributions,

$$\vec{T}_{act} = \sum_{j=1}^{m}\vec{t}_j. \tag{7.38}$$

Thus, the accelerations created by active muscle contributions are,

$$\ddot{\vec{Q}}_{t,act} = M^{-1}\left[\vec{T}_{act}\right] \tag{7.39}$$

$$= M^{-1}\left[\sum_{j=1}^{m}\vec{t}_j\right]. \tag{7.40}$$

The *accelerations created by individual muscle contractions* are simply,

$$\ddot{\vec{Q}}_j = M^{-1}\vec{t}_j . \tag{7.41}$$

$\ddot{\vec{Q}}_j$ are called the *muscle induced accelerations* for muscles $j = 1, \ldots, m$.

It is important to note that the vectors $\vec{t}_j$ are sparse (*i.e.,* mostly zero) in high degree of freedom systems because only the elements corresponding to the degrees of freedom directly affected by muscle j's attachment are nonzero. Usually, this means only the degrees of freedom for segments directly attached to or spanned by muscle j are nonzero.[6] However, just because $\vec{t}_j$ is sparse does not mean that $\ddot{\vec{Q}}_j$ will be sparse. Except by extreme coincidence M^{-1} is nonsparse and thus most, if not all, of the n elements of $\ddot{\vec{Q}}_j$ will be nonzero and non-negligible.

This implies that *each muscle has global actions, and accelerates every degree of freedom,* not just the degrees of freedom for the joints spanned by the muscle. Because the mass matrix M is dependent upon the instantaneous mass configuration of the system, the muscle induced accelerations are also configuration dependent and can be quantified to yield dynamic interpretations of muscle function.

Despite the form of Equation 7.41 of this section, the reader should note the inefficiency in computing the inverse mass matrix M^{-1}, and multiplying it by $\vec{t}_j$ on the right to obtain $\ddot{\vec{Q}}_j$. The objective of these matrix-vector multiplications is to obtain the accelerations. This can always be accomplished without computing M^{-1}, because the equations can be solved more efficiently using LU Decomposition. This is explained more fully in the next section.

7.5.3 PROPERTIES OF THE PSEUDOINVERSE

The third factor concerns the Moore-Penrose "pseudoinverse" of a nonsquare matrix A (Golub and Van Loan, 1983; Strang, 1976) for the linear system described by,

$$A\vec{x} = \vec{b}, \tag{7.42}$$

where A has dimension $n \times m$, $\vec{x}$ has dimension $m \times 1$, and $\vec{b}$ has dimension $n \times 1$. The choice of letters n and m is not arbitrary, as n will refer to degrees of freedom and m will refer to the number of musculotendon actuators of the system. In most biomechanical structures, $m \gg n$. In Section 7.1.1 the number of linearly independent rows n was considered to be equal to the number of linearly independent columns m, and there was one and only one solution. In the case where $m > n$ and matrices A and $\vec{b}$ are known, there are an infinite

[6] See Section 6.7 for examples of $\vec{t}_j$ for uniarticular, biarticular, and multiarticular muscles.

number of possible solutions $\vec{x}$ which satisfy Equation 7.42 because the number of variables being solved for is greater than the number of equations. This is the "underconstrained" case, and commonly uses optimization to find the best, or optimal solution according to some mathematical criterion.[7]

If a *right pseudoinverse* A^+ exists for matrix A, such that $AA^+ = I$, where I is the identity matrix, then Equation 7.42 can be rewritten as,

$$A\vec{x} = (AA^+)\,\vec{b} \quad (7.43)$$

$$= A\left(A^+\vec{b}\right). \quad (7.44)$$

Factoring out the A multiplying both sides of the equation from the left yields a solution,

$$\vec{x} = A^+\vec{b}. \quad (7.45)$$

Among the infinite number of possible solutions, the solution delivered above happens to be the solution having *minimum norm* $|\vec{x}|$ (Strang, 1976). This is extremely fortuitous, as it provides a way to obtain an optimal solution *without searching* the solution space. Many of the difficulties associated with trajectory search algorithms, including the computation of gradients, convergence to local minima, and long computation times can be avoided using the pseudoinverse solution if the control problem can be cast into a compatible form.

It is also fortunate that A^+ is easy to compute. For row independent matrices A of rank n, the pseudoinverse is found by performing a transpose, two matrix multiplications, and one inversion,

$$A^+ = A^T\left(AA^T\right)^{-1}. \quad (7.46)$$

Note that because A is $n \times m$ and A^T is $m \times n$, the matrix inversion is applied to the smaller product AA^T which is $n \times n$ and not $m \times m$.

7.5.4 THE PSEUDOINVERSE ALGORITHM

As a reminder, let the goal of this section be restated.

> *"Given a complex, dynamic musculoskeletal model and a well defined set of desired motion pathways for the model to follow, develop a computationally efficient way to determine the optimal set of muscle forces that simultaneously achieves the motion trajectory and minimizes the muscles' metabolic energy expenditure."*

[7]There is also another case, where $m < n$ representing an "overconstrained" solution because there are more equations than unknowns. The solution $\vec{x}$ in this latter case will represent the best fit of the data to the equations, using a least squares approach. The overconstrained case requires a left pseudoinverse, and should not be confused with the underconstrained case described in this section.

The idea is to combine the three factors described in Sections 7.5.1, 7.5.2, and 7.5.3 into a single methdology, casting the musculoskeletal control problem into a rectangular form that can be optimally solved using the properties of the pseudoinverse rather than by trajectory search algorithms. This is accomplished in the following way.

From the current time t and system configuration, compute the muscle induced accelerations created by a single muscle j contracting with a unit amount of stress ($1.0\, N/cm^2$ for instance).[8] It is most convenient to utilize the LU Decomposition method described at the end of Section 7.1.1 to compute the muscle induced accelerations, because Equation 7.41 can be recast into the linear form,

$$M\ddot{\vec{Q}}_j \;=\; \vec{t}_j\,, \tag{7.47}$$

and solved for the accelerations $\ddot{\vec{Q}}_j$. LU Decomposition is good for this step because for a system having m muscles, the muscle induced accelerations must be repeated m times starting with index $j = 1$ and ending with $j = m$. Because all m computations must be performed at a single time instant, matrix M only needs to be decomposed once into the product of lower (L) and upper (U) triangular matrices, $M = LU$. Muscles of different size and with different geometrical pathways will develop different torque inputs, $\vec{t}_j$, and these can be reloaded and solved for $\ddot{\vec{Q}}_j$ extremely quickly using the same LU Decomposition. Each vector $\ddot{\vec{Q}}_j$ of an n degree of freedom system will have dimension $n \times 1$ because each muscle accelerates all n freedom degrees.

Next, the muscle induced accelerations $\ddot{\vec{Q}}_j$ are loaded into the rectangular A matrix in Equation 7.42, such that $\ddot{\vec{Q}}_j$ forms the jth column of A,

$$A = \left[[\ddot{\vec{Q}}_1]\; [\ddot{\vec{Q}}_2]\; [\ddot{\vec{Q}}_3] \ldots [\ddot{\vec{Q}}_m] \right] . \tag{7.48}$$

Because each muscle generates one column, the dimension of A is $n \times m$.

At this point, A has been generated, but not the vector $\vec{b}$ on the right hand side of Equation 7.42. In order to do this, a distinction is made between the accelerations that are desired, $\ddot{\vec{Q}}_{des}$, and the actual accelerations $\ddot{\vec{Q}}$ of the system. $\ddot{\vec{Q}}_{des}$ can be formulated from an instantaneous calculation, but it is simpler in practice to form each term contained in $\ddot{\vec{Q}}_{des}$ from a finite difference of velocity over time interval Δt. Letting $\ddot{q}_{i,des}(t)$ be the ith term of vector

[8] The actual amount of stress assigned to each muscle is unimportant. However, the optimal muscle stresses will be delivered as multiples of the amounts initially applied, so the user should use a convenient measure.

$\ddot{\vec{Q}}_{des}$ $(i = 1, 2, \ldots, n)$,

$$\ddot{q}_{i,des}(t) = \frac{\dot{q}_{i,des}(t + \Delta t) - \dot{q}_i(t)}{\Delta t}. \tag{7.49}$$

In the equation, the speed $\dot{q}_{i,des}(t + \Delta t)$ is the magnitude of the ith *desired* velocity at time $t + \Delta t$ and $\dot{q}_i(t)$ is the magnitude of the ith *actual* velocity at time t.

Using the finite velocity difference is advantageous for two reasons. Using the current velocities and trying to achieve the desired velocities a small time Δt later allows trajectory errors that have accumulated between the actual and desired states at time t to be compensated for during succeeding time intervals. This greatly reduces the buildup of trajectory errors as time proceeds from $t = 0$ to the final time, and is referred to as *real-time* or *on-line error correction*. The second advantage of the finite difference approach is that calculations are performed over *time intervals*. This is philosophically attractive, as it means that Bellman's Principle of Optimality can be applied to argue that optimizing over subintervals of time creates a movement that is *dynamically optimized* over the entire duration of movement (see Section 7.3). This is superior to quasistatic optimizations which are performed at unrelated instants of time.

Referring to Equations 7.35, 7.36, and 7.37, and replacing $\ddot{\vec{Q}}$ with $\ddot{\vec{Q}}_{des}$, the vector $\vec{b}$ is loaded with the *accelerations desired from the muscle set,*

$$\vec{b} = \ddot{\vec{Q}}_{des} - M^{-1}\left[\vec{T}_{pass} + \vec{V} + \vec{G} + \vec{E}\right]. \tag{7.50}$$

The accelerations eliminated are functions of the system state, and cannot be immediately controlled except through future changes in the positions and velocities.

The solution, then, is achieved via the pseudoinverse,

$$\vec{x} = A^T\left(AA^T\right)^{-1}\vec{b} \tag{7.51}$$

$$= \begin{bmatrix} \sigma_1 \\ \sigma_2 \\ \vdots \\ \sigma_m \end{bmatrix}. \tag{7.52}$$

The elements of $\vec{x}$ are the scalar multipliers σ_j which superpose the columns of A to create the accelerations desired from the muscles, $\vec{b}$ (Figure 7.3). Superposition holds because the dynamic equations of motion are linear in terms of the system accelerations. Therefore, the accelerations created by unit stress contractions can be multiplied and added together to form $\vec{b}$. *Each*

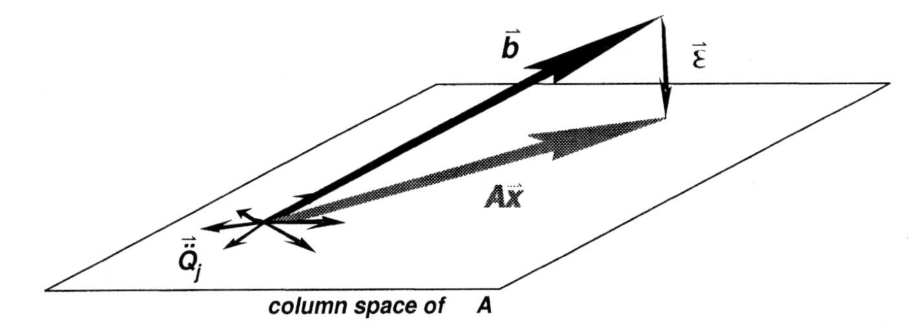

Figure 7.3. Graphical interpretation of the pseudoinverse solution. The m muscle induced acceleration vectors created by unit stress contractions are represented as the small vectors $\ddot{\vec{Q}}_j$ ($j = 1, \ldots, m$) on the lower left hand side. In actuality, each muscle induced acceleration has dimension n (which cannot be drawn). These are depicted as lying in a plane which represents the n dimensional space containing all possible linear combinations of the muscle induced accelerations, which is called the column space of A. Vector $\vec{b}$ is the desired acceleration vector, which may not lie within the column space (shown). If $\vec{b}$ cannot be attained by linear superposition of vectors $\ddot{\vec{Q}}_j$, an error $\vec{\varepsilon} = A\vec{x} - \vec{b}$ results. $A\vec{x}$ is the linear combination of the $\ddot{\vec{Q}}_j$'s that simultaneously minimizes the error and has minimum norm $|\vec{x}|$. $A\vec{x}$ can be thought of as the shadow of $\vec{b}$ cast by a light source perpendicular to the plane containing the column space of A.

individual multiplier σ_j is the actual stress within muscle j ($j = 1, 2, \ldots, m$). This is because it is numerically equivalent to compute the muscle induced accelerations due to a unit stress contraction, and then multiply the result $\vec{t}_j$ by σ_j, or to simply compute the muscle induced accelerations with the muscle stress assigned a value of σ_j.

Remembering that the pseudoinverse automatically computes the solution with minimum norm,

$$min\ |\vec{x}| = \sqrt{\sigma_1^2 + \sigma_2^2 + \ldots + \sigma_m^2} \quad (7.53)$$

$$= \sqrt{\sum_{j=1}^{m} \sigma_j^2} \quad (7.54)$$

it is easy to see that the method delivers the solution which minimizes the quadratic form of the Crowninshield-Brand criterion function (Equation 7.30 with exponent $p = 2$).

One difficulty that is not yet solved satisfactorily is that the pseudoinverse delivers muscle forces that are unconstrained. Negative muscle forces are usually delivered in about $m/3$ of the m muscles, which means that the optimal controller would assign compressive, or "pushing" stresses to the muscles if this were not prevented. Sequential elimination of the negative stresses can be accomplished by post-processing to remove the largest negative stress, called the "worst offending negative stress," at every time step.

If this largest negative stress was $-\sigma_j$, it can be eliminated by dividing the jth column of the A matrix by a large number (say $Z = 1000$). This modifies the A matrix slightly without changing its rank. Re-solving the pseudoinverse with the modified A matrix delivers a solution which is unlikely to use muscle j, because using that particular muscle would be expensive. If it did choose to use muscle j, its stress would likely become $-Z\sigma_j$, which would enlarge the cost function significantly. The next largest negative stress is identified and eliminated, and the process is repeated until all compressive muscle stresses are reduced to insignificant levels. When compressive muscles stresses are completely eliminated, the cost C increases by about 30%.

Muscle force saturation is another problem, and is defined when a muscle stress exceeds the maximal stress σ_{max} that the muscle can achieve under the current conditions of muscle length, velocity, and activation. This time, the "worst offending positive stress" σ_j is identified and replaced with σ_{max}. The accelerative effects of muscle j being replaced by a lower, but maximal stress σ_{max} can be computed as the product, $\vec{\ddot{Q}}_{j,max} = \sigma_{max}\vec{t}_j$, where $\vec{t}_j$ is the muscle induced acceleration vector produced by muscle j. Subtracting $\vec{\ddot{Q}}_{j,max}$ from the vector of accelerations ($\vec{b}$) desired from the muscle set (Equation 7.50) effectively accounts for the action of muscle j contracting maximally. Finally, the jth column of the A matrix is divided by large number Z to remove muscle j from further consideration, and the algorithm is repeated.

While the post-processing methodology reduces the optimal solution to a suboptimal control, Si, Moran, and Yamaguchi (2000) have developed an alternative *recurrent neural network* (RNN) algorithm that is a constrained form of the pseudoinverse. The RNN delivers a globally optimal solution, but takes an order of magnitude more computer time to run.[9] In addition, the RNN appears to work best when initialized with the pseudoinverse solution. Comparisons between the two algorithms suggest the pseudoinverse control solution is near optimal, easily applied, and workable.

Performance of the pseudoinverse has been good for high degree of freedom systems where minimizing energy expenditure is important. Upper extremity models of humans and primates have controlled five to seven angular trajectories with 30 to 39 muscle pathways (Kakavand, 1997; Movafagh, 2000). Trajectory fits have been excellent, maintaining an RMS averaged error across the system to less than one degree for up to three second durations. However, upper extremity models have had limited success in predicting on/off patterns

[9]For a seven degree of freedom model having 39 muscles, the pseudoinverse runs in about 3 hours on a personal computer, and the RNN runs in about a day.

of muscle EMG, because the upper extremity reaching motions are not energetically taxing. On the other hand, Carhart (2000) successfully controlled an 18 degree-of-freedom normal human walking model having 86 muscle pathways, and also simulated a compensatory stepping motion using the same model and technique. His model definitions are given in the appendices. Comparison of the predicted force patterns to the timing of the muscle EMG was excellent.

While these results are promising, the stepping study brought out one of the major weaknesses of the pseudoinverse method. Because the method superposes muscle-induced accelerations to match a desired set of accelerations, the method requires a well defined trajectory. It is therefore difficult to allow the system to find the best trajectory on its own. Also, the pseudoinverse delivers muscle-tendon stresses (forces), not muscle activations. For many applications, it would be preferable to compute muscle activation levels. At the present time, this must be performed one muscle at a time, incorporating muscle contraction dynamics with the known musculotendon length and velocity information, to convert the muscle stresses back to activations.

In closing, there are many, many ways of applying optimal control techniques to effect a coordinated biomechanical movement. At present, there is no "best" way, as all of the methods have trade-offs. None of the methods developed to date have successfully emulated the control process employed by the central nervous system, and perhaps none ever will. However, as long as forward progress is being made, we can continue to strive for the ultimate goal of predicting motor actions and movements before they happen. Perhaps then we will be able to adequately restore the limbs of paraplegics and quadriplegics, understand why people fall and develop ways to prevent falling, and find uses for our models that will make a difference in the lives of people.

Now that we can model the musculoskeletal system well, and control our models to some degree, the next step will be to compare predicted musculotendon forces to forces measured *in-vivo*. This has never been done satisfactorily on a large scale. Until this is done, and researchers collect raw data with which to compare muscle force distribution solutions, there is little point to developing additional control algorithms. Concentrating our efforts on new algorithms will be like building a house on the sand – with the weight of each additional brick adding still greater pressures on a shaky foundation. Let us figure out how to collect *in-vivo* force data, share it, and strengthen the foundation that was never firmly established.

7.6 SUMMARY

An introduction to inverse dynamic methods and forward dynamic methods was given in this chapter. Inverse dynamics takes motion trajectory data as inputs, and provides torques as outputs. The torques can be further decomposed into musculotendon forces distributed among a set of redundant muscles. This process is a quasistatic process in which events and solutions at one time step are unrelated to each other. Because they are unrelated, it is essential that trajectories be smoothed. If the angles, angular velocities, and angular acceleration data change little between adjacent time steps, the torque and muscle force solutions will have a reduced tendency to change too rapidly.

A variety of other methods were mentioned to perform forward dynamic optimizations. All are compromises between computer resource requirements, accuracy, computation time, and relevance to musculoskeletal physiology. Dynamic optimizations are superior to static optimizations because the controls are optimized over the entire duration of the movement, and not only at unrelated instants of time.

The pseudoinverse method was offered as a capable way of performing a dynamic optimization very quickly, using a minimum of computational resources. The algorithm can be implemented using nothing more than a linear system solver, and is extremely efficient using LU decomposition. Most Runge-Kutta forward dynamic simulation programs are packaged together with either Gaussian Elimination or LU decomposition routines. This makes the pseudoinverse suitable as an introductory method of controlling classroom projects.

Future developments in computational hardware and software should deliver continual improvements in price and performance. We can now model the musculoskeletal system with a high degree of accuracy, and this will no doubt be enhanced. What we cannot do is predict how the central nervous systems and the brain interact with our bodies to effect coordinated, well controlled movements. New ways of emulating biomechanical and neuromusculoskeletal control systems are needed, and new control algorithms need to be developed. First, however, raw data needs to be gathered so that any new controller or algorithm can be objectively evaluated. In the meantime, the reader should keep one eye out for developments in the field of *motor control*, and another watching the field of *automatic control*, to see what the future will bring.

7.7 EXERCISES

1. *A forward and inverse dynamic solution.*

   Using Gordon's Method, develop the dynamic equations for a three link planar linkage on a commercially available spreadsheet program with plotting capabilities, or using a computer program.

   (a) Start the system from rest, and apply torques which vary with respect to time. Create a column of data which represents time sampled angles q_1 to q_3 versus time. Use time steps of $\Delta t = 0.1$ second and an overall duration of 1 second.

   (b) Smooth the angular trajectories using a smoothing function or cubic spline interpolation to deliver the angular velocities and accelerations at the $\Delta t = 0.02$ second intervals of time (Section 7.1.1). Graph the original data points and the interpolated data points on the same plot. Are the angular values at 0.1 second intervals changed by the interpolation process? Are the angular velocities and angular accelerations smooth and continuous?

(c) Reformat the dynamic equations found in Part (a) to do an inverse dynamic computation. Apply your values obtained in Part (b) as inputs to your reformatted equations. Solve for the torques versus time, and compare these to the input torques you applied in Part (a).

2. *An inverse and forward dynamic solution.* Develop the dynamic equations for a three or four degree of freedom model of your choice, using Gordon's Method. You will need both inverse and forward dynamic equations. It is preferable to write one or two computer programs for this, but solution by a commercially available spreadsheet is possible.

   Allow a vertical endpoint force to be applied when the ground is contacted, similar to the "soft constraints" described in Section 5.3.1. It is not important what the parameter values b and k of the soft constraint are so long as the endpoint does not bounce violently upon contact.

   (a) Draw a set of desired angular trajectories for your model to follow. The angles should create an external endpoint contact during the motion.

   (b) Digitize and smooth the angular trajectories using a smoothing function or cubic spline interpolation to deliver the angular velocities and accelerations. Then use the inverse dynamic equations to solve for the torques as a function of time. Using the endpoint position and velocity, and Equation 5.17, compute the vertical endpoint force.

   (c) Next, perform a forward dynamic simulation using the torques found in Part (b) as inputs. Try a "sample and hold" algorithm, which resets the torques at each time increment, and holds them until the next time. At the appropriate times, apply the endpoint forces obtained in Part (b) in addition to the torque inputs.

   Compare your motion outputs to your desired trajectories. Comment on why the simulations do or do not match the desired motions.

3. *Inverse Dynamic Project.* Using Gordon's Method, create a planar model having four to six degrees of freedom to explore a question of biomechanical interest. Reformulate the equations into an inverse dynamic format. The equations may be input into a commercially available spreadsheet, or written into a computer code.

(a) Using motion tracking equipment, obtain planar segmental angles as a function of time. The student will likely find it easier to place markers at approximate joint locations.

(b) Use the measured angles to define smoothly varying angular velocities and accelerations.

(c) Perform an inverse dynamic analysis to obtain smoothly varying joint torques.

(d) At one joint, perform a muscle force distribution for an appropriate number of muscles spanning the joint, using moment arm estimates and muscle physiological cross-sectional areas obtained from either Appendix B or the literature.

4. *Forward Dynamic Project.* Using Kane's Method, develop the dynamic equations for a three dimensional model of your choice. If the model is of the lower extremity, Appendices A and B can be used to provide segmental and muscle parameters.

The model should be formulated in response to a biomechanical question of interest. It should have $n = 3$ degrees of freedom if derived by hand, or $n = 4$ to 6 degrees of freedom if developed with the aid of a computer program. At least one degree of freedom must create a motion that is non-coplanar with the other degrees of freedom. Only forward dynamic equations are needed. It is preferable to write one or two computer programs for this, but solution by a spreadsheet is possible.

(a) Define a desired motion trajectory. Either a sampled motion trajectory, or a hypothetical trajectory can be used. If a hypothetical trajectory is defined, it should be physically realizable for the model to perform the motion.

(b) Check the model for errors by summing the potential (*PE*) and kinetic energies (*KE*) during a forward dynamic simulation. Use the following expression for *KE*,

$$KE = \sum_{i=1}^{\mu} \left[\frac{1}{2} m_{B_i} \left| {}^N\vec{v}^{B_i^*} \right|^2 + \frac{1}{2} \sum_{j=1}^{3} I_{B_{ij}} \left({}^N\vec{\omega}^{B_i} \cdot \hat{x}_j \right)^2 \right]. \qquad (7.55)$$

B_i $(i = 1, \ldots, \mu)$ are the bodies with masses m_{B_i} and central principal moments of inertia $I_{B_{ij}}$ defined about principal axes $\hat{x}_j$. Set all torques and forces to zero, and temporarily suspend passive joint moments and damping in the system.

(c) Implement a pseudoinverse solution, using a minimum of $m > 2n$ muscles. To compute unit stress moments and muscle induced accelerations, use muscle moment arm estimates, pathways, and physiological cross-sectional areas obtained from Appendix B or the literature.

References

Allard, Paul, Stokes, Ian A.F., and Blanchi, Jean-Pierre (eds.) (1995) *Three Dimensional Analysis of Human Movement.* Human Kinetics, Champaign, IL.

Alexander, R. McN., and Vernon, A. (1975) "The dimensions of knee and ankle muscles and the forces they exert." *J. Human Movement Studies* V. 1: 115-123.

Bellman, R. E. (1957) *Dynamic programming.* Princeton University Press, Princeton, NJ.

Bremermann, H. (1970) "A method of unconstrained optimization." *Math. Biosci.*, V. 9, pp. 1-5.

Carhart, Michael R. (2000) *Biomechanical Analysis of Compensatory Stepping: Implications for Paraplegics Standing via FNS.* Ph.D. Dissertation, Department of Bioengineering, Arizona State University, Tempe, AZ.

Chao, E. Y., and An, K. N. (1978) "Graphical interpretation of the solution to the redundant problem in biomechanics." *Journal of Biomechanical Engineering*, V. 100, pp. 159-167.

Crowninshield, R. D. (1978) "Use of optimization techniques to predict muscle forces." *Journal of Biomechanical Engineering*, V. 100, pp. 88-92.

Crowninshield, R. D., and Brand, R. A. (1981) "A physiologically based criterion of muscle force prediction in locomotion." *Journal of Biomechanics*, V. 14, n. 11, pp. 793-801.

Dantzig, G. (1963) *Linear Programming and Extensions.* Princeton University Press, Princeton, NJ.

Gerritsen, K. G. M., Van den Bogert, A. J., Hulliger, M., and Zernicke, R. F. (1998) "Intrinsic muscle properties facilitate motor control - A computer simulation study." *Motor Control*, V. 2, pp. 206-220.

Golub, Gene H. and Van Loan, Charles F. (1983) *Matrix Computations.* The Johns Hopkins University Press, Baltimore, MD 21218.

Gordon, Michael E. (1990) "Planar N-Link Open Chain." Appendix 8 of *An Analysis of the Biomechanics and Muscular Synergies of Human Standing.* Ph.D. Dissertation, Department of Mechanical Engineering, Stanford University, Stanford, CA, pp. 223-239.

Hardt, D. E. (1978) "Determining muscle forces in the leg during normal human walking – An application and evaluation of optimization methods." *Journal of Biomechanical Engineering*, V. 100, n. 4, pp. 72-78.

Hatze, H. (1981) "A comprehensive model for human motion simulation and its application to the take-off phase of the long jump." *Journal of Biomechanics* V. 14, , n. 3, pp. 135-143.

Hatze, H. (1981) "The use of optimally regularized Fourier series for estimating higher-order derivatives of noisy biomechanical data." *Journal of Biomechanics*, V. 14, n. 1, pp. 13-18.

Herzog, W. (1992) "Sensitivity of muscle force estimations to changes in muscle input parameters using nonlinear optimization approaches." *Journal of Biomechanical Engineering*, V. 114, n. 2, pp. 267-268.

Herzog, W., and Leonard, T. R. (1991) "Validation of optimization models that estimate the forces exerted by synergistic muscles." *Journal of Biomechanics*, V. 24, suppl. 1, p. 31-39.

Kailath, Thomas (1980) *Linear Systems*. Prentice-Hall, Inc., Englewood Cliffs, NJ.

Kakavand, A. (1997) *Single Cell Cortical Facilitation of Muscle Activity During Volitional Arm Movement: Dynamic Modeling and Analysis.* Ph.D. Dissertation, Department of Chemical, Bio, and Materials Engineering, Arizona State University, Tempe, AZ.

Kirk, D. E. (1970) *Optimal Control Theory – An Introduction.* R. W. Newcomb (ed.), Prentice-Hall, Englewood Cliffs, NJ.

Movafagh, May (2000) Unpublished work performed at Arizona State University.

Pandy, M.G., Anderson, F.C., and Hull, D.G. (1992) "A parameter optimization approach for the optimal control of large-scale musculoskeletal systems." *Journal of Biomechanical Engineering*, V114, n.4, pp. 450-460.

Patriarco, A. G., Mann, R. W., Simon, S. R., and Mansour, J. M. (1981) "An evaluation of the approaches of optimization models in the prediction of muscle forces during human gait." *Journal of Biomechanics* V. 14, , n. 8, pp. 513-525.

Pedotti, A., Krishnan, V. V., and Stark, L. (1978) "Optimization of muscle-force sequencing in human locomotion." *Math. Biosci.*, V. 38, pp. 57-76.

Pierrynowski, M. R., and Morrison, J. B. (1985) "Estimating the muscle forces generated in the human lower extremity when walking: A physiological solution." *Mathematical Biosciences* V. 75, pp. 43-68.

Seireg, A., and Arvikar, R. J. (1975) "The prediction of muscular load sharing and joint forces in the lower extremities during walking." *Journal of Biomechanics* V. 8, pp. 89-102.

Si, J., Moran, D. W., and Yamaguchi, G. T. (2000) "Arm movement joint angle trajectory and muscle force generation using recurrent neural networks." Unpublished manuscript of work performed at Arizona State University.

Strang, G. (1976) *Linear Algebra and its Applications.* Academic Press, New York, NY.

Tözeren, Aydin (2000) *Human Body Dynamics – Classical Mechanics and Human Movement*. Springer-Verlag, New York, NY.

Woltring, H. J. (1986) "A FORTRAN package for generalized, cross-validatory spline smoothing and differentiation." *Adv. Engineering Software*, V. 8, n. 2, pp. 104-113.

Yamaguchi, Gary T., and Zajac, Felix E. (1990) "Restoring unassisted natural gait to paraplegics with functional neuromuscular stimulation: A computer simulation study." *IEEE Transactions on Biomedical Engineering*, V. 37, n. 9, pp. 886-902.

Yamaguchi, Gary T., Moran, Daniel W., and Si, Jennie (1995) "A computationally efficient method for solving the redundant problem in biomechanics." *Journal of Biomechanics*, V. 28, n. 8, pp. 999-1005.

Zajac, Felix E., and Gordon, Michael E. (1989) "Determining muscle's force and action in multi-articular movement," *Exercise and Sport Science Reviews*, V. 17, pp. 187-230.

Appendix A
Human Lower Extremity Segmental Parameters

Acknowledgment – These data sets and figure are provided courtesy of Michael R. Carhart, Ph.D. (2000).

Table A.1. Lengths of body segments and positions of their centers of mass.

Description	*Parameter Name*	*Length in m*
Location of the foot center of mass relative to the ankle	ρ_{a1}, ρ_{g1}	0.0461
	ρ_{a2}, ρ_{g2}	0.0284
Shank length	ℓ_b, ℓ_f	0.4432
Shank center of mass	ρ_b, ρ_f	0.1976
Thigh length	ℓ_c, ℓ_e	0.4312
Thigh center of mass	ρ_c, ρ_e	0.1766
Lateral distance between femoral heads	ℓ_d	0.1650
Location of the pelvis center of mass	ρ_{d1}	0.0000
	ρ_{d2}	0.0578
	ρ_{d3}	0.0825
Location of the torso origin relative to the pelvis	ℓ_{h1}	0.0000
	ℓ_{h2}	0.1488
	ℓ_{h3}	0.0825
Location of the head-arms-trunk center of mass †	ρ_{h1}	0.0000
	ρ_{h2}	0.2869
	ρ_{h3}	0.0825
Origin of the pelvis reference frame for muscle definitions	ℓ_{orig1}	0.0698
(midpoint of the line joining the anterior superior illiac	ℓ_{orig2}	0.0653
spines as measured from the left hip centroid (point D_o)).	ℓ_{orig3}	0.0825

†Center of mass location estimated with the arms folded across the chest.

Table A.2. Body segment masses and principal moments of inertia.

Body Segment	*Mass in kg*	(I_x, I_y, I_z) *in* $kg - m^2$
Foot	1.0688	(0.0011, 0.0045, 0.0049)
Shank	3.3794	(0.0412, 0.0070, 0.0432)
Thigh	11.1836	(0.2250, 0.0462, 0.2250)
Pelvis	8.8101	(0.0592, 0.0672, 0.0738)
Head, Arms and Upper Torso †	37.8264	(1.0686, 0.4960, 1.2343)

†Moments of inertia estimated with the arms folded across the chest.

Body segment lengths and inertial parameters are provided for a human male, 1.78 m tall (5' 10") and weighing 77.9 kilos (171 lb.). Parameters definitions are

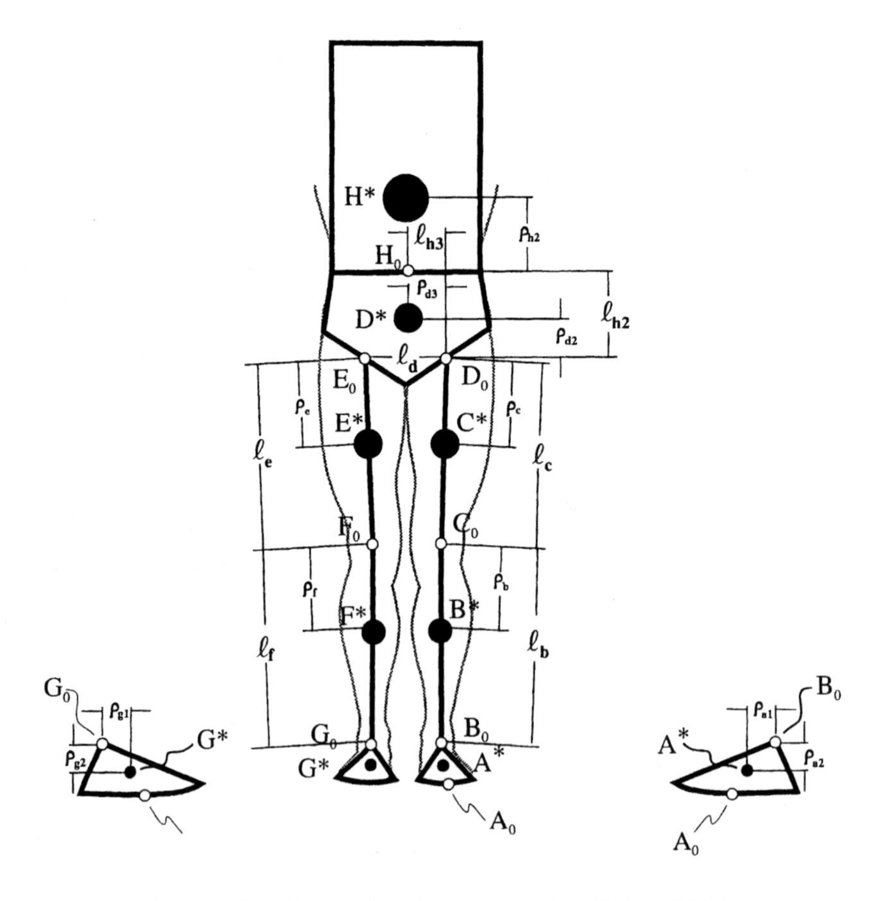

Figure A.1. Segment length parameters from Carhart (2000).

shown in Figure A.1. Specification of the inertial parameters corresponding to the segment lengths, masses, distances to mass centers, and segment moments of inertia was based on the anthropomorphic study of Zatsiorsky and Seluyanov (1983), incorporating the adjustments for joint center based segment definitions suggested by de Leva (1996). In cases where segment definitions used in the model did not match those given by de Leva (1996), the values given by Winter were used (1990).

Moments of inertia are central principal moments about the $\hat{x}$, $\hat{y}$, $\hat{z}$ axes, where $\hat{x}$ is the anterior-posterior axis, $\hat{y}$ is the superior-inferior axis, and $\hat{z}$ is the medial-lateral axis. Additionally, some dimensions are indexed to the "1", "2", and "3" directions, which are defined as anterior, superior, and lateral right, respectively.

References

Carhart, Michael R. (2000) *Biomechanical Analysis of Compensatory Stepping: Implications for Paraplegics Standing Via FNS.* Ph.D. Dissertation, Department of Bioengineering, Arizona State University, Tempe, AZ.

de Leva, P. (1996) "Adjustments to Zatsiorski-Seluyanov's segment inertia parameters." *Journal of Biomechanics*, V. 29, n. 9, pp. 1223-1230.

Winter, D. A. (1990) *Biomechanics and Motor Control of Human Movement.* John Wiley & Sons, New York, NY.

Zatsiorski, V.,and Seluyanov, V. (1983) "The mass and inertial characteristics of the main segments of the human body." *Biomechanics VIII-B*, H. Matsui and K. Kobayashi (eds.), Human Kinetics, Champaign, IL, pp. 1152-1159.

Appendix B
Human Lower Extremity Muscle Parameters

Acknowledgment – These data sets are provided courtesy of Michael R. Carhart, Ph.D. (2000).

Muscle data is provided for the right leg of a human male, 1.78 m tall (5' 10") and weighing 77.9 kilos (171 lb.). Much of the data and reference frames are based on the model reported by Delp (1990). See Figure B.1 for an integrated picture of the skeletal model of Appendix A together with the musculotendon data contained in this appendix.

The specification of the lower extremity muscular architecture was based on the origin (O) and insertion (I) coordinate data compiled by Delp (1990). The data set is a modification of the data originally reported by Brand *et al.* (1982) and Freiderich and Brand (1990). Three dimensional musculotendon pathways were modeled as a series of straight line segments extending from the origin to insertion. Where necessary (*e.g.*, to wrapping the muscle around the surface of a bone), the musculotendon pathway was subsectioned into a number of straight-line segments by introducing via points (V). Forty-three musculotendons in each leg were included in the model for a total of 86 lower extremity musculotendon actuators. For other muscles in different parts of the body, please refer to the collections of data provided in Van der Helm and Yamaguchi (2000), or Yamaguchi *et al.* (1990).

Optimal muscle forces reported for each muscle may be converted to *physiological cross sectional area*, or *PCSA* by dividing F_o^M by the specific muscle strength of 31.39 N/cm^2. Note that PCSAs obtained this way tend to be too small, which is probably explained by the characteristics of the specimens used in the Brand *et al.* (1982) and Freiderich and Brand (1990) studies. Narici (1992), Fukunaga *et al.* (1992), and Fukunaga *et al.* (1996) have performed studies on living subjects which deliver muscle PCSAs that are routinely 30% to 50% larger than those from the earlier studies. Alternatively, the PCSA

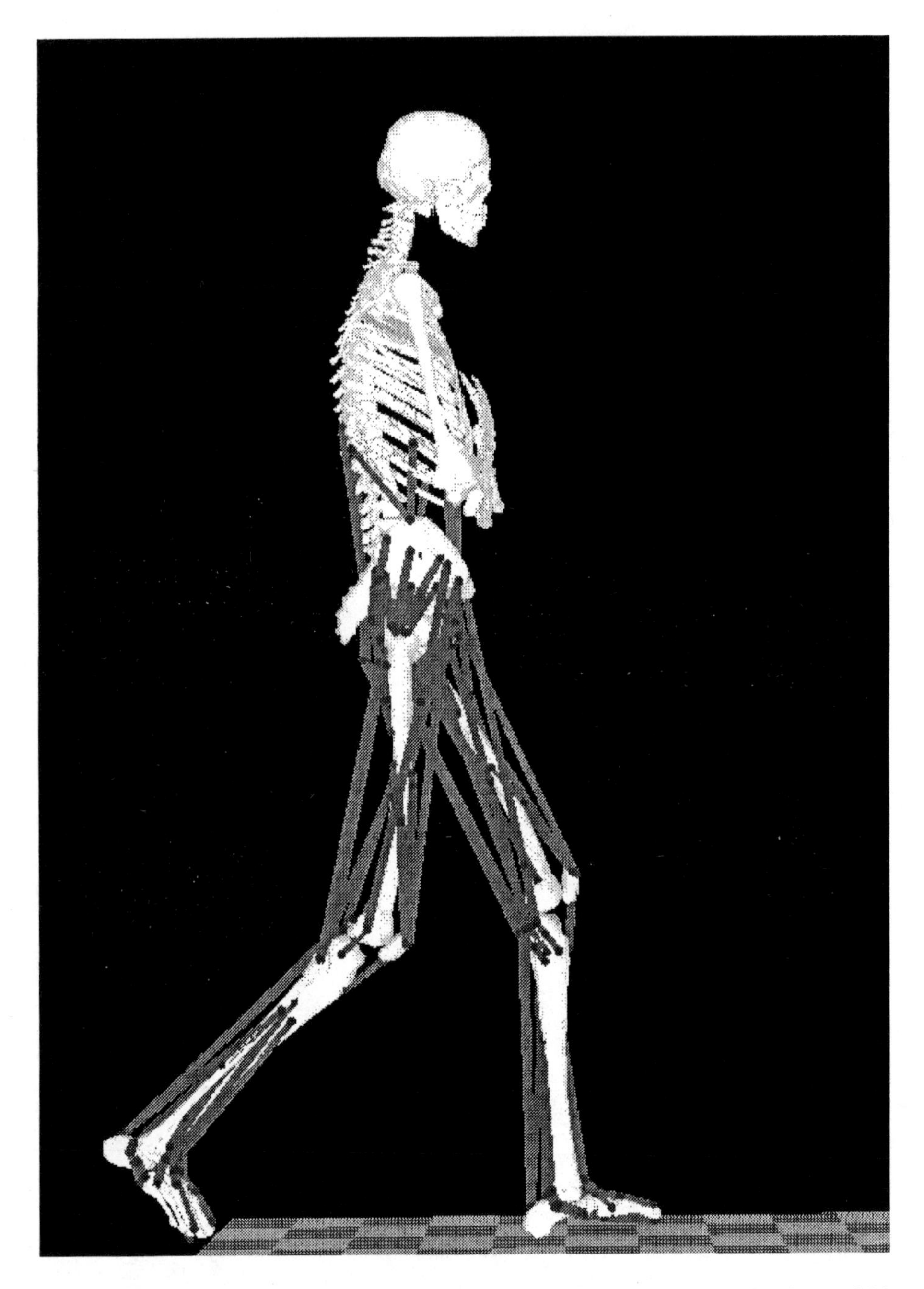

Figure B.1. Walking model of Carhart (2000). The model has 18 degrees of freedom and 96 musculotendon pathways.

parameters may be scaled based on strength data, as done by Anderson and Pandy (1999). For more athletic subjects, it is generally safe to *double* the PCSAs that would be obtained from these tables.

Key:

RF = Reference Frame

Type = O, I, or V O is a muscle origin, I is a muscle insertion, and V is a via point through which the muscle line of action must pass.

P = Pelvic Reference Frame The origin of the P reference frame is defined at the midpoint of the line joining the anterior superior illiac spines.

F = Femoral Reference Frame The origin of the F reference frame is defined as the centroid of the right femoral head.

T = Tibial Reference Frame The origin of the T reference frame is defined as the midpoint of line joining the medial and lateral femoral condyles.

ft = Foot Reference Frame The origin of the *ft* reference frame is the midpoint of a line joining the apeces of the medial and lateral malleoli.

$\hat{x}$**= anterior** The point coordinates in the $\hat{x}$ direction are reported as the distance anterior to the origin of the respective reference frame.

$\hat{y}$**= superior** The point coordinates in the $\hat{y}$ direction are reported as the distance above the origin of the respective reference frame.

$\hat{z}$**= laterally** The point coordinates in the $\hat{z}$ direction are reported as the distance to the right side of the origin of the respective reference frame.

References

Anderson, F. C., and Pandy, M. G. (1999) "A dynamic optimization solution for vertical jumping in three dimensions." *Comput. Methods Biomech. Biomed. Engin.*, V. 2, pp. 201-231.

Brand, R. A., Crowninshield, R. D., Wittstock, C. E., Pedersen, D. R., Clark, C. R., and Van Krieken, F. M. (1982) "A model of lower extremity muscular anatomy." *J. Biomechanical Engineering*, V. 104, pp. 304-310.

Carhart, Michael R. (2000) *Biomechanical Analysis of Compensatory Stepping: Implications for Paraplegics Standing Via FNS.* Ph.D. Dissertation, Department of Bioengineering, Arizona State University, Tempe, AZ.

Delp, Scott L. (1990) *Surgery simulation: A computer graphics system to analyze and design musculoskeletal reconstructions of the lower limb.* Ph.D.

Dissertation, Department of Mechanical Engineering, Stanford University, Palo Alto, CA.

de Leva, P. (1996) "Adjustments to Zatsiorski-Seluyanov's segment inertia parameters." *Journal of Biomechanics*, V. 29, n. 9, pp. 1223-1230.

Friederich, J. A. and Brand, R. A. (1990) "Muscle fiber architecture in the human lower limb." *Technical Note, Journal of Biomechanics*, V. 23, pp. 91-95.

Fukunaga, T., Roy, R. R., Shellock, F. G., Hodgson, J. A., Day, M. K., *et al.* (1992) "Physiological cross-sectional area of human leg muscles based on magnetic resonance imaging." *J. Orthop. Res.*, V. 10, n. 6, pp. 928-934.

Fukunaga, T., Roy, R. R., Shellock, F. G., Hodgson, J. A., and Edgerton, V. R. (1996) "Specific tension of human plantar flexors and dorsiflexors." *J. Appl. Physiol.*, V. 80, n. 1, pp. 158-165.

Narici, M. V., Landoni, L., and Minetti, A. E. (1992) "Assessment of human knee extensor muscles stress from *in-vivo* physiological cross-sectional area and strength measurements." *Eur. J. Appl. Physiol.*, V. 65, n. 5, pp. 438-434.

Van der Helm, Frans C. T., and Yamaguchi, G. T. (2000) "Morphological data for the development of musculoskeletal models: An update." Appendix 1 in *Biomechanics and Neural Control of Posture and Movement*,, pp. 717-773.

Yamaguchi, G. T., Sawa, A. G. U., Moran, D. W., Fessler, M. J., and Winters, J. M. (1990) "A survey of human musculotendon actuator parameters," Appendix in *Multiple Muscle Systems: Biomechanics and Movement Organization*, J. M. Winters and S. L.-Y. Woo (eds.), Springer-Verlag, New York.

Zatsiorski, V.,and Seluyanov, V. (1983) "The mass and inertial characteristics of the main segments of the human body." *Biomechanics VIII-B*. H. Matsui and K. Kobayashi. Human Kinetics, Champaign, IL, pp. 1152-1159.

Muscle Name	F_o^M *(N)*	α *(deg)*	ℓ_o^M *(m)*	ℓ_r^T *(m)*	*RF : Coordinates* $(\hat{x}, \hat{y}, \hat{z})$ *(m, m, m)*	Type
Gluteus Medius	546.0	8.0	0.0535	0.0780	P : (-0.0403, 0.0300, 0.1195)	O
(anterior)					F : (-0.0235, -0.0126, 0.0599)	I
Gluteus Medius	382.0	0.0	0.0845	0.0530	P : (-0.0845, 0.0440, 0.0757)	O
(middle)					F : (-0.0279, -0.0063, 0.0569)	I
Gluteus Medius	435.0	19.0	0.0646	0.0530	P : (-0.1208, 0.0104, 0.0640)	O
(posterior)					F : (-0.0334, -0.0051, 0.0560)	I
Gluteus Minimus	180.0	10.0	0.0680	0.0160	P : (-0.0461, -0.0079, 0.1043)	O
(anterior)					F : (-0.0078, -0.0112, 0.0605)	I
Gluteus Minimus	190.0	0.0	0.0560	0.0260	P : (-0.0625, -0.0064, 0.0979)	O
(middle)					F : (-0.0104, -0.0112, 0.0605)	I
Gluteus Minimus	215.0	21.0	0.0380	0.0510	P : (-0.0824, -0.0062, 0.0846)	O
(posterior)					F : (-0.0146, -0.0090, 0.0594)	I
Gluteus Maximus	382.0	5.0	0.1420	0.1250	P : (-0.1181, 0.0605, 0.0692)	O
(superior)					P : (-0.1276, 0.0012, 0.0875)	V
					F : (-0.0494, -0.0268, 0.0423)	V
					F : (-0.0299, -0.0611, 0.0508)	I
Gluteus Maximus	546.0	0.0	0.1470	0.1270	P : (-0.1333, 0.0174, 0.0556)	O
(middle)					P : (-0.1360, -0.0514, 0.0903)	V
					F : (-0.0460, -0.0572, 0.0316)	V
					F : (-0.0169, -0.1097, 0.0453)	I
Gluteus Maximus	368.0	5.0	0.1440	0.1450	P : (-0.1537, -0.0310, 0.0057)	O
(inferior)					P : (-0.1511, -0.1039, 0.0398)	V
					F : (-0.0323, -0.1126, 0.0146)	V
					F : (-0.0065, -0.1534, 0.0444)	I
Adductor Longus	418.0	6.0	0.1380	0.1100	P : (-0.0312, -0.0826, 0.0167)	O
					F : (0.0054, -0.2281, 0.0253)	I
Adductor Brevis	286.0	0.0	0.1330	0.0200	P : (-0.0580, -0.0904, 0.0162)	O
					F : (0.0010, -0.1292, 0.0318)	I
Adductor Magnus	346.0	5.0	0.0870	0.0600	P : (-0.0723, -0.1160, 0.0252)	O
(superior)					F : (-0.0049, -0.1309, 0.0366)	I
Adductor Magnus	312.0	3.0	0.1210	0.1300	P : (-0.0821, -0.1178, 0.0304)	O
(middle)					F : (0.0058, -0.2469, 0.0245)	I
Adductor Magnus	444.0	5.0	0.1310	0.2600	P : (-0.0762, -0.1167, 0.0273)	O
(inferior)					F : (0.0076, -0.4147, -0.0287)	I
Tensor Fasciae	155.0	3.0	0.0950	0.4250	P : (-0.0307, 0.0211, 0.1226)	O
Latae					F : (0.0318, -0.1075, 0.0645)	V
					F : (0.0058, -0.4376, 0.0386)	V
					T : (0.0062, -0.0502, 0.0306)	I
Pectineus	177.0	0.0	0.1330	0.0010	P : (-0.0426, -0.0759, 0.0446)	O
					F : (-0.0132, -0.0888, 0.0273)	I
Illiacus	429.0	7.0	0.1000	0.0900	P : (-0.0666, 0.0361, 0.0844)	O
					P : (-0.0215, -0.0543, 0.0841)	V
					P : (-0.0291, -0.0800, 0.0845)	V
					F : (0.0018, -0.0587, 0.0062)	V
					F : (-0.0208, -0.0671, 0.0139)	I

Muscle Name	F_o^M *(N)*	α *(deg)*	ℓ_o^M *(m)*	ℓ_r^T *(m)*	*RF : Coordinates* ($\hat{x}$, $\hat{y}$, $\hat{z}$) *(m, m, m)*	Type
Psoas	371.0	8.0	0.1040	0.1300	P : (-0.0639, 0.0876, 0.0286)	O
					P : (-0.0235, -0.0563, 0.0750)	V
					P : (-0.0289, -0.0795, 0.0838)	V
					F : (0.0017, -0.0548, 0.0041)	V
					F : (-0.0203, -0.0645, 0.0112)	I
Quadratus Femoris	254.0	0.0	0.0540	0.0240	P : (-0.1129, -0.1137, 0.0514)	O
					F : (-0.0412, -0.0388, 0.0395)	I
Gemelli	109.0	0.0	0.0240	0.0390	P : (-0.1119, -0.0810, 0.0705)	O
					F : (-0.0153, -0.0036, 0.0479)	I
Piriformis	296.0	10.0	0.0260	0.1150	P : (-0.1379, 0.0003, 0.0232)	O
					P : (-0.1179, -0.0273, 0.0649)	V
					F : (-0.0160, -0.0039, 0.0472)	I
Semimembranosus	1030.0	15.0	0.0800	0.3590	P : (-0.1178, -0.1003, 0.0687)	O
					T : (-0.0250, -0.0553, -0.0200)	I
Semitendinosus	328.0	5.0	0.2010	0.2620	P : (-0.1222, -0.1031, 0.0596)	O
					T : (-0.0324, -0.0562, -0.0150)	V
					T : (-0.0116, -0.0769, -0.0253)	V
					T : (0.0028, -0.0985, -0.0199)	I
Biceps Femoris (long head)	717.0	0.0	0.1090	0.3410	P : (-0.1229, -0.0989, 0.0658)	O
					T : (-0.0083, -0.0751, 0.0436)	I
Biceps Femoris (short head)	402.0	23.0	0.1730	0.1000	F : (0.0054, -0.2281, 0.0253)	O
					T : (-0.0104, -0.0747, 0.0419)	I
Sartorius	104.0	0.0	0.5790	0.0400	P : (-0.0151, -0.0013, 0.1227)	O
					F : (-0.0032, -0.3855, -0.0455)	V
					T : (-0.0058, -0.0432, -0.0411)	V
					T : (0.0062, -0.0607, -0.0395)	V
					T : (0.0250, -0.0866, -0.0260)	I
Gracilis	108.0	3.0	0.35200	0.1400	P : (-0.0556, -0.1026, 0.0078)	O
					T : (-0.0159, -0.0490, -0.0369)	V
					T : (0.0062, -0.0862, -0.0235)	I
Rectus Femoris	779.0	5.0	0.0840	0.3460	P : (-0.0291, -0.0307, 0.0956)	O
					F : (0.0392, -0.4340, 0.0028)	V†
					T : (0.0609, 0.0224, 0.0035)	V
					T : (0.0506, -0.0211, 0.0026)	V
					T : (0.0403, -0.0847, 0.0000)	I
Vastus Medialis	1294.0	5.0	0.0890	0.1260	F : (0.0151, -0.2268, 0.0203)	O
					F : (0.0385, -0.2992, 0.0010)	V
					F : (0.0441, -0.4382, -0.0129)	V†
					T : (0.0549, 0.0232, -0.0150)	V
					T : (0.0506, -0.0211, 0.0026)	V
					T : (0.0403, -0.0847, 0.0000)	I

†Indicates a via point that is dependent upon the configuration of adjacent joints.

Muscle Name	F_o^M *(N)*	α *(deg)*	ℓ_o^M *(m)*	ℓ_r^T *(m)*	*RF : Coordinates* ($\hat{x}$, $\hat{y}$, $\hat{z}$) *(m, m, m)*	Type
Vastus Intermedius	1365.0	3.0	0.08700	0.1360	F : (0.0313, -0.2079, 0.0335)	O
					F : (0.0362, -0.2252, 0.0308)	V
					F : (0.0382, -0.4339, 0.0073)	V†
					T : (0.0544, 0.0268, 0.0019)	V
					T : (0.0506, -0.0211, 0.0026)	V
					T : (0.0403, -0.0847, 0.0000)	I
Vastus Lateralis	1871.0	5.0	0.0840	0.15700	F : (0.0052, -0.2004, 0.0377)	O
					F : (0.0291, -0.2800, 0.0442)	V
					F : (0.0414, -0.4366, 0.0238)	V†
					T : (0.0591, 0.0209, 0.0170)	V
					T : (0.0506, -0.0211, 0.0026)	V
					T : (0.0403, -0.0847, 0.0000)	V
Gastrocnemius (medial)	1113.0	17.0	0.0450	0.4080	F : (-0.0137, -0.4246, -0.0254)	O
					F : (-0.0273, -0.4410, -0.0281)	V†
					T : (-0.0224, -0.0502, -0.0304)	V†
Gastrocnemius (lateral)	488.0	8.0	0.0640	0.3850	F : (-0.0167, -0.4264, 0.0294)	O
					F : (-0.0303, -0.4447, 0.0290)	V†
					T : (-0.0249, -0.0496, 0.0242)	V†
					ft : (-0.0439, -0.0108, 0.0026)	I
Soleus	2839.0	25.0	0.0300	0.2680	T : (-0.0025, -0.1580, 0.0073)	O
					ft : (-0.0439, -0.0108, 0.0026)	I
Tibialis Posterior	1270.0	12.0	0.0310	0.3100	T : (-0.0097, -0.1390, 0.0020)	O
					T : (-0.0148, -0.4177, -0.0236)	V
					ft : (-0.0070, -0.0084, -0.0205)	V
					ft : (0.0281, -0.0257, -0.0200)	I
Flexor Digitorum Longus	310.0	7.0	0.0340	0.4000	T : (-0.0086, -0.2109, -0.0019)	O
					T : (-0.0159, -0.4177, -0.0202)	V
					ft : (-0.0051, -0.0104, -0.0199)	V
					ft : (0.0218, -0.0240, -0.0182)	V
					ft : (0.1157, -0.0495, 0.0193)	I
Flexor Hallucis Longus	322.0	10.0	0.0430	0.3800	T : (-0.0081, -0.2406, 0.0252)	O
					T : (-0.0192, -0.4205, -0.0179)	V
					ft : (-0.0113, -0.0141, -0.0160)	V
					ft : (0.0544, -0.0348, -0.0175)	V
					ft : (0.1224, -0.0468, -0.0188)	I
Tibialis Anterior	603.0	5.0	0.0980	0.2230	T : (0.0185, -0.1674, 0.0119)	O
					T : (0.0339, -0.4073, -0.0182)	V
					ft : (0.0671, -0.0239, -0.0224)	I
Peroneus Brevis	348.0	5.0	0.0500	0.1610	T : (-0.0072, -0.2727, 0.0335)	O
					T : (-0.0204, -0.4313, 0.0292)	V
					T : (-0.0148, -0.4427, 0.0298)	V
					ft : (-0.0017, -0.0147, 0.0309)	V
					ft : (0.0187, -0.0198, 0.0417)	I

†Indicates a via point that is dependent upon the configuration of adjacent joints.

Muscle Name	F_o^M *(N)*	α *(deg)*	ℓ_o^M *(m)*	ℓ_r^T *(m)*	*RF : Coordinates* $(\hat{x}, \hat{y}, \hat{z})$ *(m, m, m)*	Type
Peroneus Longus	754.0	10.0	0.0490	0.3450	T : (0.0005, -0.1616, 0.0373)	O
					T : (-0.0213, -0.4334, 0.0295)	V
					T : (-0.0167, -0.4452, 0.0298)	V
					ft : (-0.0049, -0.0187, 0.0297)	V
					ft : (0.0191, -0.0310, 0.0359)	V
					ft : (0.0360, -0.0346, 0.0195)	V
					ft : (0.0707, -0.0331, -0.0104)	I
Peroneus Tertius	90.0	13.0	0.0790	0.1000	T : (0.0010, -0.2890, 0.0238)	O
					T : (0.0236, -0.4194, 0.0164)	V
					ft : (0.0365, -0.0190, 0.0374)	I
Extensor Digitorum Longus	341.0	8.0	0.1020	0.3450	T : (0.0033, -0.1424, 0.0284)	O
					T : (0.0298, -0.4130, 0.0074)	V
					ft : (0.0429, -0.0032, 0.0077)	V
					ft : (0.1116, -0.0360, 0.0207)	I
Extensor Hallucis Longus	108.0	6.0	0.1110	0.3050	T : (0.0012, -0.1821, 0.0235)	O
					T : (0.0336, -0.4108, -0.0088)	V
					ft : (0.0477, -0.0031, -0.0131)	V
					ft : (0.0796, -0.0110, -0.0176)	V
					ft : (0.1232, -0.0277, -0.0199)	I

About the Author

Dr. Gary T. Yamaguchi studied physics at Occidental College and engineering design at the California Institute of Technology, and received undergraduate degrees from both institutions. He then attended the Massachusetts Institute of Technology for a master's degree in mechanical engineering, and worked on magnetic fusion energy at the Lawrence Livermore National Laboratory before returning to graduate school at Stanford University. He has taught biomechanics, rehabilitation engineering, and engineering design at Arizona State University for 12 years. He loves to fly fish, golf, and rock climb with his wife and three children.

Index